Operator Methods
in
Quantum Mechanics

Operator Methods in Quantum Mechanics

Martin Schechter
University of California, Irvine

DOVER PUBLICATIONS, INC.
Mineola, New York

Bibliographical Note

This Dover edition, first published in 2002, is an unabridged republication of the work originally published in 1981 by Elsevier North Holland, Inc., New York.

Library of Congress Cataloging-in-Publication Data

Schechter, Martin.
Operator methods in quantum mechanics / Martin Schechter.
p. cm.
Originally published: New York : North Holland, c1981.
Includes bibliographical references and index.
ISBN 0-486-42547-9 (pbk.)
1. Operator theory. 2. Quantum theory. I. Title.

QC174.17.O63 S33 2002
515'.724—dc21

2002031300

Manufactured in the United States of America
Dover Publications, Inc., 31 East 2nd Street, Mineola, N.Y. 11501

BS"D

To the memory of my parents, Z"L,

Joshua and Rose

who left this world before their nachas was complete.

Contents

Preface

The interaction between physics and mathematics has always played an important role in the development of both sciences. The physicist who does not have the latest mathematical knowledge available to him is at a distinct disadvantage. The mathematician who shies away from physical applications will most likely miss important insights and motivations. For these reasons it has become a practice to expose science students to a course in "mathematical physics" and mathematics students to a course in "applied mathematics." Such courses usually present various mathematical techniques that are useful in science and engineering. The student is told that the mathematics is important, but he is rarely shown why it is needed. If applications are discussed, the student may find them artificially manufactured.

For several years I have been giving a course in applications of operator theory. Rather than presenting the usual mixture of theorems, I decided to pick a branch of physics and to develop the theory from the very beginning. I did it in a mathematical way without describing the experimental evidence that gave rise to the theory. I gave a few simple postulates assuming no background in physics and as little background in mathematics as possible. The students were asked to accept these few postulates and not to try to understand the reasoning that led to them. Then I gradually introduced the powerful mathematical techniques that help answer questions that are important to the physical theory. Only those mathematical methods that are needed for the physics are introduced, and they are introduced when they are required. From the very beginning it is possible to motivate the entire exposition. The student sees clearly the purpose of the method and understands the accomplishment. This book is an outgrowth of this course.

I found quantum theory to be very fruitful from a mathematical point of view. The theory gives rise to the many questions in operator theory and in the study of differential equations. I found the theory so rich that I was able to fill this entire volume analyzing only a single particle in one dimension. In fact, I was not able to cover all of the one-dimensional topics that I would have liked. It is true that most of the methods developed can be applied to systems of particles in higher dimensions, but I found that there was enough to do in the simplest case. In this way the student is introduced to the methods without being overwhelmed by details. The consideration of one situation throughout the book adds unity to the exposition.

My main thrust has been towards the understanding of the physical and mathematical principles and not towards the achieving of the strongest results. I have tried to minimize technical details wherever possible and to keep the discussion elementary. Each mathematical method is introduced as needed in the development. It is then shown how the new technique solves the problem at hand.

The book may be used at an advanced undergraduate or a beginning graduate level. The main prerequisite is advanced calculus. Elementary theory of Lebesgue integration and analytic functions of a complex variable are used unavoidably in a few sections, but the reader will have no difficulty skipping these sections if he is willing to believe certain statements. Most theorems in analysis and functional analysis are proved either on the spot or in an Appendix. There are a few theorems that I do not prove. However, the student will have no difficulty understanding their meaning, purpose, and applicability. These few theorems are standard in mathematics courses at the beginning graduate level (references are given in the Bibliography).

Theorems are designated by three numbers. The first refers to the chapter, the second to the section and the third to the order in which it appears in the section. Lemmas and propositions are similarly designated.

To my mind, the application of operator theory to quantum mechanics forms one of the most beautiful areas of knowledge. I hope this book has captured some of the beauty.

New York Martin Schechter

TVSLB'' 0

Acknowledgements

The material presented in the book is based upon results contained in various research articles. I have tried to include all of them in the Bibliography.

I wish to thank my students Merl Altabet and Alexander Gelman for reading parts of the manuscript and making important suggestions and corrections. My daughter, Sara, helped me greatly by typing parts of the manuscript, and my wife, Deborah, made invaluable contributions to the effort.

A Message to the Reader

The purpose of this volume is to show how mathematics is used to answer questions in science. We begin by assuming that the reader has a working knowledge of advanced calculus, and we present a few simple postulates describing quantum theory. The entire book is devoted to the study of a single particle moving along a straight line. We ask questions about this particle and gradually develop mathematical techniques that give the answers. Most of these techniques can be used for more complicated systems of particles, but I did not want to get involved in too many details. There are places where I need Lebesgue integration theory and complex variables, but most of the book can be understood without them. There are a few places where I ask you to believe me (of course I give references), but there should be no difficulty in understanding the statements. I hope this book can help you see how operator theory can be used as a powerful tool in the study of quantum mechanics.

List of Symbols

The page numbers indicate where the symbols are defined or are explained.

Operator Methods
in
Quantum Mechanics

1

One-Dimensional Motion

1.1. Position

We begin by studying a single particle restricted to motion along a line. The first postulate we shall use is

Postulate 1. There is a function $\psi(x, t)$ of position x and time t such that the probability that the particle is in an interval I at the time t is given by

$$\int_I |\psi(x, t)|^2 \, dx. \tag{1.1.1}$$

Note that this postulate does not tell us how to determine the position of the particle, but only how to determine the probability that it is located in an interval. The postulate's probabilistic nature represents a basic philosophical tenet of quantum mechanics. Later we shall see that it is seldom that we can determine the position (or any other quantity) precisely. For those who are unfamiliar with probability theory, it suffices at the moment to note that the probability of an event occurring is a number P, $0 \leq P \leq 1$, which, in a sense, represents the chances that the event will take place. We shall have more to say about this later.

The function ψ is called the *state function* for the particle. Usually it is complex valued, but only in the sense that it takes on complex values for each x and t. There need not be any relationship between its real and imaginary parts (i.e., it need not be analytic in the sense of complex variables). We shall not be able to justify the appearance of the absolute value and the square power in (1.1.1); although we shall see that they are highly desirable from a mathematical point of view.

Since it is certain that the particle must be somewhere along the line, we have

$$\int_{-\infty}^{\infty} |\psi(x, t)|^2 \, dx = 1 \tag{1.1.2}$$

at each time t.

1.2. Mathematical Expectation

Suppose w is a measurable quantity which can take on the values $w_1, \ldots, w_N$. Assume that one can perform an indefinite number of independent experiments (i.e., the outcome of one does not affect the outcome of another) in which w is measured. Suppose the probability that w takes on the value w_k is P_k, $1 \leq k \leq N$. Since w must take on one of the given values, we have

$$P_1 + \cdots + P_N = 1. \tag{1.2.1}$$

The quantity

$$\bar{w} = w_1 P_1 + \cdots + w_N P_N \tag{1.2.2}$$

is called the *mathematical expectation* or *average value* of w. The reason for this terminology is the following theorem from the theory of probability (cf. Feller, 1950).

Theorem 1.2.1. *Suppose a sequence of identical experiments is performed and the values $S_1, S_2, \ldots, S_n, \ldots$ are observed (the numbers S_n are among the values $w_1, \ldots, w_N$). Then the average value*

$$(S_1 + \cdots + S_n)/n \tag{1.2.3}$$

converges to $\bar{w}$ "in the sense of probability."

In order for this theorem to make sense, we must know what it means to converge "in the sense of probability." By this we mean that for each $\varepsilon > 0$ the probability that

$$\left| \frac{1}{n}(S_1 + \cdots + S_n) - \bar{w} \right| > \varepsilon \tag{1.2.4}$$

tends to 0 as $n \to \infty$. Life would be a lot simpler if (1.2.3) would converge in the usual sense, but this is not in the cards. However, it does show that the chances of the average (1.2.3) moving away from $\bar{w}$ are small.

If w can take on any value in some interval, we must assign to each subinterval I the probability P_I that the value of w lies in I. To compute the mathematical expectation of w, assume first that it is restricted to a bounded interval $[a, b]$. We divide this interval up into small subintervals

I_k and form the sum

$$\sum x_k P_{I_k}, \tag{1.2.5}$$

where x_k is an arbitrary point of I_k. If these sums converge to a limit as the maximum length of the intervals I_k tends to 0 independently of the manner in which the intervals I_k were chosen and independently of the choice of the points $x_k \in I_k$, then we call the limit $\bar{w}$ the mathematical expectation of w. If w is not restricted to a bounded interval, we compute first the limit as above for a bounded interval $[a, b]$ and then we take the limit as the length of the interval increases, assuming that this limit is independent of the choice of the interval and the manner in which it grows. If this latter limit exists, we call it the mathematical expectation $\bar{w}$ of w.

To illustrate this concept, consider the probability (1.1.1) that a particle is in an interval I. Let I be a bounded interval, and subdivide it into smaller intervals I_k. If x_k is an arbitrary point in I_k, (1.2.5) becomes

$$\sum x_k \int_{I_k} |\psi(x, t)|^2 \, dx. \tag{1.2.6}$$

The following is true:

Lemma 1.2.1. *If $\psi(x, t)$ is continuous with respect to x in I, then* (1.2.6) *converges to*

$$\int_I x|\psi(x, t)|^2 \, dx \tag{1.2.7}$$

as the maximum length of the intervals I_k tends to 0 *independently of the choice of the I_k and x_k.*

PROOF. We have

$$\left| \int_I x|\psi(x, t)|^2 \, dx - \sum x_k \int_{I_k} |\psi(x, t)|^2 \, dx \right| \le \sum \int_{I_k} |x - x_k| \, |\psi(x, t)|^2 \, dx. \tag{1.2.8}$$

Let $\varepsilon > 0$ be given, and take the length of each $I_k < \varepsilon$. Then $|x - x_k| < \varepsilon$ in I_k. Thus the right-hand side of (1.2.8) is less than

$$\varepsilon \int_I |\psi(x, t)|^2 \, dx. \qquad \square$$

The definition and Lemma 1.2.1 lead immediately to

Theorem 1.2.2. *If $\psi(x, t)$ is continuous with respect to x and*

$$\int_{-\infty}^{\infty} |x| \, |\psi(x, t)|^2 \, dx < \infty, \tag{1.2.9}$$

then the mathematical expectation of position is given by

$$\bar{x} = \int_{-\infty}^{\infty} x|\psi(x, t)|^2\,dx. \tag{1.2.10}$$

Now we give a proof of Theorem 1.2.1 for the case of the position of a particle. Suppose the positions of n identical particles are measured at a time t and that they all have the same state function. Suppose the observed values are $x_1, \ldots, x_n$. If x_0 and $a > 0$ are given, then the probability that $|x_0 - \Sigma x_k| > a$ is

$$\int \cdots \int_{|x_0 - \Sigma x_k| > a} |\psi(x_1, t) \cdots \psi(x_n, t)|^2\,dx_1 \cdots dx_n.$$

This is less than

$$a^{-2}\int \cdots \int |x_0 - \sum x_k|^2 |\psi(x_1, t) \cdots \psi(x_n, t)|^2\,dx_1 \cdots dx_n. \tag{1.2.11}$$

Set

$$(u, v) = \int_{-\infty}^{\infty} u(x)v(x)^*\,dx$$

and

$$\|u\|^2 = (u, u),$$

where the asterisk denotes complex conjugation. Since

$$|x_0 - \sum x_k|^2 = x_0^2 - 2x_0 \sum x_k + \sum x_j x_k,$$

(1.2.11) is equal to

$$a^{-2}\big(n\|x\psi\|^2 + n(n-1)(x\psi, \psi)^2 + x_0^2 - 2nx_0(x\psi, \psi)\big)$$
$$= a^{-2}\big([x_0 - n(x\psi, \psi)]^2 + n[\|x\psi\|^2 - (x\psi, \psi)^2]\big).$$

Thus the probability that $|n\bar{x} - \Sigma x_k| > n\varepsilon$ is bounded by

$$\varepsilon^{-2}[\bar{x} - (x\psi, \psi)]^2 + \varepsilon^{-2}n^{-1}\big(\|x\psi\|^2 - (x\psi, \psi)^2\big).$$

By (1.2.10), $\bar{x} = (x\psi, \psi)$, while the second term tends to 0 as $n \to \infty$.

Theorem 1.2.3. *If $f(x)$ is a continuous function satisfying*

$$\int_{-\infty}^{\infty} |f(x)|\,|\psi(x, t)|^2\,dx < \infty, \tag{1.2.12}$$

then the mathematical expectation of $f(x)$ is given by

$$\overline{f(x)} = \int_{-\infty}^{\infty} f(x)|\psi(x, t)|^2\,dx. \tag{1.2.13}$$

The proof of Theorem 1.2.3 is simple. In fact, let I be a bounded interval, and let $I_1, \ldots, I_N$ be a partition of I into smaller intervals with maximum length δ. If x_k is an arbitrary point in I_k, we have

$$\left|\int_I f(x)|\psi(x, t)|^2\,dx - \sum f(x_k)\int_{I_k}|\psi(x, t)|^2\,dx\right|$$

$$\le \sum \int_{I_k} |f(x) - f(x_k)|\,|\psi(x, t)|^2\,dx. \tag{1.2.14}$$

Since $f(x)$ is uniformly continuous in I, we can make $|f(x) - f(x_k)| < \varepsilon$ in I_k by taking δ sufficiently small. For such a δ the left-hand side of (1.2.14) is less than

$$\varepsilon \int_I |\psi(x, t)|^2\,dx \le \varepsilon.$$

Once we have established this, we merely note that

$$\int_{-\infty}^{\infty} f(x)|\psi(x, t)|^2\,dx = \lim_{R\to\infty}\int_{-R}^{R} f(x)|\psi(x, t)|^2\,dx.$$

1.3. Momentum

In classical physics the momentum of a particle is defined as

$$p = m\frac{dx}{dt} = \text{Mass} \times \text{Velocity}.$$

The second postulate we shall make concerns momentum. We state it as follows:

Postulate 2. The probability that the momentum p of the particle is contained in the interval I is given by

$$\frac{1}{\hbar}\int_I \left|\hat{\psi}\left(\frac{p}{\hbar}, t\right)\right|^2 dp,$$

where $\hbar$ is Planck's constant (a quantity physicists will have no difficulty remembering and mathematicians will have no difficulty forgetting) and $\hat{\psi}$ is the Fourier transform of ψ with respect to x defined by

$$\hat{\psi}(k, t) = \frac{1}{\sqrt{2\pi}}\int_{-\infty}^{\infty} e^{-ikx}\psi(x, t)\,dx. \tag{1.3.1}$$

(We apologize to the mathematicians for using k as a variable.) As in the case of position, the average value of p turns out to be

$$\bar{p} = \frac{1}{\hbar}\int_{-\infty}^{\infty} p\left|\hat{\psi}\left(\frac{p}{\hbar}, t\right)\right|^2 dp = \hbar\int_{-\infty}^{\infty} k|\hat{\psi}(k, t)|^2\,dk. \tag{1.3.2}$$

As in Section 1.2, we have

Theorem 1.3.1. *If $g(p)$ is continuous and satisfies*

$$\int_{-\infty}^{\infty} |g(\hbar k)|\,|\hat{\psi}(k, t)|^2\, dk < \infty, \tag{1.3.3}$$

then

$$\overline{g(p)} = \int_{-\infty}^{\infty} g(\hbar k)|\hat{\psi}(k, t)|^2\, dk. \tag{1.3.4}$$

The proof of Theorem 1.3.1 is the same as that of Theorem 1.2.3.

In order to deal with momentum, we shall need to know certain properties of the Fourier transform. We state them here without hypotheses. Precise statements and proofs will be given in Appendix A.

a. The inverse Fourier transform

$$\psi(x, t) = \frac{1}{\sqrt{2\pi}} \int_{-\infty}^{\infty} e^{ikx}\hat{\psi}(k, t)\, dk. \tag{1.3.5}$$

b. Parseval's identity

$$\int_{-\infty}^{\infty} \varphi(x, t)\psi(x, t)^*\, dx = \int \hat{\varphi}(k, t)\hat{\psi}(k, t)^*\, dk \tag{1.3.6}$$

(The asterisk denotes the complex conjugate.)

c. $\partial\hat{\psi}(k, t)/\partial k = -i[x\psi(x, t)]^\wedge$.

d. $\partial\psi(x, t)/\partial x = i[k\hat{\psi}(k, t)]^\vee$. (Here $^\vee$ denotes the inverse Fourier transform.)

e. $[\partial\psi(x, t)/\partial x]^\wedge = ik\hat{\psi}(k, t)$.

f. $[\partial\hat{\psi}(k, t)/\partial k]^\vee = -ix\psi(x, t)$.

Using these properties we can show how one can express $\bar{p}$ without using the Fourier transform. For by (1.3.2)

$$\bar{p} = \hbar \int_{-\infty}^{\infty} k\hat{\psi}(k, t)\hat{\psi}(k, t)^*\, dk$$

$$= -i\hbar \int_{-\infty}^{\infty} [\partial\psi/\partial x]^\wedge \hat{\psi}^*\, dk = -i\hbar \int_{-\infty}^{\infty} (\partial\psi/\partial x)\psi^*\, dx.$$

Thus

$$\bar{p} = \int_{-\infty}^{\infty} (L\psi)\psi^*\, dx, \tag{1.3.7}$$

where

$$L\psi = -i\hbar\, \partial\psi/\partial x. \tag{1.3.8}$$

This is called the *momentum operator*. Note that (1.3.7) is free of Fourier transforms, but the penalty paid is the introduction of the partial differential operator L.

Repeated applications of (e) give

$$[\partial^n\psi(x, t)/\partial x^n]^\wedge = (ik)^n\hat{\psi}(k, t). \tag{1.3.9}$$

Consequently, we have by (1.3.4)

$$\overline{p^n} = (L^n\psi, \psi), \tag{1.3.10}$$

where we have used the notation

$$(\varphi, \psi) = \int_{-\infty}^{\infty} \varphi(x, t)\psi(x, t)^* \, dx. \tag{1.3.11}$$

1.4. Energy

In classical physics the kinetic energy T of the particle is given by

$$T = p^2/2m. \tag{1.4.1}$$

By (1.3.10) the expectation of T is

$$\overline{T} = \frac{1}{2m}(L^2\psi, \psi). \tag{1.4.2}$$

The potential energy is given by a real-valued function of position $V(x)$. The total energy is given by

$$E = \frac{p^2}{2m} + V. \tag{1.4.3}$$

If $V(x)$ is continuous and satisfies

$$\int_{-\infty}^{\infty} |V(x)| \, |\psi(x, t)|^2 \, dx < \infty, \tag{1.4.4}$$

then Theorem 1.2.3 gives as the expectation value of the potential energy

$$\overline{V} = \int_{-\infty}^{\infty} V(x)|\psi(x, t)|^2 \, dx. \tag{1.4.5}$$

How can we compute the average of the total energy? It turns out to be a simple matter if we use

Theorem 1.4.1. *The mathematical expectation of a sum is equal to the sum of the mathematical expectations.*

Thus, if we add (1.4.2) and (1.4.5), we get

$$\overline{E} = (H\psi, \psi),$$

where

$$H = \frac{1}{2m}L^2 + V \tag{1.4.6}$$

is the *energy operator* or *Hamiltonian*.

It should be noted that Theorem 1.4.1 holds whether or not the quantities are independent. We shall give the proof only for the case of discrete variables.

Let u be a measurable quantity that can only take on the values $u_1, \ldots, u_M$ with probabilities $P_1, \ldots, P_M$, and let v be a measurable quantity that can take on only the values $v_1, \ldots, v_N$ with probabilities $Q_1, \ldots, Q_N$. Let R_{ij} be the probability that $u = u_i$ and that $v = v_j$. (If the quantities are independent, then $R_{ij} = P_i Q_j$. Otherwise, this need not be so.) Thus

$$P_i = \sum_j R_{ij}, \qquad Q_j = \sum_i R_{ij}.$$

Now the mathematical expectation of $u + v$ is

$$\overline{u + v} = \sum_{i,j} (u_i + v_j) R_{ij} = \sum u_i R_{ij} + v_j R_{ij} = \sum u_i P_i + \sum v_j R_j = \bar{u} + \bar{v}.$$

This proves the theorem for the discrete case.

Although we have "proved" (1.4.5) only for continuous functions $V(x)$, there are many potentials of physical interest which possess discontinuities or even singularities. For such functions we shall take (1.4.5) as the definition of the expectation value.

1.5. Observables

Any quantity that can be measured is called an *observable*. We have discussed three, namely, position, momentum, and energy. In each of these cases, we have noticed that there corresponds an operator in the sense that if a denotes the variable and $\bar{a}$ is its average value, then there is an operator A such that

$$\bar{a} = (A\psi, \psi). \tag{1.5.1}$$

We can make a table:

a	A
x	x
p	$L = -i\hbar\, \partial/\partial x$
E	$H = \dfrac{1}{2m} L^2 + V$

In each of the cases mentioned the observable took on real values only. In such cases the average value must be real as well. This requires that $(A\psi, \psi)$ be real for any state function ψ. This leads to the question, what operators A have the property that $(A\psi, \psi)$ is real for every state function ψ? Fortunately the answer is simple. An operator A is called *Hermitian* if

$$(A\psi, \varphi) = (\psi, A\varphi) \tag{1.5.2}$$

holds for all ψ, φ. The answer to our question is given by

Lemma 1.5.1. *An operator A is Hermitian if and only if $(A\psi, \psi)$ is real for all ψ.*

Before we give the proof of Lemma 1.5.1, we had better clarify certain concepts. An *operator* is a mapping which takes a set $\mathcal{A}$ into a set $\mathcal{B}$ (which may coincide with $\mathcal{A}$). It is always assumed single valued, that is, it takes an element of $\mathcal{A}$ into only one element of $\mathcal{B}$. The set of elements on which it acts is called its *domain*. An operator A is called *linear* if (1) $\alpha x + \beta y$ is in its domain $D(A)$ whenever x, y are in its domain and α, β are scalars, and (2) $A(\alpha x + \beta y) = \alpha Ax + \beta Ay$. We shall take our scalars to be complex, and when we use the word *operator*, we shall mean a linear operator.

The expression (1.3.11) is called a *scalar product*. It has the following properties:

$$(\alpha\varphi, \psi) = \alpha(\varphi, \psi) \qquad \text{for any scalar } \alpha; \tag{1.5.3}$$

$$(\varphi + \sigma, \psi) = (\varphi, \psi) + (\sigma, \psi); \tag{1.5.4}$$

$$(\varphi, \psi) = (\psi, \varphi)^*; \tag{1.5.5}$$

$$\|\psi\|^2 = (\psi, \psi) > 0 \qquad \text{unless} \quad \psi = 0. \tag{1.5.6}$$

Technically speaking, statement (1.5.6) is not quite correct. It should say $\psi = 0$ almost everywhere (a.e.). However, we shall identify functions which agree almost everywhere and write $\psi = 0$ when we mean $\psi = 0$ a.e. This leads us to another question we have been avoiding, and that is, what functions are acceptable as state functions? Since they just satisfy (1.1.2), we must insist that they be square integrable on $(-\infty, \infty)$. Thus they must belong to the well-known space $L^2 = L^2(-\infty, \infty)$. Those of you who are unfamiliar with Lebesgue integration theory, do not despair. You can consider the integrations in the sense of Riemann without serious misgivings. However, you should keep in mind that Lebesgue integration has the following important advantage over that of Riemann. If there is a sequence $\{\psi_k\}$ of functions in L^2 such that

$$\|\psi_j - \psi_k\| \to 0 \qquad \text{as} \quad j, k \to \infty, \tag{1.5.7}$$

then there is a $\psi \in L^2$ such that

$$\|\psi_k - \psi\| \to 0 \qquad \text{as} \quad k \to \infty. \tag{1.5.8}$$

This property is called *completeness*. A sequence satisfying (1.5.7) is called a *Cauchy sequence*, while one satisfying (1.5.8) is said to *converge* to ψ. Completeness says that every Cauchy sequence in L^2 converges to some element in L^2.

We shall say that ψ_k converges to ψ in L^2 and write $\psi_k \to \psi$ when (1.5.8) holds. This is called *strong convergence* or *convergence in norm*. We shall say that ψ_k *converges weakly* to ψ if

$$(\psi_k, \varphi) \to (\psi, \varphi), \qquad \varphi \in L^2. \tag{1.5.9}$$

One should note that strong convergence implies weak convergence. This follows easily from the Schwarz inequality

$$|(f, g)| \leq \|f\| \, \|g\|. \tag{1.5.10}$$

In fact one has by (1.5.8) and (1.5.10)

$$|(\psi_k - \psi, \varphi)| \leq \|\psi_k - \psi\| \, \|\varphi\| \to 0.$$

Sometimes we shall write $\psi_k \rightharpoonup \psi$ to mean that (1.5.9) holds. A proof of (1.5.10) is simple. It is certainly true if $f = 0$. If not, set $\alpha = -(f, g)^*/\|f\|^2$. Then

$$\|f\|^2\|\alpha f + g\|^2 = \|f\|^2\|g\|^2 - |(f, g)|^2. \tag{1.5.11}$$

Since this is nonnegative, (1.5.10) must hold.

An important consequence of (1.5.10) is the *triangle inequality*

$$\|f + g\| \leq \|f\| + \|g\|. \tag{1.5.12}$$

It is proved by expanding $\|f + g\|^2$ by means of (1.5.4)–(1.5.6) and using (1.5.10). From (1.5.12) it follows that $\|\psi_k\| \to \|\psi\|$ whenever $\psi_k \to \psi$ strongly. This need not be the case when the convergence is weak. A subset S of L^2 is called *closed* if for every sequence of functions in S which converges in L^2, the limit is also in S.

A subset S of L^2 will be called *dense* if for every $\varepsilon > 0$ and every $\psi \in L^2$, there is a $\varphi \in S$ such that $\|\psi - \varphi\| < \varepsilon$. The following lemma will be useful later.

Lemma 1.5.2. *If S is dense in L^2 and*

$$(\psi, \varphi) = 0, \qquad \varphi \in S, \tag{1.5.13}$$

then $\psi = 0$.

The proof of Lemma 1.5.2 is simple; we leave it as an exercise. We end this section with the

PROOF OF LEMMA 1.5.1. Clearly (1.5.2) implies

$$\text{Im}(A\psi, \psi) = 0. \tag{1.5.14}$$

For by (1.5.5), $(A\psi, \psi) = (\psi, A\psi) = (A\psi, \psi)^*$. Conversely, (1.5.14) implies

$$\text{Re}(A\varphi, \psi) = \text{Re}(\varphi, A\psi). \tag{1.5.15}$$

In fact we have

$$(A[i\varphi + \psi], i\varphi + \psi) = (A\varphi, \varphi) + i(A\varphi, \psi) - i(A\psi, \varphi) + (A\psi, \psi).$$

Taking imaginary parts, we obtain (1.5.15). Next we note that by (1.5.15)

$$\begin{aligned} \text{Im}(A\varphi, \psi) &= \text{Im}(-i)(Ai\varphi, \psi) = -\text{Re}(Ai\varphi, \psi) \\ &= -\text{Re}(i\varphi, A\psi) = -\text{Re}\, i(\varphi, A\psi) = \text{Im}(\varphi, A\psi). \end{aligned}$$

Combining this with (1.5.15), we obtain (1.5.2). □

1.6. Operators

As we noticed in the beginning of Section 1.5 for each of the observables a which we have discussed so far, there is an operator A such that (1.5.1) holds. We wish to examine this a bit more carefully. Let us take, for example, the case of momentum. In this case $A = -i\hbar\,\partial/\partial x$ and

$$(A\psi, \psi) = -i\hbar\int_{-\infty}^{\infty} \psi_x(x, t)\psi(x, t)^*\,dx. \tag{1.6.1}$$

This integrand makes sense only if ψ is differentiable with respect to x and the expression is finite only if the integral exists. In other words, $D(A)$ does not consist of all $\psi \in L^2$. This is typical in quantum theory. However, for the theory to be reasonable, the number of state functions for which the observable can be measured must not be too small. This is indeed the case for all known observables. In fact the domains of the corresponding operators are dense in L^2. This will be assumed in the theory.

Another consideration deserves attention. The expression (1.6.1) makes sense and is finite for some ψ which are not differentiable everywhere. The restrictions on ψ necessary to ensure the existence and finiteness of (1.6.1) are not difficult to determine. We shall not be involved with this particular question at the moment, but we should note that it is important that the domain of A be as large as possible. Thus we shall assume that $D(A)$ is the largest domain for which (1.5.1) holds.

We summarize the ideas presented so far in this section by

Postulate 3. To every observable a there corresponds an operator A with dense domain such that

$$\bar{a} = (A\psi, \psi) \tag{1.6.2}$$

holds for each $\psi \in D(A)$. If B is a Hermitian operator such that $D(A) \subset D(B)$ and

$$\bar{a} = (B\psi, \psi)$$

whenever $\psi \in D(A)$, then $B = A$.

Note that $\bar{a}$ depends on ψ. Strictly speaking we should write it as $\bar{a}(\psi)$ or $\bar{a}_\psi$. When we write $B = A$ we mean that $D(B) = D(A)$ and $B\psi = A\psi$ there. As a consequence of Lemma 1.5.1 we have

Theorem 1.6.1. *If the observable a is real valued, then the corresponding operator is Hermitian.*

We should verify that the operators x, L, and H are indeed Hermitian. This is a simple matter, and we leave it as an exercise.

Let us make a few observations concerning domains of operators. Let A be an operator with a given domain $D(A)$. One simple way of attempting to extend A to a larger domain is as follows. Suppose that ψ is not in $D(A)$, but that there exists a sequence $\{\psi_n\} \subset D(A)$ such that $\psi_n \to \psi$ and $A\psi_n \to f$. Then we can define $A\psi$ to be f. Of course, this definition will make sense only if f does not depend on the sequence $\{\psi_n\}$. For if $\{\psi_n'\} \subset D(A)$ is such that $\psi_n' \to \psi$ and $A\psi_n' \to g$, we must have $g = f$. Or to put it another way, if $\{\varphi_n\} \subset D(A)$ satisfies $\varphi_n \to 0$ and $A\varphi_n \to w$, we must have $w = 0$ in order for this method to work. An operator having this property is called *closable* or *preclosed*. The extended operator is called its *closure* and is sometimes denoted by $\bar{A}$. It has the following property.

Lemma 1.6.1. *If $\{\psi_n\} \subset D(\bar{A})$, $\psi_n \to \psi$, $\bar{A}\psi_n \to f$, then $\psi \in D(\bar{A})$ and $\bar{A}\psi = f$.*

Thus $\bar{A}$ cannot be extended any further by the method outlined above. The proof of Lemma 1.6.1 is simple and is left as an exercise. An operator having the property described in the lemma is called *closed*. Thus the closure is the smallest closed extension of a closable operator and is uniquely determined by it.

Not all operators are closable, but we have

Lemma 1.6.2. *A Hermitian operator with dense domain is closable.*

PROOF. If $\varphi_n \in D(A)$, $\varphi_n \to 0$, and $A\varphi_n \to w$, then for each $h \in D(A)$,

$$(A\varphi_n, h) = (\varphi_n, Ah).$$

Taking the limit (see Section 1.5), we get $(w, h) = 0$ for all $h \in D(A)$. We now apply Lemma 1.5.2 to conclude that $w = 0$. □

This discussion now gives us

Theorem 1.6.2. *The operator A corresponding to a real-valued observable a is closed.*

PROOF. A is densely defined by Postulate 3 and Hermitian by Theorem 1.6.1. Thus it is closable by Lemma 1.6.2. $\bar{A}$ is Hermitian. For if $\{\psi_n\}$, $\{\varphi_n\} \subset D(A)$, $\psi_n \to \psi$, $\varphi_n \to \varphi$, $A\psi_n \to \bar{A}\psi$, $A\varphi_n \to \bar{A}\varphi$, then

$$(\bar{A}\psi, \varphi) = \lim(A\psi_n, \varphi_n) = \lim(\psi_n, A\varphi_n) = (\psi, \bar{A}\varphi).$$

We take $B = \bar{A}$ in Postulate 3 to conclude that $A = \bar{A}$. Hence it is closed by Lemma 1.6.1. □

1.7. Functions of Observables

By definition, an observable is any quantity that can be measured in physics. However, it is rare that any quantity is measured directly. It is usually determined by measuring other quantities which are related to it. Thus, when we use the term *measure* we mean "determined by means of measurement." Hence any function of an observable is also an observable. In particular a^2 is an observable for each observable a. If A is the operator corresponding to a, what operator should correspond to a^2? In view of (1.2.13) and (1.3.10) we expect it to be A^2. We state this as a postulate.

Postulate 4. A function of an observable is an observable. If A is the operator corresponding to a, then A^2 corresponds to a^2.

We should take care in defining A^2. We say that $\psi \in D(A^2)$ if ψ and $A\psi$ are both in $D(A)$. We define $A^2\psi$ to be $A(A\psi)$.

Now we come to a very important theoretical question. Let a be an observable that takes on real values, and let A be its corresponding Hermitian operator. Then (1.5.1) holds. As mentioned in Section 1.1 it may not be possible to determine certain things concerning the observable a. However, it is considered imperative in physics that for each interval I we should be able to compute the probability $P(a \in I)$ that the observable a be in the interval I. Moreover, this information should be determined by ψ and A. (An example of this is given by Postulate 1.) We now show how one can go about obtaining this information.

Let $\chi_I(\lambda)$ be the characteristic function of the interval I, that is,

$$\chi_I(\lambda) = \begin{cases} 1, & \lambda \in I \\ 0, & \lambda \notin I. \end{cases} \tag{1.7.1}$$

Since a is either in I or it is not, $\chi_I(a)$ can take on only the values 0 and 1. The probability that it takes on the value 0 is $P(a \notin I)$ and the probability that it takes on the value 1 is $P(a \in I)$. Since $\chi_I(a)$ is an observable, we have

$$\overline{\chi_I(a)} = 0 \cdot P(a \notin I) + 1 \cdot P(a \in I), \tag{1.7.2}$$

and consequently

$$P(a \in I) = \overline{\chi_I(a)}\,. \tag{1.7.3}$$

Let E_I denote the Hermitian operator corresponding to $\chi_I(a)$. Then

$$\overline{\chi_I(a)} = (E_I\psi, \psi). \tag{1.7.4}$$

Thus

$$P(a \in I) = (E_I\psi, \psi). \tag{1.7.5}$$

In a sense, this answers our question. The only problem is that we do not know what the operator E_I is, how it can be constructed, or even how it is related to A.

In order to gain some insight into the matter, let us examine the operator E_I and try to discover some of its properties. The first thing we note is

Lemma 1.7.1. $\|E_I\varphi\| \leq \|\varphi\|$, $\varphi \in D(E_I^2)$.

PROOF. We have by Postulate 4

$$\overline{\chi_I(a)^2} = (E_I^2\psi, \psi) = \|E_I\psi\|^2, \tag{1.7.6}$$

which shows that

$$\|E_I\psi\| \leq 1 \quad \text{for} \quad \|\psi\| = 1. \tag{1.7.7}$$

If $\varphi \neq 0$, set $\psi = \varphi/\|\varphi\|$. Then $\|\psi\| = 1$ and consequently $\|E_I\psi\| \leq 1$. This leads to the conclusion. □

We also note

Lemma 1.7.2. $D(E_I) = L^2$.

PROOF. We note first that $D(E_I^2)$ is dense since E_I^2 corresponds to an observable (Postulate 3). E_I is closed because it corresponds to a real observable (Theorem 1.6.2). Now let ψ be any element of L^2. Then there is a sequence $\{\psi_n\} \subset D(E_I^2)$ such that $\psi_n \to \psi$. By Lemma 1.7.1, $\|E_I(\psi_m - \psi_n)\| \leq \|\psi_m - \psi_n\| \to 0$. Thus there is an element $f \in L^2$ such that $E_I\psi_m \to f$ (completeness of L^2; see Section 1.5). By Lemma 1.6.1, $\psi \in D(E_I)$ and $E_I\psi = f$. □

Lemma 1.7.3. $E_I^2 = E_I$.

PROOF. Since we have $\chi_I(a)^2 = \chi_I(a)$, it follows from Postulate 4 that

$$(E_I^2\psi, \psi) = (E_I\psi, \psi), \quad \psi \in L^2. \tag{1.7.8}$$

This implies our result in view of the following lemma, which will be proved at the end of this section. □

Lemma 1.7.4. *If A is a densely defined Hermitian operator and*

$$(A\psi, \psi) = 0, \quad \psi \in D(A), \tag{1.7.9}$$

then $A = 0$, that is, $A\psi = 0$ for each $\psi \in D(A)$.

Another application of Lemma 1.7.4 gives

Theorem 1.7.1. *For each I, J,*

$$E_{I \cup J} = E_I + E_J - E_{I \cap J}. \tag{1.7.10}$$

PROOF. First we note that

$$\chi_{I\cup J}(a) = \chi_I(a) + \chi_J(a) - \chi_{I\cap J}(a). \tag{1.7.11}$$

In view of Theorem 1.4.1 this implies

$$(E_{I\cup J}\psi, \psi) = (E_I\psi, \psi) + (E_J\psi, \psi) - (E_{I\cap J}\psi, \psi).$$

Now we apply Lemma 1.7.4. □

An operator A on a Hilbert space is called *bounded* if there is a constant C such that

$$\|Au\| \leq C\|u\|, \qquad u \in D(A).$$

The smallest constant C that works is called the *norm* of A and is denoted by $\|A\|$. Lemma 1.7.1 states that E_I is bounded with norm ≤ 1. Note that bounded operators are closable. Another important property of E_I is

Theorem 1.7.2. *For each I, J,*

$$E_{I\cap J} = E_I E_J. \tag{1.7.12}$$

PROOF. Since

$$[\chi_I(a) + \chi_J(a)]^2 = \chi_I(a) + \chi_J(a) + 2\chi_{I\cap J}(a), \tag{1.7.13}$$

we have by Postulate 4

$$(E_I + E_J)^2 = E_I + E_J + 2E_{I\cap J}. \tag{1.7.14}$$

But the left-hand side of (1.7.14) equals $E_I + E_J + E_I E_J + E_J E_I$. Thus we have

$$2E_{I\cap J} = E_I E_J + E_J E_I. \tag{1.7.15}$$

Apply $E_I + E_J$ to both sides. This gives

$$2(E_I + E_J)E_{I\cap J} = 2E_{I\cap J} + E_J E_I E_J + E_I E_J E_I. \tag{1.7.16}$$

Assume for the moment that $I \cap J$ is empty. Then

$$E_J E_I E_J + E_I E_J E_I = 0. \tag{1.7.17}$$

But

$$([E_J E_I E_J + E_I E_J E_I]\psi, \psi) = \|E_I E_J \psi\|^2 + \|E_J E_I \psi\|^2.$$

Hence $E_I E_J = E_J E_I = 0$ when $I \cap J = \varnothing$ (the empty set). Next let I, J be arbitrary. Then we can write $I = (I \cap J) \cup I_1$, $J = (I \cap J) \cup J_1$, where $I_1 \cap J = J_1 \cap I = \varnothing$. Thus

$$E_I E_J = (E_{I\cap J} + E_{I_1})(E_{I\cap J} + E_{J_1}) = E_{I\cap J}^2,$$

which gives (1.7.12). □

If $I_1, \ldots, I_N$ are nonoverlapping bounded intervals and $c_1, \ldots, c_N$ are complex constants, we call

$$f(\lambda) = \sum c_k \chi_{I_k}(\lambda) \tag{1.7.18}$$

a *step function*. For such a function we define

$$f(A) = \sum c_k E_{I_k}. \tag{1.7.19}$$

First we note

Lemma 1.7.5. *If f, g are step functions and $h(\lambda) = f(\lambda)g(\lambda)$, then $h(A) = f(A)g(A)$.*

PROOF. If

$$f(\lambda) = \sum b_j \chi_{I_j}(\lambda), \qquad g(\lambda) = \sum c_k \chi_{J_k}(\lambda),$$

then

$$h(\lambda) = \sum b_j c_k \chi_{I_j \cap J_k}(\lambda).$$

Thus

$$f(A) = \sum b_j E_{I_j}, \qquad g(A) = \sum c_k E_{J_k}, \qquad h(A) = \sum b_j c_k E_{I_j \cap J_k}.$$

The result now follows from Theorem 1.7.2. □

By $I \to R$ we shall mean that I is a finite interval with lower endpoint tending to $-\infty$ and upper endpoint tending to ∞. For later considerations we shall need

Lemma 1.7.6. *If $I \to R$, then*

$$\overline{[1 - \chi_I(a)]} \to 0, \tag{1.7.20}$$

$$\overline{[1 - \chi_I(a)]^2} \to 0, \tag{1.7.21}$$

$$\overline{\chi_I(a)} \to 1, \tag{1.7.22}$$

$$\overline{a\chi_I(a)} \to \bar{a}, \tag{1.7.23}$$

$$\overline{a^2\chi_I(a)} \to \overline{a^2}, \tag{1.7.24}$$

$$\overline{[a^2 - a^2\chi_I(a)]} \to 0, \tag{1.7.25}$$

$$\overline{[a - a\chi_I(a)]^2} \to 0. \tag{1.7.26}$$

PROOF. If $I \supset [-N, N]$, then

$$N^2[1 - \chi_I(a)]^2 \le a^2. \tag{1.7.27}$$

Thus, if $\psi \in D(A)$,

$$N^2\|(1 - E_I)\psi\|^2 \le \|A\psi\|^2. \tag{1.7.28}$$

Letting $N \to \infty$, we get (1.7.20) for such ψ. Let ψ be any element of L^2 and let $\varepsilon > 0$ be given. Since $D(A)$ is dense in L^2, there is a $\psi_1 \in D(A)$ such that $\|\psi - \psi_1\| < \varepsilon/4$. Thus,

$$\|(1 - E_I)\psi\| \le \|(1 - E_I)(\psi - \psi_1)\| + \|(1 - E_I)\psi_1\|. \tag{1.7.29}$$

The first term on the right is less than $\frac{1}{2}\varepsilon$ while the second tends to 0 as $I \to R$. Since this is true for every ε, we have

$$\|(1 - E_I)\psi\| \to 0 \qquad \text{as} \quad I \to R \tag{1.7.30}$$

for each $\psi \in L^2$. This gives (1.7.20). Since $1 - \chi_I(a)$ is a characteristic function, (1.7.21) follows. Theorem 1.4.1 gives (1.7.22) as well. Next we note that (1.7.23) is merely the definition of $\bar{a}$. To verify (1.7.24), let $I = [-N, N]$ and $J = [-N^2, N^2]$. Then

$$\overline{a^2\chi_I(a)} = \overline{a^2\chi_J(a^2)} \to \overline{a^2}$$

by (1.7.23). The remaining two statements follow from Theorem 1.4.1. □

We have discovered several properties of the projections E_I. We shall continue our investigation in the next section. Meanwhile we end this section by giving the

PROOF OF LEMMA 1.7.4. First we claim that

$$(A\psi, \varphi) = 0 \qquad \text{for all} \quad \psi, \varphi \in D(A). \tag{1.7.31}$$

To see this note that

$$(A[\psi + \varphi], \psi + \varphi) = (A\psi, \psi) + (A\varphi, \varphi) + 2\,\mathrm{Re}(A\psi, \varphi).$$

This shows that $\mathrm{Re}(A\psi, \varphi) = 0$ for every φ, ψ. The same is true for $\mathrm{Re}(Ai\psi, \varphi) = -\mathrm{Im}(A\varphi, \psi)$. This gives (1.7.31). Once we know this, we can apply Lemma 1.5.2. □

1.8. Self-Adjoint Operators

In Section 1.7 we let E_I denote the operator corresponding to the function $\chi_I(a)$, where a is an observable with corresponding operator A. We determined some of the properties of the operators E_I, but we still do not know how to define them in terms of A. Moreover, we do not even know whether they can be defined for every densely defined closed Hermitian operator A.

In order to answer these questions we draw some more consequences from our postulates.

We shall prove

Theorem 1.8.1. *If A is an operator corresponding to a real observable a, then for each $f \in L^2$ we can find a $u \in D(A^2)$ such that $u + A^2u = f$.*

In other words the operator $1 + A^2$ is onto. We should note that for any two operators A, B, $D(A + B) = D(A) \cap D(B)$ with the obvious definition for $A + B$ on $D(A + B)$. Thus $D(1 + A^2) = D(A^2)$.

Theorem 1.8.1 hints that not every densely defined closed Hermitian operator can serve as the operator corresponding to an observable. We note some consequences.

Let f be any function in L^2. Theorem 1.8.1 tells us that there is a $u \in D(A^2)$ such that

$$(A - i)(A + i)u = (A + i)(A - i)u = f. \tag{1.8.1}$$

Since $(A \pm i)u \in D(A)$, we have

Corollary 1.8.1. *For each $f \in L^2$, there are $v, w \in D(A)$ such that*

$$(A - i)v = (A + i)w = f. \tag{1.8.2}$$

This leads us to the important

Theorem 1.8.2. *Let a be a real-valued observable with corresponding operator A. Suppose that $\psi, f \in L^2$ are such that*

$$(\psi, A\varphi) = (f, \varphi), \qquad \varphi \in D(A). \tag{1.8.3}$$

Then $\psi \in D(A)$ and $A\psi = f$.

PROOF. By Corollary 1.8.1, there is a $w \in D(A)$ such that $(A - i)w = f - i\psi$. Thus, for $\varphi \in D(A)$

$$(\psi, (A + i)\varphi) = (f - i\psi, \varphi) = ([A - i]w, \varphi) = (w, [A + i]\varphi).$$

Again by Corollary 1.8.1 there is a $\varphi \in D(A)$ such that $(A + i)\varphi = \psi - w$. This shows that $\psi = w \in D(A)$. The rest follows from the fact that A is Hermitian and densely defined. □

The property of A described in Theorem 1.8.2 is not shared by all densely defined closed Hermitian operators. Such operators having this property are called *self-adjoint*. The reason for this terminology is as follows. For any densely defined operator A on L^2 we can define an operator A^* by stipulating that $\psi \in D(A^*)$ and $A^*\psi = f$ if (1.8.3) holds. This definition makes sense only because $D(A)$ is dense. Otherwise, there might be more than one function ψ satisfying (1.8.3). Clearly, A^* is a linear operator. Moreover, when A is Hermitian, $D(A) \subset D(A^*)$ and A^* coincides with A on $D(A)$. For if $\psi \in D(A)$, it satisfies (1.8.3) with $f = A\psi$. When A is self-adjoint, we have $A^* = A$. This means that their domains are the same and they coincide on their common domain. This is precisely the property described in Theorem 1.8.2. Thus we can rephrase this theorem as

Theorem 1.8.3. *If a is a real-valued observable, then the corresponding operator is self-adjoint.*

Now we give an example of a densely defined closed Hermitian operator that is not self-adjoint. Let B be the operator defined by

$$B\psi = i\, d\psi/dx,$$

with $D(B)$ consisting of continuously differentiable functions which vanish when $|x|$ is large and $x = 0$. Since

$$\int_{-\infty}^{\infty} \psi(x)[i\varphi'(x)]^* \, dx = \int_{-\infty}^{\infty} i\psi'(x)\varphi(x)^* \, dx \tag{1.8.4}$$

for $\psi, \varphi \in D(B)$, we see that B is Hermitian. We shall show in Section 7.8 that $D(B)$ is dense in L^2. Let $\bar{B}$ denote its closure (see Section 1.6). Then $\bar{B}$ is a densely defined closed Hermitian operator. It will now be shown that continuous functions in $D(\bar{B})$ must vanish for $x = 0$. If this is so, then $\bar{B}$ cannot be self-adjoint. For if ψ is any continuously differentiable function vanishing for $|x|$ large, then (1.8.4) holds for all $\varphi \in D(B)$. This implies

$$(\psi, \bar{B}\varphi) = (i\psi', \varphi), \qquad \varphi \in D(\bar{B}). \tag{1.8.5}$$

Thus $\psi \in D(\bar{B}^*)$ whether or not it vanishes at $x = 0$. But it cannot be in $D(\bar{B})$ unless it vanishes there. Thus $\bar{B}^* \neq \bar{B}$.

The assertion that functions in $D(\bar{B})$ must vanish at $x = 0$ follows from the inequality

$$|\psi(x)|^2 \leq \|\psi'\|^2 + 2\|\psi\|^2, \qquad -\infty < x < \infty, \tag{1.8.6}$$

which holds for all continuously differentiable ψ which vanish for large $|x|$. In fact, if $\psi \in D(\bar{B})$, there is a sequence $\{\psi_n\} \subset D(B)$ such that $\psi_n \to \psi$, $B\psi_n \to \bar{B}\psi$ in L^2. Thus by (1.8.6),

$$|\psi_n(x) - \psi_m(x)|^2 \leq \|B(\psi_n - \psi_m)\|^2 + 2\|\psi_n - \psi_m\|^2 \to 0.$$

Thus the ψ_n converge uniformly. The limit must coincide with ψ. Since each of the ψ_n vanishes at $x = 0$, the same must be true of ψ.

It remains to prove (1.8.6). It is a consequence of

Lemma 1.8.1. *Let I be an interval of length* 1, *and let $u(x)$ be continuously differentiable in I. Then for every $\varepsilon > 0$*

$$|u(x)|^2 \leq \varepsilon \int_I |u'(y)|^2 \, dy + (1 + \varepsilon^{-1}) \int_I |u(y)|^2 \, dy, \qquad x \in I. \tag{1.8.7}$$

PROOF. By separating out real and imaginary parts, we need only consider the case when u is real valued. Let x, x' be any two points in I. Then

$$u(x)^2 - u(x')^2 = 2\int_{x'}^{x} u(y)u'(y)\,dy$$

$$\leq \varepsilon\int_I u'(y)^2\,dy + \varepsilon^{-1}\int_I u(y)^2\,dy, \tag{1.8.8}$$

where we have used the inequality $2ab \leq \varepsilon a^2 + \varepsilon^{-1}b^2$. We pick x' so that

$$u(x')^2 = \int_I u(y)^2\,dy. \tag{1.8.9}$$

We can do this by the theorem of the mean. Inequality (1.8.7) now follows from (1.8.8) and (1.8.9). □

1.9. Hilbert Space

A Hilbert space is a vector space having a scalar product satisfying (1.5.3)–(1.5.6) and which is complete with respect to the norm given by (1.5.6). A very important example is the space L^2 described in Section 1.5. Many of the theorems we shall prove are true in general Hilbert spaces, and it will be convenient to state them in such a context. On the other hand, anyone who prefers to consider everything within the framework of L^2 can do so without any difficulty.

A subset V of a Hilbert space $\mathcal{H}$ is called a subspace if $\alpha u + \beta v$ are in V for all $u, v \in V$ and all scalars α, β. For an arbitrary operator A we let the *range* $R(A)$ of A be the set of all elements of the form Au, where $u \in D(A)$. If A is a linear operator, then both $D(A)$ and $R(A)$ are subspaces (see Section 1.5).

In this section we shall prove Theorem 1.8.1. Our proof will be based upon a representation theorem due to F. Riesz. We describe this theorem here; a proof of it will be given in Appendix B. An operator F from a Hilbert space $\mathcal{H}$ to the scalars (complex numbers) is called a *functional*. We call it *bounded* if

$$|Fu| \leq C\|u\|, \qquad u \in \mathcal{H}, \tag{1.9.1}$$

holds for some constant C. The smallest such constant (there is one) is called the *norm* of F and is denoted by $\|F\|$. The representation theorem of Riesz can be stated as follows.

Theorem 1.9.1. *Let F be a bounded linear functional defined everywhere on a Hilbert space $\mathcal{H}$. Then there exists an element $f \in \mathcal{H}$ such that $\|f\| = \|F\|$ and*

$$Fu = (u, f), \qquad u \in \mathcal{H}. \tag{1.9.2}$$

In other words, the theorem states that every bounded linear functional on a Hilbert space is merely a scalar product. We shall use this theorem to prove.

Theorem 1.9.2. *Let B be a densely defined operator on a Hilbert space such that*

$$\|u\|^2 \le (Bu, u), \qquad u \in D(B). \tag{1.9.3}$$

Then there is an extension C of B such that

$$\|u\|^2 \le (Cu, u), \qquad u \in D(C), \tag{1.9.4}$$

and $R(C) = \mathcal{H}$.

PROOF. By Lemma 1.5.1, B is a Hermitian operator. Thus if we set

$$b(u, v) = (Bu, v), \qquad b(u) = b(u, u), \qquad u, v \in D(B), \tag{1.9.5}$$

we see that $b(u, v)$ is a scalar product on $D(B)$ [i.e., it satisfies (1.5.3)–(1.5.6)]. We shall extend b to a larger subspace of $\mathcal{H}$. Let M be the set of those $u \in \mathcal{H}$ for which there exists a sequence $\{u_n\}$ of elements of $D(B)$ such that

$$u_n \to u \text{ in } \mathcal{H} \qquad \text{and} \qquad b(u_n - u_m) \to 0 \text{ as } m, n \to \infty. \tag{1.9.6}$$

Clearly, M is a subspace. For each $u \in M$ we call a sequence $\{u_n\} \subset D(B)$ satisfying (1.9.6) an *admitting sequence* for u. Note that $b(u_n)$ is bounded for an admitting sequence. Clearly, each element of M may have more than one admitting sequence. But if $u, v \in M$ have admitting sequences $\{u_n\}, \{v_n\}$, then

$$c(u, v) = \lim_{n\to\infty} b(u_n, v_n) \tag{1.9.7}$$

exists and is independent of the particular admitting sequences. To show that the limit exists, note that

$$b(u_n, v_n) - b(u_m, v_m) = b(u_n, v_n - v_m) + b(u_n - u_m, v_m)$$
$$\le b(u_n)^{1/2} b(v_n - v_m)^{1/2} + b(u_n - u_m)^{1/2} b(v_m)^{1/2} \to 0.$$

To show that the limits are independent of the admitting sequences, suppose $\{u_n'\}, \{v_n'\}$ are others for u, v. Set $g_n = u_n - u_n'$, $h_n = v_n - v_n'$. Then there is a constant K such that

$$b(g_n) + b(h_n) \le K^2.$$

Moreover,

$$b(g_n - g_m) + b(h_n - h_m) \to 0.$$

Let $\varepsilon > 0$ be given. Then there is an N so large that

$$b(g_n - g_m) \le \varepsilon^2 / K^2, \qquad m, n > N.$$

Now

$$b(g_n) = b(g_n, g_n - g_m) + (Bg_n, g_m)$$
$$\leq b(g_n)^{1/2} b(g_n - g_m)^{1/2} + \|Bg_n\| \, \|g_m\|$$
$$\leq \varepsilon + \|Bg_n\| \, \|g_m\|$$

for $m, n > N$. Let $m \to \infty$. This gives

$$b(g_n) \leq \varepsilon \qquad \text{for} \quad n > N.$$

Since ε was arbitrary, we see that $b(g_n) \to 0$ as $n \to \infty$. Similarly, $b(h_n) \to 0$. Thus we have

$$b(u_n, v_n) - b(u_n', v_n') = b(g_n, v_n) + b(u_n', h_n)$$
$$\leq b(g_n)^{1/2} b(v_n)^{1/2} + b(u_n')^{1/2} b(h_n)^{1/2} \to 0.$$

Next we note that $c(u, v)$ defined by (1.9.7) is a scalar product on M. This follows from the fact that $b(u, v)$ is a scalar product. Note that

$$c(u) = \lim b(u_n) \geq \lim \|u_n\|^2 = \|u\|^2. \tag{1.9.8}$$

Thus $c(u) = 0$ implies $u = 0$. Also we note that if $u \in M$, then it has an admitting sequence $\{u_n\}$ such that

$$\|u_n - u\| < 1/n, \qquad b(u_n - u_m) \leq (1/n) + (1/m). \tag{1.9.9}$$

Such a sequence can be obtained by taking a subsequence of any admitting sequence for u. For a sequence satisfying (1.9.9) we have

$$c(u_n - u) = \lim_{m\to\infty} b(u_n - u_m) \leq 1/n. \tag{1.9.10}$$

Finally, we claim that M is a Hilbert space with scalar product $c(u, v)$. Everything but completeness has been verified. To show this, suppose $\{u_j\}$ is a sequence in M such that $c(u_j - u_k) \to 0$. By (1.9.8), $\{u_k\}$ is a Cauchy sequence in $\mathcal{H}$, and hence it converges to an element u in $\mathcal{H}$. By (1.9.10) there is a $w_j \in D(B)$ such that $c(w_j - u_j) \leq 1/j$. Thus

$$\|w_j - u\| \leq \|w_j - u_j\| + \|u_j - u\| \to 0 \text{ and}$$

$$b(w_j - w_k)^{1/2} \leq c(w_j - u_j)^{1/2} + c(u_j - u_k)^{1/2} + c(u_k - w_k)^{1/2}. \tag{1.9.11}$$

This converges to 0 as $j, k \to 0$. Thus $\{w_j\}$ is an admitting sequence for u, and u is in M. Moreover, by (1.9.11)

$$c(w_j - u)^{1/2} = \lim_{k\to\infty} b(w_j - w_k)^{1/2} \leq \lim_{k\to\infty} c(u_j - u_k)^{1/2} + 1/j \to 0$$
$$\text{as} \quad j \to \infty.$$

Consequently

$$c(u_j - u)^{1/2} \leq c(u_j - w_j)^{1/2} + c(w_j - u)^{1/2} \to 0. \tag{1.9.12}$$

Now we define the operator C as follows. We shall say that $u \in D(C)$ and $Cu = f$ if $u \in M$ and

$$c(u, v) = (f, v), \qquad v \in M. \tag{1.9.13}$$

This definition makes sense since only one f can satisfy (1.9.13). For if f' were another, we would have $(f - f', v) = 0$ for all $v \in D(B)$, a dense set. Thus $f' = f$ by Lemma 1.5.2. Moreover, if $u \in D(B)$, then $c(u, v) = b(u, v) = (Bu, v)$, $v \in D(B)$. This also holds for all $v \in M$ by taking limits of admitting sequences. Thus $u \in D(C)$ and $Cu = Bu$. Hence C is an extension of B. It satisfies (1.9.4) since

$$\|u\|^2 \leq c(u) \leq (Cu, u)$$

by (1.9.8). It remains to show that $R(C) = \mathcal{H}$. Let f be any element of $\mathcal{H}$. Then $Fv = (v, f)$ is a linear functional in M. It is bounded since $|(v, f)| \leq \|v\| \, \|f\| \leq c(v)^{1/2}\|f\|$. Since M is a Hilbert space, there is a $u \in M$ such that

$$Fv = c(v, u), \qquad v \in M$$

(Theorem 1.9.1). This says that $u \in D(C)$ and $Cu = f$. Thus $f \in R(C)$, and the proof is complete. □

Now we give the

PROOF OF THEOREM 1.8.1. Put $B = 1 + A^2$. It satisfies the hypotheses of Theorem 1.9.2. Thus it has an extension C with $R(C) = L^2$ satisfying (1.9.4). In particular, C is Hermitian (Lemma 1.5.1). But $B - 1$ is an operator corresponding to an observable (Postulate 4). Hence $B - 1 = C - 1$ (Postulate 3). Thus, $B = C$. □

The following two theorems follow immediately from the proof of Theorem 1.8.2.

Theorem 1.9.3. *If A is a Hermitian densely defined operator and $R(1 + A^2)$ is the whole space, then A is self-adjoint.*

Theorem 1.9.4. *If A is a Hermitian densely defined operator and $R(A \pm i)$ are both the whole space, then A is self-adjoint.*

1.10. The Spectral Theorem

In Section 1.8 we showed that the operators corresponding to real-valued observables must be self-adjoint. This resulted from the requirement that such an operator have as large a domain as possible (Postulate 3) and that a function of an observable is an observable (Postulate 4). In particular, the operators E_I corresponding to the functions $\chi_I(a)$ should exist and all of the statements of Sections 1.7 and 1.8 should hold. At the moment we know that in order for these things to be true, A must be self-adjoint. However, as far as we know, other requirements may be necessary in order to make them true—self-adjointness may not be enough. However, it turns out that no further requirements are needed. If A is self-adjoint, there are operators with all of the properties mentioned in Sections 1.7 and 1.8 (and more). This result is far from trivial and is known as the spectral theorem. We state the theorem here. For a proof we refer to Akhiezer and Glazman (1961). An operator E is called an *orthogonal projection* if it is bounded, Hermitian, and satisfies $E^2 = E$. We have

Theorem 1.10.1 (Spectral Theorem). *Let A be a self-adjoint operator on a Hilbert space $\mathcal{H}$. Then there is a family $\{E(\lambda)\}$ of orthogonal projections depending on a real parameter λ such that:*

a. *If $\lambda_1 < \lambda_2$, then $E(\lambda_1)E(\lambda_2) = E(\lambda_1)$.*
b. *For each $\varepsilon > 0$, $u \in \mathcal{H}$, and real λ, $E(\lambda + \varepsilon)u \to E(\lambda)u$ as $\varepsilon \to 0$.*
c. *For each $u \in \mathcal{H}$, $E(\lambda)u \to 0$ as $\lambda \to -\infty$ and $E(\lambda)u \to u$ as $\lambda \to \infty$.*
d. *$u \in D(A)$ if and only if (iff) $\int_{-\infty}^{\infty} \lambda^2 \, d\|E(\lambda)u\|^2 < \infty$.*
e. *For $u \in D(A)$ and $v \in \mathcal{H}$,*

$$(Au, v) = \int_{-\infty}^{\infty} \lambda \, d(E(\lambda)u, v).$$

f. *For $f(\lambda)$ any complex-valued function, the operator*

$$f(A) = \int_{-\infty}^{\infty} f(\lambda) \, dE(\lambda)$$

is defined on the set $D[f(A)]$ consisting of those $u \in \mathcal{H}$ such that

$$\int_{-\infty}^{\infty} |f(\lambda)|^2 \, d\|E(\lambda)u\|^2 < \infty.$$

g. *For $u \in D[f(A)]$, $v \in D[g(A)]$, $(f(A)u, g(A)v) = \int_{-\infty}^{\infty} f(\lambda)g(\lambda)^* \, d(E(\lambda)u, v)$.*
h. *$f(A)^* = \int_{-\infty}^{\infty} f(\lambda)^* \, dE(\lambda)$.*
i. *If $h(\lambda) = f(\lambda) + g(\lambda)$, then $h(A)$ is an extension of $f(A) + g(A)$.*
j. *If $h(\lambda) = f(\lambda)g(\lambda)$, then $h(A)$ is an extension of $f(A)g(A)$.*
k. *If $u \in D[g(A)]$, $|f_k(\lambda)| \le g(\lambda)$, and $f_k(\lambda) \to f(\lambda)$ for each λ, then $f_k(A)u \to f(A)u$.*
l. *If I is the interval $(-\infty, \lambda]$, then $E(\lambda) = \chi_I(A)$.*

Once we have the spectral theorem we have no problem finding the operator corresponding to a function of an observable. For if a is an observable and A is its corresponding operator, then the operator corresponding to $f(a)$ is $f(A)$ given by the spectral theorem. In particular, the projection operator E_I is $\chi_I(A)$. Sometimes we shall use the notation $E(I)$

in place of E_I. It follows from the spectral theorem that the E_I have all of the properties mentioned in Sections 1.7 and 1.8. The set of operators $\{E(\lambda)\}$ is called the *spectral family* of A.

Exercises

1. Prove (1.5.11). Where did the value of α come from?

2. Prove Lemma 1.5.2.

3. Prove Lemma 1.6.1.

4. Find E_I when $A = x$.

5. Find E_I when $A = p$.

6. If $f(\lambda)$ is a step function, show that

$$\|f(A)\psi\| \leq \max|f| \, \|E_I\psi\|,$$

where I is the interval outside of which $f(\lambda)$ vanishes.

7. Show that (1.7.27) implies (1.7.28).

8. Show that (1.8.4) implies (1.8.5).

9. Show that $\|\psi_k\| \to \|\psi\|$ if $\psi_k \to \psi$ strongly. Give an example in which $\psi_k \to \psi$ weakly but $\|\psi_k\|$ does not converge to $\|\psi\|$.

10. Show that M defined in the proof of Theorem 1.9.2 is a subspace.

11. Prove (1.5.12).

12. Show that there is a smallest constant C such that (1.9.1) holds.

13. Prove Theorems 1.9.3 and 1.9.4.

2
The Spectrum

2.1. The Resolvent

In this chapter we shall apply the spectral theorem to various questions concerning observables. The first general question we shall ask is, what values can an observable assume? Is it possible for an observable to attain any value in its range of definition, or are there restrictions? Obviously, the answer to this question will depend on both the corresponding operator and on the state function.

We begin with some definitions. A complex number z is in the *resolvent set* $\rho(A)$ of a closed operator A on a Hilbert space $\mathcal{H}$ if there is a bounded operator B on $\mathcal{H}$ such that

$$(z - A)Bu = u, \quad u \in \mathcal{H}, \qquad B(z - A)v = v, \quad v \in D(A). \tag{2.1.1}$$

The operator B depends on z and is called the *resolvent operator* of A. It is sometimes denoted by $R(z)$. Points not in $\rho(A)$ are in the spectrum $\sigma(A)$ of A. Our first consequence of the spectral theorem is

Theorem 2.1.1. *If A is self-adjoint, then all nonreal numbers are in $\rho(A)$.*

PROOF. If z is not real, $g(\lambda) = (z - \lambda)^{-1}$ is a bounded continuous function on the real line. Hence $g(A)$ is a bounded operator defined everywhere. Moreover, $B = g(A)$ satisfies (2.1.1) (Theorem 1.10.1). □

Theorem 2.1.2. *If I is an open interval in the real line and $E_I = \chi_I(A) = 0$, then $I \subset \rho(A)$.*

PROOF. Let λ_0 be any point of I. Its distance from the ends of I is positive. Set $g(\lambda) = (\lambda_0 - \lambda)^{-1}$ for $\lambda \notin I$, $g(\lambda) = 0$ for $\lambda \in I$. Then $g(\lambda)$ is piecewise continuous and bounded on the real line. Thus $g(A)$ is a bounded operator defined everywhere. Now $g(\lambda)(\lambda_0 - \lambda) = 1 - \chi_I(\lambda)$. Thus

$$(\lambda_0 - A)g(A) = 1 - \chi_I(A) = 1 \qquad \text{on} \quad \mathcal{H}$$

and

$$g(A)(\lambda_0 - A) = 1 - \chi_I(A) = 1 \qquad \text{on} \quad D(A).$$

Hence $\lambda_0 \in \rho(A)$. Since λ_0 was any point in I, the result follows. □

Theorem 2.1.3. *If $\lambda_0 \in \rho(A)$, then there is an open interval I containing λ_0 such that $\chi_I(A) = 0$.*

PROOF. If not, there would be a sequence $\{I_n\}$ of open intervals with lengths tending to 0, each containing λ_0 and such that $E_{I_n} \neq 0$. Hence we can find a $u_n \in \mathcal{H}$ such that $\|u_n\| = 1$ and $E_{I_n}u_n = u_n$. Thus

$$\|(\lambda_0 - A)u_n\| = \|(\lambda_0 - A)E_{I_n}u_n\| \leq \sup_{\lambda \in I_n} |\lambda_0 - \lambda| \to 0 \qquad \text{as} \quad n \to \infty$$

by Theorem 1.10.1. This is impossible, since it would imply

$$\|u_n\| = \|R(\lambda_0)(\lambda_0 - A)u_n\| \leq \|R(\lambda_0)\| \, \|(\lambda_0 - A)u_n\| \to 0,$$

contradicting the fact that $\|u_n\| = 1$. □

Corollary 2.1.1. *$\rho(A)$ is an open set.*

PROOF. Suppose $z \in \rho(A)$. If it is not real, then it is the center of a small disk containing no real points. By Theorem 2.1.1 this disk is in $\rho(A)$. If z is real, it is the center of an interval contained in $\rho(A)$ (Theorems 2.1.2 and 2.1.3). This interval is the diameter of a disk contained in $\rho(A)$. □

Corollary 2.1.2. *$\sigma(A)$ is a closed set.*

Corollary 2.1.3. *An observable can assume values only in the spectrum of its corresponding operator.*

PROOF. Let a be an observable with corresponding operator A. If $\lambda_0 \in \rho(A)$, then there is an open interval I containing λ_0 such that $\chi_I(A) = 0$ (Theorem 2.1.3). Thus

$$P(a \in I) = (\chi_I(A)\psi, \psi) = 0$$

for all state functions ψ. Thus a cannot take on values in I for any state function. □

Corollary 2.1.3 shows why the spectrum of an operator is so important. We shall be examining operators arising in quantum theory, and one of the first things we shall want to determine is their spectra. Thus we shall need criteria for determining the spectrum of an operator, which will be treated in the next section.

2.2. Finding the Spectrum

In this section we shall prove a very useful theorem for finding the spectrum of a self-adjoint operator. In giving the proof we shall need the following lemma, the proof of which we leave as an exercise.

Lemma 2.2.1. *A closed subspace of a Hilbert space is a Hilbert space.*

The theorem is

Theorem 2.2.1. *If A is self-adjoint, then a real number λ is in $\sigma(A)$ iff there is a sequence $\{u_n\}$ of elements in $D(A)$ such that*

$$\|u_n\| = 1 \qquad \text{and} \qquad \|(\lambda - A)u_n\| \to 0 \qquad \text{as} \qquad n \to \infty. \tag{2.2.1}$$

PROOF. Suppose (2.2.1) holds. If λ were in $\rho(A)$, we would have $u_n = R(\lambda)(\lambda - A)u_n \to 0$ by the second part of (2.2.1). This contradicts the first part. Conversely, suppose (2.2.1) does not hold. Then there is a constant C such that

$$\|u\| \leq C\|(\lambda - A)u\|, \qquad u \in D(A). \tag{2.2.2}$$

For if (2.2.2) did not hold, there would be a sequence $\{v_n\} \subset D(A)$ such that

$$\|(\lambda - A)v_n\|^{-1}\|v_n\| \to \infty \qquad \text{as} \quad n \to \infty.$$

Set $u_n = v_n/\|v_n\|$. Then u_n satisfies (2.2.1). Thus (2.2.2) holds. This inequality tells us two things. The first is that for each $f \in R(\lambda - A)$ there is only one $u \in D(A)$ such that $(\lambda - A)u = f$. For if there were two, their difference v would satisfy $(\lambda - A)v = 0$. But v would have to vanish by (2.2.2). The second bit of information we can obtain from (2.2.2) is that $R(\lambda - A)$ is closed. For if $\{f_n\}$ is a sequence of elements in $R(\lambda - A)$ and $f_n \to f$ in $\mathcal{H}$, let u_n be the unique solution of $(\lambda - A)u_n = f_n$. By (2.2.2), the u_n forms a Cauchy sequence. Thus they converge to some element $u \in \mathcal{H}$. Thus if $v \in D(A)$ we have

$$(u, (\lambda - A)v) = \lim(u_n, (\lambda - A)v) = \lim(f_n, v) = (f, v).$$

Hence $u \in D(A)$ and $(\lambda - A)u = f$. Once we know this, we can show that $R(\lambda - A)$ is the whole space $\mathcal{H}$. This in turn implies that $\lambda \in \rho(A)$. For we can define $R(\lambda)f$ to be the solution of $(\lambda - A)u = f$. It clearly satisfies (2.1.1) and it is bounded by (2.2.2). Thus it remains only to show that $R(\lambda - A) = \mathcal{H}$. Let f be any element in $\mathcal{H}$, and let w be any element in $R(\lambda - A)$. Let v be the unique element of $D(A)$ such that $(\lambda - A)v = w$. Set $Fw = (v, f)$. This defines a linear functional on $R(\lambda - A)$, which is a

Hilbert space by Lemma 2.2.1 (see Section 1.9). F is also bounded. For we have by (2.2.2)

$$|Fw| \leq \|v\| \, \|f\| \leq C\|f\| \, \|w\|.$$

By the Riesz representation theorem (Theorem 1.9.1), there is a $u \in R(\lambda - A)$ such that $Fw = (w, u)$ for all $w \in R(\lambda - A)$. This gives

$$(u, (\lambda - A)v) = (f, v), \qquad v \in D(A). \tag{2.2.3}$$

Since A is self-adjoint, this implies that $u \in D(A)$ and $(\lambda - A)u = f$. Hence $f \in R(\lambda - A)$, and the proof is complete. □

An operator A is said to be *one-to-one* or *injective* if the only $u \in D(A)$ satisfying $Au = 0$ is $u = 0$. The proof of Theorem 2.2.1 also proves

Corollary 2.2.1. *Self-adjoint operators are closed.*

Corollary 2.2.2. *If A is a closed operator and*

$$\|v\| \leq C\|Av\|, \qquad v \in D(A), \tag{2.2.4}$$

then A is one-to-one and $R(A)$ is closed.

A number λ is an *eigenvalue* of A if there is a nonvanishing $u \in D(A)$ such that $(\lambda - A)u = 0$. We call u an *eigenelement* (or in the case of L^2, an *eigenfunction*) of A corresponding to λ. An immediate consequence of Theorem 2.2.1 is

Corollary 2.2.3. *An eigenvalue is in $\sigma(A)$.*

Now that we have criteria for determining the spectrum of an operator, we turn our attention to the operators that we have encountered so far in quantum theory.

2.3. The Position Operator

By Theorem 1.2.2, the operator corresponding to position is multiplication by x. However, we did not specify the domain of this operator. Needless to say, an operator is not defined unless its domain is specified. Another important point to keep in mind is that the operators corresponding to real observables are required to be self-adjoint. Even a superficial glance at the definition will convince one that the slightest change in the domain can spoil self-adjointness. Thus we must define the domain of the operators and verify that they are indeed self-adjoint. For $A = x$, this is not a particularly difficult task. The simplest domain one can choose is the set of those $\psi \in L^2$ such that $x\psi \in L^2$ (this is also the largest). Clearly A is Hermitian on this domain. It is also easy to verify that it is self-adjoint if we use Theorem 1.9.4. For if f is any function in L^2, then $\psi = f/(x + i)$ is in $D(A)$ and $(A + i)\psi = f$. Thus $R(A + i) = L^2$, and similarly $R(A - i) = L^2$. Moreover, $D(A)$ is dense in L^2. For if f is any element in L^2, then $\psi_\varepsilon = f/(\varepsilon x^2 + 1)$ is in $D(A)$ for every $\varepsilon > 0$. But $|\psi_\varepsilon - f| \leq \varepsilon x^2|f|/(1 + \varepsilon x^2)$. This is sufficient to show that $\psi_\varepsilon \to f$ in L^2 as $\varepsilon \to 0$. Thus $D(A)$ is dense in L^2. We can now apply Theorem 1.9.4 to conclude that A is self-adjoint.

We turn to the question of determining the spectrum of A. The easiest points to look for are eigenvalues. Thus, we search for solutions of $(\lambda - x)\psi = 0$. Such an equation can hold only if $\psi(x) = 0$ for $x \neq \lambda$. Thus any solution would have to vanish a.e. As far as we are concerned, this means $\psi = 0$. Thus, there are no eigenvalues. We are forced to use Theorem 2.2.1. It is difficult to motivate at this time the selection of a sequence satisfying (2.2.1). Let λ be any real number. Any eigenfunction of x with respect to λ would have to vanish for $x \neq \lambda$, and therefore it could not have norm 1. We shall try to select a sequence of functions which converge to 0 for $x \neq \lambda$, but whose norms remain 1. This is easily done. But we also want $\lambda - x$ applied to these functions to converge to 0. We try the sequence

$$\psi_n(x) = c_n e^{-n^2(x-\lambda)^2/2},$$

where c_n is chosen so that $\|\psi_n\| = 1$. Thus

$$c_n = n^{1/2}\pi^{-1/4}. \tag{2.3.1}$$

Now

$$\|(x - \lambda)\psi_n\|^2 = c_n^2 n^{-3}\int_{-\infty}^{\infty} y^2 e^{-y^2}\, dy \to 0$$

as $n \to \infty$. Thus we have a sequence satisfying (2.2.1) and $\lambda \in \sigma(A)$. Since λ was any real number, we see that $\sigma(A)$ consists of the whole real line.

2.4. The Momentum Operator

The operator (1.3.8) corresponds to momentum. Again we have not yet defined its domain. Here the definition is not as clearcut as that for the position operator. It can be done directly if one uses techniques employed in the study of differential equations. These techniques are not difficult, and they do not require much in the way of preparation. However, we shall take the easy way out and avoid them. We do this in the following way. First, we note that Parseval's identity (1.3.5) shows that the Fourier transform maps L^2 onto itself in a one-to-one way with norm and scalar product being preserved. Second, we note that the Fourier transform of $L\psi$ is $\hbar k\hat{\psi}$. Thus we can define L by

$$L\psi = \hbar\left(k\hat{\psi}\right)^{\vee}, \tag{2.4.1}$$

and this is equivalent to the other definition. This has the effect of converting L to an operator similar to the position operator. In particular, we can take $D(L)$ to be the set of those $\psi \in L^2$ such that $k\hat{\psi}(k)$ is in L^2. The proof that L with this domain is self-adjoint is almost identical to the proof we gave for the position operator.

The study of the spectrum can also be handled by the methods of the last section. In fact, let λ be any real number, and set

$$\varphi_n(k) = c_n e^{-n^2(k-\lambda)^2/2},$$

where c_n is given by (2.3.1). From the calculations of Section 2.3 we have

$$\|\varphi_n\| = 1, \qquad \|(k-\lambda)\varphi_n\| \to 0. \tag{2.4.2}$$

Let $\psi_n(x)$ be the inverse Fourier transform of the function $\varphi_n(k)$. Then

$$(L-\lambda)\psi_n = \hbar[(k-\lambda)\varphi_n]^{\vee}.$$

Thus

$$\|(L-\lambda)\psi_n\| = \hbar\|(k-\lambda)\varphi_n\| \to 0$$

and $\|\psi_n\| = 1$ by Parseval's identity. This shows that $\sigma(L)$ also consists of the whole real axis.

We want to give another, somewhat more suggestive method for determining $\sigma(L)$. We start out as in Section 2.3 by searching for eigenvalues. If $(L-\lambda)\psi = 0$, we have

$$\lambda\psi + i\hbar\psi_x = 0. \tag{2.4.3}$$

A solution of this is

$$\psi_0 = e^{i\lambda x/\hbar} \tag{2.4.4}$$

and all other solutions are just constants times ψ_0. Since $|\psi_0| = 1$, ψ_0 is not in L^2, and consequently there is no solution of (2.4.3) in L^2 other than 0. However, ψ_0 misses being an eigenfunction by very little. If it were to die down like $|x|^{-(1/2)-\varepsilon}$ at ∞, it would be in L^2. This suggests the following procedure.

Set $\varphi(x) = ce^{-x^2}$, where the constant c is chosen so that

$$\int \varphi(x)^2\,dx = 1. \tag{2.4.5}$$

Define

$$\psi_n(x) = n^{-1/2}\varphi(x/n)\psi_0(x).$$

Then

$$\|\psi_n\|^2 = n^{-1}\int \varphi(x/n)^2\,dx = 1$$

by (2.4.5). Since

$$\psi_n'(x) = n^{-3/2}\varphi'(x/n)\psi_0 + i\lambda\psi_n(x)/\hbar,$$

we have

$$L\psi_n = \lambda\psi_n - i\hbar n^{-3/2}\varphi'(x/n)\psi_0.$$

Hence

$$\|(L-\lambda)\psi_n\|^2 = \hbar^2 n^{-3}\int \varphi'(x/n)^2\,dx$$

$$= \hbar^2 n^{-2}\int \varphi'(y)^2\,dy \to 0.$$

Again this shows that $\sigma(L)$ is the whole real line.

2.5. The Energy Operator

Next we consider the Hamiltonian operator given by (1.4.6). In this case the considerations are a bit more involved. Set

$$H_0 = \frac{1}{2m}L^2. \tag{2.5.1}$$

Then

$$H = H_0 + V. \tag{2.5.2}$$

As we noted previously, the domain of the sum of two operators is the intersection of their domains. Thus

$$D(H) = D(H_0) \cap D(V).$$

Thus, to examine H we must study both of the operators H_0 and V. Let us look at H_0. The domain that seems to be natural for this operator is $D(L^2)$. If we use this domain, is H_0 self-adjoint? The answer is given by

Theorem 2.5.1. *The square of a self-adjoint operator is self-adjoint.*

PROOF. We have $A^2 + 1 = (A+i)(A-i)$ [note that the domains of both these operators are $D(A^2)$]. If A is self-adjoint, then for each f there is a $u \in D(A^2)$ such that $(A^2+1)u = f$. In fact, by Theorem 2.1.1 there is a $v \in D(A)$ such that $(A-i)v = f$ and a $u \in D(A)$ such that $(A+i)u = v$. Since both u and v are in $D(A)$, the same is true of Au. Hence $u \in D(A^2)$ and $(A^2+1)u = f$. Thus $R(A^2+1)$ is the whole space, and we can apply Theorem 1.9.3. □

Now that we know that H_0 is self-adjoint, let us turn to the study of its spectrum. For each real λ

$$H_0 - \lambda = \frac{1}{2m}(L + \sqrt{2m\lambda})(L - \sqrt{2m\lambda}). \tag{2.5.3}$$

Moreover, $\pm\sqrt{2m\lambda}$ are in $\rho(L)$ unless they are real, that is, unless $\lambda \geq 0$. The proof of Theorem 2.5.1 shows us that $\lambda \in \rho(H_0)$ if λ is negative.

To examine the case when $\lambda > 0$, suppose $\lambda = \hbar^2\gamma^2/2m$. First we look for eigenvalues. Now,

$$(H_0 - \lambda)\psi = 0 \tag{2.5.4}$$

is equivalent to

$$\psi_{xx} + \gamma^2\psi = 0. \tag{2.5.5}$$

The general solution of this is

$$\psi(x) = c_+ e^{i\gamma x} + c_- e^{-i\gamma x}. \tag{2.5.6}$$

The only way this can be in L^2 is if $c_+ = c_- = 0$. Thus there are no eigenvalues. However, we can use the same method as before to find points in the spectrum. Take $\varphi(x)$ as in Section 2.4 and define

$$\psi_n(x) = n^{-1/2}\varphi(x/n)e^{i\gamma x}.$$

Then

$$\|\psi_n\|^2 = n^{-1}\int |\varphi(x/n)|^2\,dx = 1. \tag{2.5.7}$$

Moreover,

$$\psi_n' = i\gamma\psi_n + \varphi'(x/n)e^{i\gamma x}n^{-3/2}$$

and

$$\psi_n'' = -\gamma^2\psi_n + 2i\gamma n^{-3/2}\varphi'(x/n)e^{i\gamma x} + n^{-5/2}\varphi''(x/n)e^{i\gamma x}.$$

Hence

$$\begin{aligned}\|\psi_n'' + \gamma^2\psi_n\| &\le 2\gamma n^{-3/2}\|\varphi'(x/n)\| + n^{-5/2}\|\varphi''(x/n)\| \\ &= 2\gamma n^{-1/2}\|\varphi'\| + n^{-3/2}\|\varphi''\|. \end{aligned} \tag{2.5.8}$$

This shows that

$$(H_0 - \lambda)\psi_n \to 0. \tag{2.5.9}$$

Thus $\lambda \in \sigma(H_0)$ (Theorem 2.2.1). Since λ was any positive number, we see that the positive real axis is contained in $\sigma(H_0)$. Since the spectrum is a closed set (Corollary 2.1.2), 0 must also be in $\sigma(H_0)$. Thus we have proved

Theorem 2.5.2. *The spectrum of H_0 consists of the nonnegative real numbers. In symbols,*

$$\sigma(H_0) = [0, \infty). \tag{2.5.10}$$

2.6. The Potential

In determining a domain for the operator (2.5.2) we must decide on a domain for V. Again the simplest and largest domain one can choose for V is the set of those $\psi \in L^2$ such that $V\psi \in L^2$. This domain will be denoted by $D(V)$. Once this is done we know precisely the domain of H, namely $D(H_0) \cap D(V)$. Is H self-adjoint? Herein lies a tale.

So far we have not restricted $V(x)$ in any way other than taking it real valued. If we make no further restrictions on V, it will not be true, in general, that H is self-adjoint. In fact, it is rather simple to construct an example of a potential V such that $D(H)$ consists only of the function 0. (We shall construct one in Section 4.2.) This came as a terrible blow to many people, and it raised many questions. One of them is whether or not there exist potentials such that H is self-adjoint. Fortunately, there is a rather large class of potentials such that this is the case. They are described in

Theorem 2.6.1. *If there are constants $a < 1$, b such that*

$$\|V\psi\| \leq a\|H_0\psi\| + b\|\psi\|, \qquad \psi \in D(H_0), \tag{2.6.1}$$

then H is self-adjoint.

In proving Theorem 2.6.1 we shall make use of the following two theorems of general interest.

Theorem 2.6.2. *If M is a bounded linear operator on a Hilbert space $\mathcal{H}$ and $\|M\| < 1$, then there is a bounded linear operator N on $\mathcal{H}$ such that $N(1 - M) = (1 - M)N = 1$ and*

$$\|N\| \leq 1/(1 - \|M\|). \tag{2.6.2}$$

Theorem 2.6.3. *Let A be a densely defined Hermitian operator on a Hilbert space $\mathcal{H}$. Assume there is a complex number z such that $R(z - A) = R(\bar{z} - A) = \mathcal{H}$. Then A is self-adjoint.*

The operator N in Theorem 2.6.2 is called the *inverse* of $1 - M$ and is denoted by $(1 - M)^{-1}$. If $(1 - M)u = f$, then $Nf = u$. Theorem 2.6.2 is well known. For those readers unfamiliar with it, a proof will be given in Appendix B. Theorem 2.6.3 is an easy consequence of Theorem 1.9.4. It is left as an exercise.

Now we show how Theorems 2.6.2 and 2.6.3 imply Theorem 2.6.1. If λ is real and $\psi \in D(H_0)$, then

$$\|(H_0 - i\lambda)\psi\|^2 = \|H_0\psi\|^2 + 2 \operatorname{Re} i\lambda(H_0\psi, \psi) + \lambda^2\|\psi\|^2.$$

Since λ is real and H_0 is Hermitian, the middle term on the right-hand side vanishes. Since H_0 is self-adjoint (see Section 2.5), if $\lambda \neq 0$, for each $f \in L^2$ we can find a $\psi \in D(H_0)$ such that $(H_0 - i\lambda)\psi = f$ (Theorem 2.1.1). Thus

$$\|f\|^2 = \|H_0(H_0 - i\lambda)^{-1}f\|^2 + \lambda^2\|(H_0 - i\lambda)^{-1}f\|^2.$$

This implies

$$\|H_0(H_0 - i\lambda)^{-1}f\| \le \|f\| \tag{2.6.3}$$

and

$$|\lambda|\,\|(H_0 - i\lambda)^{-1}f\| \le \|f\|. \tag{2.6.4}$$

Now suppose V satisfies (2.6.1). Then

$$\begin{aligned}\|V(H_0 - i\lambda)^{-1}f\| &\le a\|H_0(H_0 - i\lambda)^{-1}f\| + b\|(H_0 - i\lambda)^{-1}f\|\\ &\le (a + b|\lambda|^{-1})\|f\|.\end{aligned}$$

Since $a < 1$, this implies that

$$\|V(H_0 - i\lambda)^{-1}\| < 1 \tag{2.6.5}$$

for $|\lambda|$ sufficiently large. Now (2.6.1) implies that $D(H_0) \subset D(V)$. Hence $D(H) = D(H_0)$. Consequently,

$$H - i\lambda = H_0 - i\lambda + V = \left[1 + V(H_0 - i\lambda)^{-1}\right](H_0 - i\lambda).$$

By (2.6.5) and Theorem 2.6.2, the first operator on the right is onto, that is, its range coincides with the whole space, for $|\lambda|$ sufficiently large. The same is true of the other operator. Hence $R(H - i\lambda) = L^2$ for λ real and $|\lambda|$ sufficiently large. The self-adjointness of H now follows from Theorem 2.6.3.

2.7. A Class of Functions

Theorem 2.6.1 gives us a very useful way of determining when H is self-adjoint. However, it requires us to know when a potential function V satisfies inequality (2.6.1) with $a < 1$. Fortunately, a simple criterion exists. It is given by

Theorem 2.7.1. *The following statements are equivalent:*

a. $D(H_0) \subset D(V)$.
b. $\|V\psi\|^2 \le C(\|H_0\psi\|^2 + \|\psi\|^2)$, $\psi \in D(H_0)$.
c. $C_0 = \sup_x \int_x^{x+1} |V(y)|^2\,dy < \infty$.
d. *For each* $\varepsilon > 0$ *there is a constant* K *such that*

$$\|V\psi\|^2 \le \varepsilon\|H_0\psi\|^2 + K\|\psi\|^2, \qquad \psi \in D(H_0). \tag{2.7.1}$$

e. *Inequality* (2.7.1) *can be replaced by*

$$\|V\psi\| \le \varepsilon\|H_0\psi\| + K\|\psi\|, \qquad \psi \in D(H_0). \tag{2.7.2}$$

In proving Theorem 2.7.1 we shall need an important theorem from functional analysis.

Theorem 2.7.2. *If A is a closed operator mapping one Hilbert space $\mathcal{H}_1$ into another and $D(A) = \mathcal{H}_1$, then A is bounded (see Section 1.7).*

Theorem 2.7.2 is called the *closed graph* theorem; a proof will be given in Appendix B. We shall use it in giving the

PROOF OF THEOREM 2.7.1. Suppose (a) holds. Put

$$[u, v] = (u, v) + (H_0u, H_0v), \qquad u, v \in D(H_0). \tag{2.7.3}$$

One checks easily that this is a scalar product. Let $|||\ |||$ denote the corresponding norm. It can be claimed that $D(H_0)$ is complete with respect to this norm. In fact, if $\{v_n\} \subset D(H_0)$ is a Cauchy sequence with respect to the norm, then $\{v_n\}$ and $\{H_0v_n\}$ are both Cauchy sequences in L^2. Thus there are functions $v, g \in L^2$ such that $v_n \to v$ and $H_0v_n \to g$. Since H_0 is a closed operator (Corollary 2.2.1), we see that $v \in D(H_0)$ and $H_0v = g$. Thus $v_n \to v$ with respect to the norm. This shows that $D(H_0)$ is a Hilbert space when (2.7.3) is the scalar product. Next we show that V is a closed operator from $D(H_0)$ to L^2. For if $v_n \to v$ in $D(H_0)$ and $Vv_n \to f$ in L^2, then $v_n \to v$ and $H_0v_n \to H_0v$ in L^2. Thus

$$(Vv_n, \varphi) = (v_n, V\varphi), \qquad \varphi \in D(H_0),$$

and consequently

$$(f, \varphi) = (v, V\varphi), \qquad \varphi \in D(H_0).$$

since $v \in D(H_0)$, this gives

$$(f - Vv, \varphi) = 0, \qquad \varphi \in D(H_0),$$

and since $D(H_0)$ is dense in L^2, we see that $Vv = f$ (Lemma 1.5.2). We are now in a position to apply Theorem 2.7.2 to conclude that V is a bounded operator from $D(H_0)$ to L^2, that is, there is a constant C such that (b) holds.

Now suppose (b) holds. Let K be such a large constant that the function

$$\varphi(x) = Ke^{-x^2} \tag{2.7.4}$$

is ≥ 1 in the interval $0 \leq x \leq 1$. For each real number λ set $\varphi_\lambda(x) = \varphi(x - \lambda)$. If (b) holds, we have

$$\int_\lambda^{\lambda+1} |V(x)|^2\, dx \leq \int_{-\infty}^{\infty} |V(x)\varphi_\lambda(x)|^2\, dx$$
$$\leq C(\|H_0\varphi_\lambda\|^2 + \|\varphi_\lambda\|^2) \leq C'(\|\varphi''\|^2 + \|\varphi\|^2). \tag{2.7.5}$$

The right-hand side of (2.7.5) is finite and independent of λ. Thus (c) holds.

Next assume that (c) is true. Suppose $u \in D(H_0)$ is continuously differentiable. By Lemma 1.8.1, for every $\varepsilon > 0$ and every interval I of length 1

$$|u(x)|^2 \le \varepsilon \int_I |u'(y)|^2\, dy + (1+\varepsilon^{-1}) \int_I |u(y)|^2\, dy, \qquad x \in I. \quad (2.7.6)$$

If C_0 is the finite expression in (c), we have

$$\int_I |V(x)u(x)|^2\, dx \le \varepsilon C_0 \int_I |u'(y)|^2\, dy + C_0(1+\varepsilon^{-1}) \int_I |u(y)|^2\, dy, \quad (2.7.7)$$

where we multiplied (2.7.6) by $|V(x)|^2$ and integrated with respect to x over the interval I. Since the real line is the union of intervals of length 1, we obtain

$$\int_{-\infty}^{\infty} |V(x)u(x)|^2\, dx \le \varepsilon C_0 \int_{-\infty}^{\infty} |u'(y)|^2\, dy + (1+\varepsilon^{-1}) C_0 \int_{-\infty}^{\infty} |u(y)|^2\, dy \quad (2.7.8)$$

by summing (2.7.7) over such intervals. Now by Parseval's formula (1.3.6),

$$\begin{aligned} \int_{-\infty}^{\infty} |u'(y)|^2\, dy &= \int_{-\infty}^{\infty} |k\hat{u}(k)|^2\, dk \\ &\le \frac{1}{2}\int_{-\infty}^{\infty} |k^2\hat{u}(k)|^2\, dk + \frac{1}{2}\int_{-\infty}^{\infty} |\hat{u}(k)|^2\, dk \\ &= \frac{2m^2}{\hbar^4}\|H_0 u\|^2 + \frac{1}{2}\|u\|^2. \end{aligned} \quad (2.7.9)$$

Substituting this into (2.7.8), we have

$$\int_{-\infty}^{\infty} |V(x)u(x)|^2\, dx \le \frac{2\varepsilon m^2 C_0}{\hbar^4}\|H_0 u\|^2 + \left(1 + \frac{1}{\varepsilon} + \frac{\varepsilon}{2}\right) C_0 \|u\|^2. \quad (2.7.10)$$

Since this is true for every $\varepsilon > 0$, we see that (2.7.1) holds for continuously differentiable functions $\psi \in D(H_0)$. But this is enough, for we shall show at the end of this section that all functions in $D(H_0)$ are indeed continuously differentiable.

That (d) implies (e) is trivial. We merely add the term $2\sqrt{\varepsilon K}\,\|H_0\psi\|\,\|\psi\|$ to the right-hand side of (2.7.1) and take the square root of both sides. This gives (2.7.2). Finally we note that (e) implies (a). For inequality (2.7.2) says that $V\psi \in L^2$ whenever $\psi \in D(H_0)$. This means that $D(H_0) \subset D(V)$. □

Theorem 2.7.1 gives a very convenient method for verifying that H is self-adjoint. Combining Theorems 2.6.1 and 2.7.1, we have

Corollary 2.7.1. *If* (c) *holds, then* H *is self-adjoint.*

We are left to show that functions in $D(H_0)$ are continuously differentiable. This follows from

Theorem 2.7.3. *If u and $k\hat{u}$ are in L^2, then u is continuous.*

PROOF. We have by the inverse Fourier transform (1.3.5)

$$u(x) - u(x_0) = \frac{1}{\sqrt{2\pi}} \int_{-\infty}^{\infty} (e^{ikx} - e^{ikx_0})\hat{u}(k)\, dk. \tag{2.7.11}$$

Thus

$$2\pi|u(x) - u(x_0)|^2 \leq \int_{-\infty}^{\infty} |e^{ikx} - e^{ikx_0}|^2(k^2+1)^{-1}\, dk \times \int_{-\infty}^{\infty} (k^2+1)|\hat{u}(k)|^2\, dk$$

by the Schwarz inequality (1.5.10). Now,

$$\int_{-\infty}^{\infty} |e^{ikx} - e^{ikx_0}|^2(k^2+1)^{-1}\, dk \to 0 \quad \text{as} \quad x \to x_0. \tag{2.7.12}$$

This shows that $u(x) \to u(x_0)$ as $x \to x_0$. □

Theorem 2.7.4. *Functions in $D(H_0)$ are continuously differentiable.*

PROOF. If $u \in D(H_0)$, then u and $k^2\hat{u}$ are in L^2. By inequality (2.7.9), $k\hat{u}$ is also in L^2. Set

$$w(x) = \frac{i}{\sqrt{2\pi}} \int_{-\infty}^{\infty} e^{ikx}k\hat{u}(k)\, dk.$$

By Theorem 2.7.3, $w(x)$ is continuous. Moreover, by (2.7.11)

$$\frac{u(x) - u(x_0)}{x - x_0} - w(x_0) = \frac{i}{\sqrt{2\pi}} \int_{-\infty}^{\infty} \left\{ \frac{e^{ikx} - e^{ikx_0}}{ik(x - x_0)} - e^{ikx_0} \right\} k\hat{u}(k)\, dk,$$

and consequently

$$\left| \frac{u(x) - u(x_0)}{x - x_0} - w(x_0) \right|^2 \leq \frac{1}{2\pi} \int_{-\infty}^{\infty} \left| \frac{e^{ikx} - e^{ikx_0}}{ik(x - x_0)} - e^{ikx_0} \right|^2 \frac{dk}{(k^2+1)} \times \int_{-\infty}^{\infty} (k^2+1)k^2|\hat{u}(k)|^2\, dk. \tag{2.7.13}$$

As before, this tends to 0 as $x \to x_0$. Thus the derivative of u exists and equals w, which is continuous. □

We should be careful with one detail. For functions in L^2, the inverse Fourier transform formula (2.7.11) holds only a.e. Thus, strictly speaking, in Theorem 2.7.3 we have only shown that u is a.e. equal to a continuous function. In other words, we can make u continuous by changing it on a set of measure 0. A similar statement is true for Theorem 2.7.4. However, since we identify functions that differ only on a set of measure 0, no harm will come from considering u itself continuous.

2.8. The Spectrum of H

Now we undertake a study of the spectrum of H. As might be expected, it depends in a substantial way on the potential V. We undertake this study in the next chapter.

Exercises

1. Prove Lemma 2.2.1.
2. Show that (2.2.3) implies that $u \in D(A)$ and $(\lambda - A)u = f$.
3. Show that ψ_ε defined in Section 2.3 converges to f in L^2.
4. Prove (2.3.1).
5. Show that the operator (2.4.1) with the prescribed domain is self-adjoint.
6. Show that all solutions of (2.4.3) are of the form $C\psi_0$, where C is a constant.
7. Show that $\lambda \in \rho(H_0)$ if $\lambda < 0$.
8. Prove that $A^2 + 1 = (A + i)(A - i)$ with identical domains.
9. Show that every solution of (2.5.5) is of the form (2.5.6).
10. If ψ is given by (2.5.6), show that it is not in L^2 unless it vanishes.
11. Prove that 0 is not an eigenvalue of H_0.
12. Prove Theorem 2.6.3.
13. Show that (2.7.3) is a scalar product.
14. Prove (2.7.5) assuming (b) of Theorem 2.7.1.
15. Prove (2.7.12).
16. Show that $E(\lambda) = 0$ if $(-\infty, \lambda] \subset \rho(A)$.

3
The Essential Spectrum

3.1. An Example

In Section 2.5 we showed that when $V = 0$ the spectrum of H consists of the nonnegative real axis. Now we want to study the spectrum of H when V does not vanish. To feel our way we begin by considering simple functions without regard to their applicability. The following two theorems are very useful in helping to determine the spectrum of H. We assume $D(H) = D(H_0)$.

Theorem 3.1.1. *If $V(x) \geq b$ for all x, then every $\lambda < b$ is in $\rho(H)$. In symbols,*

$$(-\infty, b) \subset \rho(H). \tag{3.1.1}$$

PROOF. We have

$$([H-\lambda]\psi, \psi) = \frac{\hbar^2}{2m}\|k\hat{\psi}\|^2 + ([V-\lambda]\psi, \psi), \qquad \psi \in D(H). \tag{3.1.2}$$

If $\lambda < b$, this gives

$$\|\psi\| \leq \frac{1}{b-\lambda}\|(H-\lambda)\psi\|.$$

This shows that there does not exist a sequence satisfying (2.2.1) with $A = H$. Thus $\lambda \in \rho(H)$ by Theorem 2.2.1. □

Theorem 3.1.2. *If $V(x) = b$, a constant, on an unbounded interval I, then all $\lambda \geq b$ are in $\sigma(H)$. Thus,*

$$[b, \infty) \subset \sigma(H). \tag{3.1.3}$$

PROOF. Suppose $I = [a, \infty)$. We may assume $a \geq 0$. Take any $\lambda \geq b$. We look for a solution of

$$(H - \lambda)\psi = 0 \qquad \text{in} \quad I. \tag{3.1.4}$$

This is equivalent to

$$\psi'' + \gamma^2\psi = 0, \tag{3.1.5}$$

where $\gamma^2 = 2m(\lambda - b)/\hbar^2$. A solution of this is $\psi = e^{i\gamma x}$ in I, but this is not in $L^2(I)$. However, this does not stop us. Let $\varphi(x) \not\equiv 0$ be a smooth function which vanishes for large x and outside I. By "smooth" we mean twice continuously differentiable. We shall show in Section 7.8 that such things exist. By multiplying φ by a suitable constant, we may arrange $\|\varphi\|$ to be 1. Set

$$\psi_n(x) = n^{-1/2}\varphi(x/n)e^{i\gamma x}. \tag{3.1.6}$$

Note that these functions vanish for large x and outside I. Simple computations give $\|\psi_n\| = 1$ and

$$\psi_n'' + \gamma^2\psi_n = 2i\gamma n^{-3/2}\varphi'(x/n)e^{i\gamma x} + n^{-5/2}\varphi''(x/n)e^{i\gamma x}. \tag{3.1.7}$$

Since

$$(\lambda - H)\psi_n = \frac{\hbar^2}{2m}(\psi_n'' + \gamma^2\psi_n), \tag{3.1.8}$$

we have

$$\|(\lambda - H)\psi_n\| \leq \frac{\hbar^2\gamma}{mn}\|\varphi'\| + \frac{\hbar^2}{2mn^2}\|\varphi''\|. \tag{3.1.9}$$

This converges to 0 as $n \to \infty$. The result now follows from Theorem 2.2.1. □

To illustrate how these theorems can be applied, consider the case when

$$V(x) = \begin{cases} b_1 & \text{for} \quad x < a \\ b_2 & \text{for} \quad x > a. \end{cases}$$

By Theorem 3.1.1, $(-\infty, \min(b_1, b_2)) \subset \rho(H)$. On the other hand, by Theorem 3.1.2, $[\min(b_1, b_2), \infty) \subset \sigma(H)$. Thus

$$\sigma(H) = [\min(b_1, b_2), \infty). \tag{3.1.10}$$

Next, let us consider something a bit more complicated. Suppose

$$V(x) = \begin{cases} b_1, & x < 0 \\ b_2, & 0 < x < a \\ b_3, & a < x. \end{cases} \tag{3.1.11}$$

Assume first that $b_2 \geq b_0 = \min(b_1, b_3)$. Then $(-\infty, b_0) \subset \rho(H)$ by Theorem 3.1.1, while $[b_0, \infty) \subset \sigma(H)$ by Theorem 3.1.2. Thus we have

$$\sigma(H) = [b_0, \infty). \tag{3.1.12}$$

What happens if $b_2 < b_0$? In this case we have $(-\infty, b_2) \subset \rho(H)$ by Theorem 3.1.1 while $[b_0, \infty) \subset \sigma(H)$ by Theorem 3.1.2. The theorems do not give us the answer for the interval $[b_2, b_0)$. We examine this in detail in the next section.

3.2. A Calculation

We try to determine whether there are any points in $[b_2, b_0)$ which are in $\sigma(H)$. As is our custom, we first look for eigenvalues. Suppose $b_2 \leq \lambda < b_0$. A solution of (3.1.4) satisfies

$$\hbar^2\psi'' + 2m(\lambda - b_j)\psi = 0 \quad \text{in } I_j, \quad j = 1, 2, 3, \tag{3.2.1}$$

where $I_1 = (-\infty, 0)$, $I_2 = (0, a)$, and $I_3 = (a, \infty)$. Set

$$\gamma_j^2 = 2m|\lambda - b_j|/\hbar^2, \quad j = 1, 2, 3. \tag{3.2.2}$$

Then

$$\gamma_j^2 + \gamma_2^2 = 2m(b_j - b_2)/\hbar^2 \equiv c_j^2, \quad j = 1, 3, \tag{3.2.3}$$

and

$$\psi'' - \gamma_j^2\psi = 0 \quad \text{in } I_j, \quad j = 1, 3, \tag{3.2.4}$$

$$\psi'' + \gamma_2^2\psi = 0 \quad \text{in } I_2. \tag{3.2.5}$$

The most general solution of (3.2.4) and (3.2.5) in L^2 is

$$\psi = \begin{cases} A_1 e^{\gamma_1 x} & \text{in } I_1 \\ A_2 \sin(\gamma_2 x + \alpha) & \text{in } I_2 \\ A_3 e^{-\gamma_3 x} & \text{in } I_3. \end{cases} \tag{3.2.6}$$

Since we can multiply ψ by an arbitrary constant without spoiling the fact that it satisfies (3.2.4) and (3.2.5), we may assume that $A_2 = 1$ (this simplifies the calculations). Now V satisfies (c) in Theorem 2.7.1. Hence $D(H) = D(H_0)$. Thus any eigenfunction of H must be continuously differentiable (Theorem 2.7.4). Thus, in order that (3.2.6) be an eigenfunction of H, we must choose the constants A_1, A_3, and α so that ψ is continuously differentiable. For continuity we need

$$A_1 = \sin\alpha, \qquad A_3 e^{-a\gamma_3} = \sin(\gamma_2 a + \alpha). \tag{3.2.7}$$

Since

$$\psi' = \begin{cases} \gamma_1 A_1 e^{\gamma_1 x} & \text{in } I_1 \\ \gamma_2 \cos(\gamma_2 x + \alpha) & \text{in } I_2 \\ -\gamma_3 A_3 e^{-\gamma_3 x} & \text{in } I_3, \end{cases} \tag{3.2.8}$$

we need

$$\gamma_1 A_1 = \gamma_2 \cos\alpha, \qquad \gamma_3 A_3 e^{-a\gamma_3} = -\gamma_2 \cos(\gamma_2 a + \alpha) \tag{3.2.9}$$

for continuous differentiability. Thus we must have

$$\gamma_1 = \gamma_2 \cot\alpha, \qquad \gamma_3 = -\gamma_2 \cot(\gamma_2 a + \alpha). \tag{3.2.10}$$

In particular, one must have $\cot\alpha > 0$ and $\cot(\gamma_2 a + \alpha) < 0$. Thus α must be in the first or third quadrant and $\gamma_2 a + \alpha$ must be in the second or fourth. Now, by (3.2.10)

$$\sin\alpha = \pm\gamma_2/c_1, \qquad \sin(\gamma_2 a + \alpha) = \pm\gamma_2/c_3. \tag{3.2.11}$$

Hence,

$$\alpha = \sin^{-1}(\gamma_2/c_1) + n_1\pi \tag{3.2.12}$$

and

$$\gamma_2 a + \alpha = -\sin^{-1}(\gamma_2/c_3) + n_2\pi, \tag{3.2.13}$$

where $\sin^{-1} t$ denotes the principal value (the angle between $-\pi/2$ and $\pi/2$ whose sine equals t) and n_1, n_2 are integers. [The minus sign in (3.2.13) comes from the quadrant considerations mentioned above.] Combining (3.2.12) and (3.2.13), we have

$$a\gamma_2 = n\pi - \sin^{-1}(\gamma_2/c_1) - \sin^{-1}(\gamma_2/c_3), \tag{3.2.14}$$

where $n = n_2 - n_1$. Thus we have shown that in order for (3.2.1) to have a nontrivial solution which is continuously differentiable, it is necessary for γ_2 to satisfy (3.2.14) for some integer n, where the c_j are given by (3.2.3). Conversely, if we have a solution of (3.2.14), we can retrace our steps and find a nonvanishing continuously differentiable solution (3.2.6) of (3.2.1) (the second-order derivatives could be discontinuous at $x = 0$ and $x = a$).

Two questions arise immediately. The first is whether or not the solution (3.2.6) so constructed is in $D(H)$ (otherwise, it could not be an eigenfunction). This will be answered affirmatively in Section 3.4. In the next section we shall answer the second question, namely, how many solutions does (3.2.14) have?

3.3. Finding the Eigenvalues

In searching for solutions of (3.2.14), we note that if any of the quantities $\gamma_1, \gamma_2, \gamma_3$ vanishes, then (3.2.7), (3.2.9), (3.2.10), and the fact that $\psi \in L^2$ imply $\psi \equiv 0$. Thus we are looking for positive solutions. This forces the integer n to be positive. Moreover, (3.2.3) implies

$$\gamma_2^2 \leq c_0^2 = \min(c_1^2, c_3^2) = 2m(b_0 - b_2)/\hbar^2. \tag{3.3.1}$$

Hence we are looking for solutions satisfying

$$0 < \gamma_2 < c_0 \tag{3.3.2}$$

($\gamma_2 = c_0$ would imply either $\gamma_1 = 0$ or $\gamma_3 = 0$). Consider the curves

$$y_1 = a\gamma, \tag{3.3.3}$$

$$y_2 = n\pi - \sin^{-1}(\gamma/c_1) - \sin^{-1}(\gamma/c_3). \tag{3.3.4}$$

We obtain a solution of (3.2.14) when these curves intersect. Now y_2 is a decreasing function of γ, and it satisfies

$$(n-1)\pi \le y_2 \le n\pi. \tag{3.3.5}$$

It is clear that y_1 will intersect y_2 for some γ in the interval (3.3.2) iff

$$n\pi - \sin^{-1}(c_0/c_1) - \sin^{-1}(c_0/c_3) < ac_0. \tag{3.3.6}$$

Set $c_4 = \max(c_1, c_3)$. Since either $c_0 = c_1$ and $c_4 = c_3$, or $c_0 = c_3$ and $c_4 = c_1$, (3.3.6) becomes

$$n\pi < ac_0 + \tfrac{1}{2}\pi + \sin^{-1}(c_0/c_4). \tag{3.3.7}$$

In view of (3.2.3), this becomes

$$n\pi\hbar < a\sqrt{2m(b_0 - b_2)} + \frac{1}{2}\pi\hbar + \hbar \sin^{-1}\sqrt{\frac{b_0 - b_2}{b_4 - b_2}},$$

where $b_4 = \max(b_1, b_3)$. Thus we have

Proposition 3.3.1. *Let N be the largest integer satisfying*

$$N < \frac{a}{\pi\hbar}\sqrt{2m(b_0 - b_2)} + \frac{1}{2} + \frac{1}{\pi}\sin^{-1}\sqrt{\frac{b_0 - b_2}{b_4 - b_2}}. \tag{3.3.8}$$

Then there are precisely N values of λ in the interval $[b_2, b_0)$ for which (3.2.1) *has a continuously differentiable solution in L^2 which does not vanish identically.*

Corollary 3.3.1. *The number N of such solutions satisfies*

$$N - 1 < \frac{a}{\pi\hbar}\sqrt{2m(b_0 - b_2)} < N + \frac{1}{2}. \tag{3.3.9}$$

Corollary 3.3.2. *If $b_1 = b_3$, then N is the largest integer satisfying*

$$N < \frac{a}{\pi\hbar}\sqrt{2m(b_0 - b_2)} + 1. \tag{3.3.10}$$

In particular, N is at least one and satisfies

$$N - 1 < \frac{a}{\pi\hbar}\sqrt{2m(b_0 - b_2)} < N. \tag{3.3.11}$$

We leave the proofs of Proposition 3.3.1 and Corollaries 3.3.1 and 3.3.2 as exercises.

3.4. The Domain of H

We have shown that (3.2.1) has nonvanishing continuously differentiable solutions in L^2 only for a finite number of values of λ, and that these values are in the interval (b_2, b_0). We want to show that these values of λ are eigenvalues of H. In order to do this we must prove that the solutions ψ given by (3.2.6) are in $D(H)$. We do this in the present section. Recall that $D(H) = D(H_0)$. This follows trivially from Theorem 2.7.1. Also note that ψ given by (3.2.6) not only has continuous first-order derivatives, but it has piecewise continuous second-order derivatives. Moreover, ψ and its derivatives die down exponentially as $|x| \to \infty$. To prove that $\psi \in D(H_0)$, we must show that $k\hat{\psi}$ and $k^2\hat{\psi}$ are in L^2 (see Section 2.5). Now

$$\begin{aligned}\sqrt{2\pi}\, k\hat{\psi} &= i\int_{-\infty}^{\infty} \frac{\partial(e^{-ikx})}{\partial x}\psi(x)\,dx\\ &= i\lim_{R\to\infty}\left[e^{-ikR}\psi(R) - e^{ikR}\psi(-R) - \int_{-R}^{R} e^{-ikx}\psi'(x)\,dx\right]\\ &= -i\int_{-\infty}^{\infty} e^{-ikx}\psi'(x)\,dx. \end{aligned} \tag{3.4.1}$$

Thus

$$\begin{aligned}\sqrt{2\pi}\, k^2\hat{\psi} &= \lim_{R\to\infty}\left[e^{-ikR}\psi'(R) - e^{ikR}\psi'(-R) - \int_{-R}^{0} - \int_{0}^{a}\right.\\ &\qquad \left. - \int_{a}^{R} e^{-ikx}\psi''(x)\,dx\right]\\ &= -\int_{-\infty}^{\infty} e^{-ikx}\psi''(x)\,dx. \end{aligned} \tag{3.4.2}$$

(Note that the boundary terms vanish in (3.4.2) because ψ' is continuous.) Since ψ and ψ'' are both in L^2, we see by Parseval's identity that $k\hat{\psi}$ and $k^2\hat{\psi}$ are in L^2. Thus $\psi \in D(H_0)$.

Note that we have proved

Proposition 3.4.1. *Let ψ be a continuously differentiable function such that ψ and ψ' tend to 0 as $|x| \to \infty$. If ψ'' is piecewise continuous and in L^2, then $\psi \in D(H_0)$.*

Now we can state

Theorem 3.4.1. *If V is given by (3.1.11), then H has precisely N eigenvalues, where N is the largest nonnegative integer satisfying (3.3.8). All of these eigenvalues are in the interval (b_2, b_0).*

PROOF. Since $(-\infty, b_2)$ is in the resolvent set of H, there can be no eigenvalues below b_2. It is also easily checked that (3.2.1) cannot have a nonvanishing solution for $\lambda \geq b_0$. Thus the only possible eigenvalues must be in $[b_2, b_0)$. By Proposition 3.3.1, (3.2.1) has nonvanishing continuously differentiable solutions for precisely N values of λ in this interval, and all of them are interior points. Moreover, the solutions are given by (3.2.6), and we have just shown that they are in $D(H_0)$. Consequently, the N values of λ are eigenvalues. □

There is still one more question to be answered. We know that $(-\infty, b_2)$ is in $\rho(H)$, while $[b_0, \infty)$ is in $\sigma(H)$. We also know that there are exactly N eigenvalues in (b_2, b_0). What about the rest of the points in $[b_2, b_0)$? To answer this question we shall need a bit more Hilbert space theory. We discuss this in the next section.

3.5. Back to Hilbert Space

In order to answer further questions concerning the spectrum of H we shall need several more notions from Hilbert space. A subspace N of a Hilbert space $\mathcal{H}$ is called *closed* if $u \in N$ whenever there is a sequence u_k of elements of N converging to u in $\mathcal{H}$. First we state

Lemma 3.5.1. *Let N be a closed subspace of $\mathcal{H}$ and suppose u_0 is not in N. Let M be the set of all elements of the form $u = \alpha u^0 + w$, where α is a scalar and $w \in N$. Then M is a closed subspace of $\mathcal{H}$ and there is a constant C_0 depending only on u_0 and N such that*

$$|\alpha| \leq C_0\|u\|, \qquad u \in M. \tag{3.5.1}$$

PROOF. Clearly, M is a subspace. If (3.5.1) did not hold, there would be a sequence $u_n = \alpha_n u_0 + w_n$ of elements in M such that

$$|\alpha_n| = 1, \qquad u_n \to 0. \tag{3.5.2}$$

The α_n have a subsequence (also denoted by α_n) which converges to a number α (why?). The corresponding $w_n = u_n - \alpha_n u_0$ converge to $-\alpha u_0$. Since N is closed, we have $\alpha u_0 \in N$. Since $u_0 \notin N$, the only way this can happen is if $\alpha = 0$. But this is impossible since $|\alpha| = \lim|\alpha_n| = 1$. Thus (3.5.1) holds. To see that M is closed, suppose $u_n = \alpha_n u_0 + w_n \to v$ in $\mathcal{H}$, where $u_n \in M$. In particular, there is a constant C such that $\|u_n\| \leq C$ (why?). By (3.5.1), $|\alpha_n| \leq C_0 C$. Thus $\{\alpha_n\}$ has a subsequence (again denoted by $\{\alpha_n\}$) which converges to some number α. Thus $w_n = u_n - \alpha_n u_0 \to v - \alpha u_0 = w$. Since N is closed, $w \in N$. Hence $v = \alpha u_0 + w \in M$. □

Vectors $v_1, \ldots, v_n$ are called *linearly independent* if the only scalars α_k which satisfy

$$\alpha_1 v_1 + \cdots + \alpha_n v_n = 0 \tag{3.5.3}$$

are $\alpha_1 = \alpha_2 = \cdots = \alpha_n = 0$. They are said to *span* a vector space V if every element of V is a linear combination of them:

$$v = \beta_1 v_1 + \cdots + \beta_n v_n. \tag{3.5.4}$$

The *dimension* of V is n if it is spanned by n linearly independent vectors. We have the following corollaries to Lemma 3.5.1.

Corollary 3.5.1. *Finite-dimensional subspaces are closed.*

PROOF. We use induction on n. It is certainly true if $n = 1$, for then the subspace consists of vectors of the form αu_0. Now suppose the corollary is true for all subspaces of dimension $n - 1$. Suppose M is a subspace of dimension n. Let $u_1, \ldots, u_n$ be linearly independent vectors which span M. Let N be the set of all linear combinations of $u_1, \ldots, u_{n-1}$. Clearly N is a subspace and its dimension is $n - 1$. By induction hypothesis, N is closed. Moreover, M consists of all u of the form $\alpha u_n + w$, where $w \in N$. Thus the hypotheses of Lemma 3.5.1 are satisfied. □

Corollary 3.5.2. *In a finite-dimensional subspace, every bounded sequence has a convergent subsequence.*

PROOF. Again we use induction on n. Suppose N is a closed subspace having this property, and M is defined as in Lemma 3.5.1. Let u_n be a bounded sequence in M. By Lemma 3.5.1, the α_n are bounded, and consequently there is a subsequence converging to some α. Thus the corresponding w_n are bounded. Since N has the desired property, there is a subsequence which converges. Hence u_n has a convergent subsequence. □

We shall say that u is *orthogonal* to v and write $u \perp v$ if $(u, v) = 0$. We shall say that u is orthogonal to a set S if $u \perp v$ for all $v \in S$. The following is called the *projection theorem*. A proof is given in Appendix B.

Theorem 3.5.1. *For each closed subspace N of $\mathcal{H}$ and each $u \notin N$ there is a unique $u' \in N$ such that $u - u' \perp N$.*

Corollary 3.5.3. *If N is a closed subspace of $\mathcal{H}$ but is not the whole space, then there is a vector $u_0 \perp N$ such that $\|u_0\| = 1$.*

For a closed subspace N we can define an operator P by $Pu = u'$, where u' is the element given by Theorem 3.5.1. One checks easily that P is a linear operator and that $P^2 = P$. It is called the *orthogonal projection* onto N. Note also that

$$\|Pu\|^2 + \|(1 - P)u\|^2 = \|u\|^2. \tag{3.5.5}$$

A sequence $\{\varphi_n\}$ is called *orthonormal* if it satisfies

$$(\varphi_j, \varphi_k) = \delta_{jk} \qquad (= 0 \text{ for } j \neq k, \; = 1 \text{ for } j = k) \tag{3.5.6}$$

We shall need

Lemma 3.5.2. *Every infinite-dimensional subspace S contains an orthonormal sequence.*

PROOF. We show that if S contains n vectors satisfying (3.5.6), then it contains $n + 1$. Clearly, this gives the lemma. If S contains n such vectors, let T be the set of linear combinations of $\varphi_1, \ldots, \varphi_n$. Since the φ_k are linearly independent (why?), the dimension of T is n. Hence T is closed (Corollary 3.5.1). By Corollary 3.5.3 there is a $\varphi_{n+1} \perp T$ such that $\|\varphi_{n+1}\| = 1$. Clearly, $\varphi_1, \ldots, \varphi_{n+1}$ satisfy (3.5.6). □

3.6. Compact Operators

In this section we shall give some general theorems which will help not only with the interval $[b_2, b_0)$ but also with many other situations as well.

An operator K mapping a Hilbert space $\mathcal{H}_1$ into a Hilbert space $\mathcal{H}_2$ is called *compact* if it takes any bounded sequence in $\mathcal{H}_1$ into a sequence having a convergent subsequence, that is, if $\|u_n\| \leq C$ implies that $\{Ku_n\}$ has a convergent subsequence. By Corollary 3.5.2 we have

Lemma 3.6.1. *A bounded operator with finite-dimensional range is compact.*

Let A be a self-adjoint operator on a Hilbert space. We define the *essential spectrum* of A as

$$\sigma_e(A) = \bigcap_K \sigma(A + K), \tag{3.6.1}$$

where the intersection is taken over all compact Hermitian operators K. Thus $\lambda \notin \sigma_e(A)$ iff there is a compact Hermitian operator K such that $\lambda \in \rho(A + K)$. The definition (3.6.1) is not easy to verify. Therefore it is desirable to obtain other criteria for determining the essential spectrum of an operator. First we have

Theorem 3.6.1. *A point λ is in $\sigma_e(A)$ iff there is a sequence $\{\psi_n\}$ of elements in $D(A)$ such that*

a. $\|\psi_n\| = 1$,
b. $\{\psi_n\}$ *has no convergent subsequence, and*
c. $(A - \lambda)\psi_n \to 0$.

PROOF. Suppose $\lambda \notin \sigma_e(A)$. Then there is a compact operator K such that $\lambda \in \rho(A + K)$. In particular, we have

$$\|\psi\| \le C\|(A + K - \lambda)\psi\|, \qquad \psi \in D(A) \tag{3.6.2}$$

[cf. (2.2.2)]. Since K is compact and the sequence $\{\psi_n\}$ is bounded by (a), there is a subsequence (also denoted by $\{\psi_n\}$) such that $\{K\psi_n\}$ converges. Hence by (3.6.2) and (c)

$$\begin{aligned}\|\psi_n - \psi_m\| &\le C\|(A + K - \lambda)(\psi_n - \psi_m)\| \\ &\le C(\|(A - \lambda)\psi_n\| + \|(A - \lambda)\psi_m\| + \|K(\psi_n - \psi_m)\|) \to 0\end{aligned}$$

as $m, n \to \infty$. This contradicts (b).

Conversely, suppose $\lambda \in \sigma_e(A)$. If $\dim N(A - \lambda) = \infty$, we can find an orthonormal sequence $\{\varphi_n\}$ such that $(A - \lambda)\varphi_n = 0$ (Lemma 3.5.2). Note that an orthonormal sequence has no convergent subsequence since

$$\|\varphi_n - \varphi_m\|^2 = 2, \qquad n \neq m. \tag{3.6.3}$$

Hence the sequence satisfies (a)–(c). Now suppose $N(A - \lambda)$ is finite dimensional. Then it can be claimed that there is a sequence $\{\psi_n\}$ of elements in $D(A)$ such that

$$\psi_n \perp N(A - \lambda), \qquad \|\psi_n\| = 1, \qquad (A - \lambda)\psi_n \to 0. \tag{3.6.4}$$

If this is true, then $\{\psi_n\}$ can have no convergent subsequence. For if $\psi_n \to \psi$, then $\psi \perp N(A - \lambda)$ and $\|\psi\| = 1$. But $\psi \in N(A - \lambda)$ since A is a closed operator (Corollary 2.2.1). Hence $\psi = 0$, which is a contradiction. Once we know that $\{\psi_n\}$ cannot have a convergent subsequence, we see that it satisfies (a)–(c).

Therefore, it remains only to prove (3.6.4). Suppose it did not hold. Let P be the orthogonal projection onto $N(A - \lambda)$. Then P is a compact operator (Lemma 3.6.1). If (3.6.4) did not hold, there would be a constant C such that

$$\|\psi\| \le C\|(A - \lambda)\psi\|, \qquad \psi \perp N(A - \lambda). \tag{3.6.5}$$

For arbitrary ψ this implies

$$\begin{aligned}\|\psi\|^2 &= \|(1 - P)\psi\|^2 + \|P\psi\|^2 \le C^2(\|(A - \lambda)\psi\|^2 + \|P\psi\|^2) \\ &= C^2\|(A + P - \lambda)\psi\|^2\end{aligned} \tag{3.6.6}$$

since

$$([A - \lambda]\psi, P\psi) = 0.$$

Inequality (3.6.6) implies that $\lambda \in \rho(A + P - \lambda)$, and consequently that $\lambda \notin \sigma_e(A)$. This contradiction completes the proof. □

Another useful criterion is given by

Theorem 3.6.2. *If $\lambda_0 \in \sigma(A)$ is not an isolated point of $\sigma(A)$, then λ_0 is in $\sigma_e(A)$.*

PROOF. If λ_0 is not isolated, there is a sequence $\{\lambda_n\}$ of points of $\sigma(A)$ such that $\lambda_n \neq \lambda_0$ and $\lambda_n \to \lambda_0$. For each n there is a $\psi_n \in D(A)$ such that

$$\|\psi_n\| = 1, \qquad \|(A - \lambda_n)\psi_n\| < |\lambda_0 - \lambda_n|/n. \tag{3.6.7}$$

Clearly $(A - \lambda_0)\psi_n \to 0$. I claim that $\{\psi_n\}$ has no convergent subsequence. For if $\psi_n \to \psi$, then we have $\|\psi\| = 1$ and $\psi \in N(A - \lambda_0)$. On the other hand,

$$\begin{aligned}(\lambda_0 - \lambda_n)(\psi_n, \psi) &= (\psi_n, \lambda_0\psi) - \lambda_n(\psi_n, \psi)\\ &= (\psi_n, A\psi) - \lambda_n(\psi_n, \psi) = ([A - \lambda_n]\psi_n, \psi).\end{aligned}$$

Hence, by (3.6.7)

$$|\lambda_0 - \lambda_n|\,|(\psi_n, \psi)| < |\lambda_0 - \lambda_n|/n.$$

Since $\lambda_n \neq \lambda_0$, this implies $(\psi_n, \psi) \to 0$. But $\psi_n \to \psi$, and consequently we know that $(\psi_n, \psi) \to \|\psi\|^2 = 1$. This contradiction proves the theorem. □

Corollary 3.6.1. *If $\lambda_0 \in \sigma(A)$ is not in $\sigma_e(A)$, then λ_0 is isolated.*

A sequence $\{u_n\}$ of elements of a Hilbert space $\mathcal{H}$ is said to converge *weakly* to a limit u if

$$(u_n - u, v) \to 0, \qquad v \in \mathcal{H}. \tag{3.6.8}$$

This is in contrast with convergence in norm, which is called *strong* convergence (cf. Section 1.5). We shall need the following important theorem, which will be proved in Appendix B.

Theorem 3.6.3. *Every bounded sequence in a Hilbert space has a weakly convergent subsequence.*

In contrast with Theorem 3.6.1 we have

Theorem 3.6.4. *A point λ is in $\sigma_e(A)$ iff there is a sequence $\{\psi_n\}$ of elements in $D(A)$ such that*

a. $\|\psi_n\| = 1$,
b. *ψ_n converges to 0 weakly, and*
c. $(A - \lambda)\psi_n \to 0$.

PROOF. In one direction this theorem is a simple consequence of Theorem 3.6.1. To see this note that a sequence satisfying (a) and (b) of our theorem cannot have a convergent subsequence. For if ψ_n satisfies (a) and (b) and $\psi_n \to \psi$, then $\|\psi\| = 1$ by (a) and $\psi = 0$ by (b). Thus (a)–(c) of the present theorem imply the conditions (a)–(c) of Theorem 3.6.2. Consequently they imply $\lambda \in \sigma_e(A)$. Conversely, suppose $\lambda \in \sigma_e(A)$. Then there is a sequence $\{\psi_n\}$ satisfying (a)–(c) of Theorem 3.6.1. By Theorem 3.6.3 there is a subsequence (also denoted by $\{\psi_n\}$) such that $\psi_n \to \psi$ weakly. Since $\{\psi_n\}$ has no convergent subsequence, there is a positive number δ such that

$$\|\psi_n - \psi\| \geq \delta, \qquad n = 1, 2, \ldots . \tag{3.6.9}$$

Since

$$(\psi_n, [A - \lambda]\varphi) = ([A - \lambda]\psi_n, \varphi), \qquad \varphi \in D(A),$$

we have

$$(\psi, (A - \lambda)\varphi) = 0, \qquad \varphi \in D(A).$$

Since A is self-adjoint, we see that $\psi \in N(A - \lambda)$. Set

$$\varphi_n = (\psi_n - \psi)/\|\psi_n - \psi\|.$$

Then $\|\varphi_n\| = 1$ and

$$(\varphi_n, \varphi) = (\psi_n - \psi, \varphi)/\|\psi_n - \psi\| \to 0$$

and

$$(A - \lambda)\varphi_n = (A - \lambda)\psi_n/\|\psi_n - \psi\| \to 0$$

in view of (3.6.9). Thus $\{\varphi_n\}$ satisfies (a)–(c) of the present theorem. □

We shall also need

Lemma 3.6.2. *If $\lambda \in \sigma(A)$ but $\lambda \notin \sigma_e(A)$, then λ is an eigenvalue of A.*

PROOF. Since $\lambda \in \sigma(A)$, there is a sequence $\{\psi_n\} \subset D(A)$ such that $\|\psi_n\| = 1$ and $(A - \lambda)\psi_n \to 0$ (Theorem 2.2.1). This sequence must have a convergent subsequence, for otherwise we would have $\lambda \in \sigma_e(A)$ (Theorem 3.6.1). Let $\{\psi_n\}$ denote the subsequence as well. Then $\psi_n \to \psi$. Since A is closed, $\psi \in N(A - \lambda)$. Moreover, $\psi \neq 0$ since $\|\psi\| = 1$. Hence λ is an eigenvalue of A. □

3.7. Relative Compactness

When V is given by (3.1.11) and $b_2 < b_0$, we showed in Section 3.1 that $(-\infty, b_2) \subset \rho(H)$, while $[b_0, \infty) \subset \sigma(H)$. In Sections 3.2–3.4 we showed that H had precisely N eigenvalues, where N is given in Theorem 3.4.1, and that all of them are in (b_2, b_0). We are now concerned with the remaining points of $[b_2, b_0)$. Our aim is to show that except for the N eigenvalues, there are no other points of $\sigma(H)$ in this interval. We shall do this by proving

Theorem 3.7.1. $\sigma_e(H) = [b_0, \infty)$.

Once this is known, it follows that any points in $[b_2, b_0)$ which are in $\sigma(H)$ must be isolated eigenvalues (Corollary 3.6.1 and Lemma 3.6.2). Thus there can be no points in $\sigma(H) \cap [b_2, b_0)$ other than the N eigenvalues. This gives the complete picture of $\sigma(H)$.

To prove Theorem 3.7.1 we need a bit more preparation. Let A be a closed operator. An operator B is called *compact relative to A* or *A-compact* if $D(A) \subset D(B)$ and

$$\|u_n\| + \|Au_n\| \leq C \tag{3.7.1}$$

implies that $\{Bu_n\}$ has a convergent subsequence. We have

Theorem 3.7.2. *If B is A-compact, then*

a. $\|Bu\| \leq C(\|u\| + \|Au\|) \leq C'(\|u\| + \|(A + B)u\|)$ *and*
b. *B is $(A + B)$-compact, and*
c. *If B is also closable, then for each $\varepsilon > 0$ there is a constant K_ε such that*

$$\|Bu\| \leq \varepsilon\|Au\| + K_\varepsilon\|u\|.$$

PROOF. If the first inequality in (a) did not hold, then there would exist a sequence $\{u_n\}$ such that

$$\|Bu_n\| \to \infty, \qquad \|u_n\| + \|Au_n\| = 1. \tag{3.7.2}$$

This is impossible because the u_n satisfy (3.7.1), while $\{Bu_n\}$ has no convergent subsequence. If the second inequality in (a) did not hold, there would be a sequence $\{u_n\}$ such that

$$\|u_n\| + \|Au_n\| = 1, \qquad \|u_n\| + \|(A + B)u_n\| \to 0. \tag{3.7.3}$$

Since the u_n satisfy (3.7.1), $\{Bu_n\}$ has a convergent subsequence (also denoted by $\{Bu_n\}$). Thus $Bu_n \to f$, and consequently $Au_n \to -f$. Since $u_n \to 0$ and A is closed, we must have $f = 0$. This contradicts the first part of (3.7.3). To prove (b) suppose

$$\|u_n\| + \|(A + B)u_n\| \leq C.$$

Then (3.7.1) holds with some other constant by (a). Hence $\{Bu_n\}$ has a convergent subsequence. If (c) did not hold, there would be an $\varepsilon > 0$ and a sequence $\{u_n\}$ such that

$$\|u_n\| + \|Au_n\| = 1, \qquad \|Bu_n\| > \varepsilon\|Au_n\| + n\|u_n\| \tag{3.7.4}$$

[if the first part of (3.7.4) does not hold, we can make it true by dividing u_n by $\|u_n\| + \|Au_n\|$]. By (a), $\|Bu_n\| \leq C$. Thus $u_n \to 0$, and consequently $\|Au_n\| \to 1$. By hypothesis, $\{Bu_n\}$ has a convergent subsequence, so we may assume $Bu_n \to w$. Since B is closable, w must vanish. But

$$\|w\| = \lim\|Bu_n\| \geq \varepsilon \lim\|Au_n\| = \varepsilon.$$

This contradiction shows that (c) holds. □

Theorem 3.7.3. *If A is self-adjoint and B is Hermitian and A-compact, then*

a. *$A + B$ is self-adjoint and*
b. $\sigma_e(A + B) = \sigma_e(A)$.

PROOF. In view of Theorem 2.6.1, (a) follows from (c) of Theorem 3.7.2. To prove (b), suppose $\lambda \in \sigma_e(A)$. Then there is a sequence $\{\psi_n\}$ satisfying (a)–(c) of Theorem 3.6.4. Since B is A-compact, $\{B\psi_n\}$ has a convergent subsequence, so we may assume $B\psi_n \to w$. Thus, for $v \in D(A)$

$$(w, v) = \lim(B\psi_n, v) = \lim(\psi_n, Bv) = 0.$$

Since $D(A)$ is dense, this implies $w = 0$. Hence $(A + B - \lambda)\psi_n \to 0$. Hence $\lambda \in \sigma_e(A + B)$ by Theorem 3.6.4. Conversely, assume that $\lambda \in \sigma_e(A + B)$. Since the operator $-B$ is $(A + B)$-compact (Theorem 3.7.2), it follows from what we have just proved that $\lambda \in \sigma_e(A + B - B) = \sigma_e(A)$. □

Theorem 3.7.4. *Suppose A is self-adjoint and B, C are Hermitian operators such that $D(A) \subset D(B)$, C is A-compact, and*

$$\|Bu\| \leq a\|Au\| + b\|u\|, \qquad u \in D(A), \tag{3.7.5}$$

with $a < 1$. Then

a. *C is $(A + B)$-compact and*
b. $\sigma_e(A + B + C) = \sigma_e(A + B)$.

PROOF. We have

$$\|Au\| \leq \|(A + B)u\| + \|Bu\| \leq \|(A + B)u\| + a\|Au\| + b\|u\|.$$

Hence

$$(1 - a)\|Au\| \leq \|(A + B)u\| + b\|u\|.$$

If

$$\|(A + B)u_n\| + \|u_n\| \leq C,$$

then (3.7.1) holds. Consequently $\{Cu_n\}$ has a convergent subsequence. This proves (a), and (b) follows from Theorem 3.7.3. □

The final step in proving Theorem 3.7.1 is

Theorem 3.7.5. *A potential $V(x)$ is H_0-compact iff it is locally square integrable and*

$$\int_a^{a+1} |V(x)|^2\, dx \to 0 \qquad \text{as} \quad |a| \to \infty. \tag{3.7.6}$$

We shall prove Theorem 3.7.5 in the next section. Meanwhile we shall show how it can be used to prove Theorem 3.7.1. Let V be given by

(3.1.11) and set

$$V_1(x) = \begin{cases} b_1, & x < a \\ b_3, & x > a, \end{cases}$$

$$V_2(x) = \begin{cases} 0, & x < 0 \\ b_2 - b_1, & 0 < x < a \\ 0, & x > a. \end{cases}$$

Then $V = V_1 + V_2$. Moreover, V_2 is H_0-compact by Theorem 3.7.5. V_1 satisfies an inequality of the type (3.7.5) with $a < 1$ (Theorem 2.7.1). Thus all of the hypotheses of Theorem 3.7.4 are satisfied. Hence we can conclude that

$$\sigma_e(H_0 + V) = \sigma_e(H_0 + V_1). \tag{3.7.7}$$

However, we showed in Section 3.1 that

$$\sigma(H_0 + V_1) = [b_0, \infty). \tag{3.7.8}$$

Moreover, since none of the points of this interval is isolated, Corollary 3.6.1 tells us that

$$\sigma_e(H_0 + V_1) = [b_0, \infty). \tag{3.7.9}$$

Theorem 3.7.1 follows from (3.7.7) and (3.7.9). □

3.8. Proof of Theorem 3.7.5

Suppose (3.7.6) does not hold. Then there exists a sequence $\{a_n\}$ of points and a $\delta > 0$ such that $|a_n| \to \infty$ and

$$\int_{a_n}^{a_n+1} |V(x)|^2\, dx \ge \delta, \qquad \text{each} \quad n. \tag{3.8.1}$$

Let $\varphi(x)$ be a twice continuously differentiable function such that

$$\varphi(x) \begin{cases} \ge 1, & 0 < x < 1 \\ = 0, & |x| > 2 \end{cases}$$

(the existence of such a function will be proved in Section 7.8). Set $\psi_n(x) = \varphi(x - a_n)$. Then

$$\|\psi_n\| + \|H_0\psi_n\| = \|\varphi\| + \|H_0\varphi\| \equiv C_0. \tag{3.8.2}$$

If V were H_0-compact, there would be a subsequence of $\{\psi_n\}$ such that $V\psi_n$ converges. Thus we may assume that $V\psi_n \to f$. Since

$$\int_{-\infty}^{\infty} |V\psi_n|^2\, dx \ge \int_{a_n}^{a_{n+1}} |V(x)|^2\, dx \ge \delta, \tag{3.8.3}$$

we have

$$\int_{-\infty}^{\infty} |f(x)|^2\, dx \ge \delta. \tag{3.8.4}$$

Let I be any bounded interval. For n sufficiently large, $\psi_n(x)$ vanishes in I. Since

$$\int_I |f(x)|^2\, dx \le 2\int_I |f - V\psi_n|^2\, dx + 2\int_I |V\psi_n|^2\, dx,$$

we see that

$$\int_I |f(x)|^2\, dx = 0.$$

Since I is arbitrary, we have

$$\int_{-\infty}^{\infty} |f(x)|^2\, dx = 0,$$

contradicting (3.8.4). Hence (3.8.1) implies that V is not H_0-compact.

Now we show that (3.7.6) implies that V is H_0-compact. Let $\varepsilon > 0$ be given. Take b so large that

$$\int_a^{a+1} |V(x)|^2\, dx < \varepsilon, \qquad |a| > b.$$

Set

$$V_b(x) = \begin{cases} V(x), & |x| \le b \\ 0, & |x| > b. \end{cases}$$

Then by (2.7.10),

$$\|(V - V_b)\psi\|^2 \le \frac{2\varepsilon m^2}{\hbar^4}\|H_0\psi\|^2 + \frac{5\varepsilon}{2}\|\psi\|^2. \tag{3.8.5}$$

Since

$$\psi(x) = \frac{1}{\sqrt{2\pi}}\int_{-\infty}^{\infty} e^{ikx}\hat{\psi}(k)\, dk \tag{3.8.6}$$

and

$$\psi'(x) = \frac{i}{\sqrt{2\pi}}\int_{-\infty}^{\infty} e^{ikx}\hat{\psi}(k)k\, dk, \tag{3.8.7}$$

we have

$$|\psi(x)|^2 \le \frac{1}{2\pi}\int_{-\infty}^{\infty}(1 + k^2)|\hat{\psi}(k)|^2\, dk \int_{-\infty}^{\infty}(1 + k^2)^{-1}\, dk \tag{3.8.8}$$

and

$$|\psi'(x)|^2 \le \frac{1}{2\pi}\int_{-\infty}^{\infty}(1 + k^2)k^2|\hat{\psi}(k)|^2\, dk \int_{-\infty}^{\infty}(1 + k^2)^{-1}\, dk. \tag{3.8.9}$$

Moreover, since

$$\int_{-\infty}^{\infty} k^2|\hat{\psi}(k)|^2\, dk \le \frac{1}{2}\int_{-\infty}^{\infty}|\hat{\psi}(k)|^2\, dk + \frac{1}{2}\int_{-\infty}^{\infty} k^4|\hat{\psi}(k)|^2\, dk, \tag{3.8.10}$$

we have

$$|\psi(x)|^2 + |\psi'(x)|^2 \le C(\|H_0\psi\|^2 + \|\psi\|^2), \qquad \psi \in D(H_0). \quad (3.8.11)$$

Now suppose $\{\psi_n\} \subset D(H_0)$ is a sequence such that

$$\|H_0\psi_n\| + \|\psi_n\| \le C_1. \quad (3.8.12)$$

Then (3.8.11) shows that $\psi_n(x)$ and $\psi_n'(x)$ are uniformly bounded. By the Arzela–Ascoli theorem, there is a subsequence of $\{\psi_n\}$ which converges uniformly on any bounded interval. If $\{\psi_n\}$ denotes this subsequence as well, we have

$$\|V_b(\psi_n - \psi_m)\|^2 = \int_{-b}^{b} |V(x)[\psi_n(x) - \psi_m(x)]|^2\, dx$$

$$\le \max_{|x|\le b} |\psi_n(x) - \psi_m(x)|^2 \int_{-b}^{b} |V(y)|^2\, dy \to 0.$$

Thus we may take this $< \varepsilon$ by taking m and n sufficiently large. On the other hand, (3.8.5) and (3.8.12) imply

$$\|(V - V_b)(\psi_n - \psi_m)\|^2 \le 2\varepsilon C_1^2 \max\left(\frac{m^2}{\hbar^4}, \frac{5}{2}\right) = \varepsilon C_2.$$

Hence we have $\|V(\psi_n - \psi_m)\|^2 \le 2\varepsilon(1 + C_2)$ for m, n sufficiently large. Since ε was arbitrary, we see that $V\psi_n$ converges. Thus V is H_0-compact.

Combining Theorems 3.7.3 and 3.7.5 we have

Theorem 3.8.1. *If $V(x)$ is locally square integrable and satisfies* (3.7.6), *then* $\sigma_e(H_0 + V) = [0, \infty)$.

More generally, we have by Theorem 3.7.4

Theorem 3.8.2. *Let V_1, V_2 be locally square-integrable functions such that*

$$\sup_a \int_a^{a+1} |V_1(x)|^2\, dx < \infty \quad (3.8.13)$$

and

$$\int_a^{a+1} |V_2(x)|^2\, dx \to 0 \quad \text{as} \quad |a| \to \infty. \quad (3.8.14)$$

Then

$$\sigma_e(H_0 + V_1 + V_2) = \sigma_e(H_0 + V_1). \quad (3.8.15)$$

To illustrate Theorems 3.8.1 and 3.8.2, let V_1 be given by (3.1.11) and let V_2 be any bounded function which tends to 0 as $|x| \to \infty$. We have $\sigma_e(H_0 + V_1 + V_2) = [b_0, \infty)$. The same holds true if

$$V_2(x) = c|x|^{-\alpha}, \qquad 0 < \alpha < \tfrac{1}{2}.$$

For then V_2 satisfies (3.8.14). It should be noticed that (3.8.14) does not require $V_2(x) \to 0$ as $|x| \to \infty$. For example, let $\{a_n\}$ be a sequence of positive numbers tending to 0. Let $V_2(x)$ equal 1 in the intervals

$$(0, a_1), (1 + a_1, 1 + a_1 + a_2), (2 + a_1 + a_2, 2 + a_1 + a_2 + a_3), \dots,$$

$$\left(n + \sum_{m=1}^{n} a_m, n + \sum_{m=1}^{n+1} a_m\right), \dots,$$

and vanish elsewhere. The widths of these intervals tend to 0 and the distance between two consecutive intervals is one. Although V_2 does not converge to 0, it is easily checked that (3.8.14) holds.

Exercises

1. Prove (3.1.7), (3.1.8), and (3.1.9).

2. Show that (3.2.6) is the most general solution of (3.2.4) and (3.2.5) in L^2.

3. Prove (3.2.11).

4. Prove (3.2.12) and (3.2.13).

5. Suppose ψ given by (3.2.8) is in L^2. If (3.2.7), (3.2.9), and (3.2.10) hold and one of the quantities $\gamma_1, \gamma_2, \gamma_3$ vanishes, then $\psi \equiv 0$.

6. For $n = 1, \dots, N$ draw the graphs of the equations (3.3.3) and (3.3.4) and prove Proposition 3.3.1.

7. Prove Corollaries 3.3.1 and 3.3.2.

8. What happens to the eigenvalues and eigenfunctions of H when b_1 and $b_3 \to \infty$ in (3.1.11)?

9. Can you approximate the eigenvalues of H when $b_1 = b_3$ and $a\sqrt{2m(b_0 - b_2)}\,/\pi\hbar$ is very small?

10. Show that (3.2.1) cannot have a nonvanishing solution in L^2 for $\lambda \geq b_0$.

11. Prove Corollary 3.5.3.

12. Prove the properties of an orthogonal projection described in Section 3.5.

13. If (3.6.4) does not hold, then (3.6.5) must hold.

14. Let A be a self-adjoint operator and suppose there is a compact operator K such that $\lambda \in \rho(A + K)$. Show that there is a compact Hermitian operator K_1 such that $\lambda \in \rho(A + K_1)$.

15. Prove (3.6.7).

16. If $\psi_n \to \psi$ weakly and $\psi_n \to \varphi$ strongly, show that $\psi = \varphi$.

17. Prove (3.7.2) and (3.7.3).

18. Prove (3.7.4).

19. Without using Corollary 3.6.1, show that the conclusion of Theorem 3.1.2 can be strengthened to read $[b, \infty) \subset \sigma_e(H)$.

20. Prove (3.8.2), (3.8.3), and (3.8.4).

21. Prove Theorem 3.8.2.

22. Show that the example given at the end of Section 3.8 satisfies (3.8.14).

4

The Negative Eigenvalues

4.1. The Possibilities

In Section 3.8 we gave a sufficient condition on V so that

$$\sigma_e(H_0 + V) = [0, \infty) \tag{4.1.1}$$

holds. When this happens we know that any negative spectrum must consist of isolated eigenvalues (Theorem 3.6.2 and Lemma 3.6.2). However, we do not know how many negative eigenvalues there are. There may be

a. none,
b. a finite number,
c. an infinite number.

The purpose of the present chapter is to give criteria for determining which of these possibilities occurs.

However, before we begin, we want to weaken the hypotheses on V that were made in Chapter 3. This is to allow more general potentials than those allowed there. However, we shall have to pay a price. The theory is not so simple, and we shall be required to use the theory of Lebesgue integration at one point. For those of you who are unfamiliar with the theory, we suggest you either believe that the theorems quoted are true (if you trust us) or skip the first three sections completely. Those choosing the latter course need only assume that all the potentials mentioned in this chapter are locally square integrable and satisfy (4.7.6). They can then begin with Section 4.4 without missing anything. The next two sections discuss potentials which do not satisfy these conditions.

4.2. Forms Extensions

If $V(x)$ is not locally square integrable, the intersection of $D(H_0)$ with $D(V)$ need not be very large. In fact, one can give simple examples where this intersection consists of only the function 0. To construct one, set

$$g(x) = \begin{cases} |x|^{-1/2}, & |x| \leq 1 \\ 0, & |x| > 1, \end{cases}$$

and let $\{\alpha_n\}$ be a sequence of all the rational numbers (positive and negative). Let

$$V(x) = \sum_{n=1}^{\infty} g(x - \alpha_n)/2^n.$$

Note that

$$\int_{-\infty}^{\infty} V(x)\, dx \leq \int_{-\infty}^{\infty} g(x)\, dx \sum_{n=1}^{\infty} 2^{-n},$$

but that V is not locally square integrable. Now it can be claimed that any continuous function in $D(V)$ must vanish identically. For, suppose $\varphi(x)$ is continuous and in $D(V)$. If it does not vanish everywhere, there must be an interval I with $|I| < 1$ such that $|\varphi(x)| \geq c > 0$ in I. Moreover, some α_n is in I. Thus,

$$V(x) \geq g(x - \alpha_n)/2^n,$$

and consequently,

$$\int |V(x)\varphi(x)|^2\, dx \geq c^2 2^{-2n} \int_I |x - \alpha_n|^{-1}\, dx = \infty.$$

On the other hand, every function in $D(H_0)$ is continuous (Theorem 2.7.4). Thus, $D(H_0) \cap D(V) = \{0\}$. It is obvious, therefore, that the operator

$$H_1 = H_0 + V \tag{4.2.1}$$

is not self-adjoint in such cases.

There are potentials of physical interest which are not locally square integrable and for which the operator (4.2.1) is not self-adjoint. Yet we know that the total energy is an observable. What is the corresponding operator?

To examine this problem, recall that (1.4.6) was a consequence of the fact that the expectation of a sum is the sum of the expectations (Theorem 1.4.1). Now the expectation of the kinetic energy is

$$\bar{T} = \frac{1}{2m}(L^2\psi, \psi) = \frac{1}{2m}(L\psi, L\psi) \tag{4.2.2}$$

and that of the potential energy is

$$\bar{V} = (V\psi, \psi). \tag{4.2.3}$$

Let C_0^∞ denote the set of *test functions*, that is, infinitely differentiable functions which vanish for $|x|$ large. Both $\bar{T}$ and $\bar{V}$ are defined on C_0^∞, provided $V(x)$ is locally integrable. Thus

$$\bar{E} = \frac{1}{2m}(L\psi, L\psi) + (V\psi, \psi) \tag{4.2.4}$$

is defined on the test functions for such V. By (4.2.1) this implies

$$\bar{E} = (H_1\psi, \psi). \tag{4.2.5}$$

But (4.2.5) holds only for $\psi \in D(H_1) = D(H_0) \cap D(V)$, which may be a very small set (in particular, it consists of only one function for the example just mentioned). Thus (4.2.5) may not say very much. But it does show that the self-adjoint operator corresponding to E is an extension of H_1. How does one find it?

This leads to several mathematical questions. Does H_1 have self-adjoint extensions, and if so, how many? And if there are more than one, which one do you choose? This last question makes physicists feel uneasy because the self-adjoint operator corresponding to an observable should be completely determined by the physical theory and should not be ambiguous (enough indeterminism is already allowed).

We are happy to report that H_1 *always* has self-adjoint extensions (we shall not bother to prove this fact since we shall not use it). But this does not satisfy anyone. At first, physicists and mathematicians sought conditions on V which would guarantee that H_1 have only one self-adjoint extension. In this case H_1 is called *essentially self-adjoint*. The literature abounds with research articles devoted to this question (mostly in higher dimensions). We shall not give any theorems on this topic because the best results known do not cover many potentials of physical interest. Instead, we shall take a different approach which is more physically significant. We shall look for a procedure which will find a well-determined self-adjoint operator from (4.2.4). Since the operator determined by this procedure will depend uniquely on L and V, it will meet the requirements of physics.

Let

$$h(u, v) = \frac{1}{2m}(Lu, Lv) + (Vu, v). \tag{4.2.6}$$

If V is locally integrable, $h(u, v)$ is defined for $u, v \in C_0^\infty$. It is linear in u and conjugate linear in v—just like a scalar product. For this reason, it is called a *bilinear form*. The expression $h(u) = h(u, u)$ is called the corresponding *quadratic form*. We want to extend the definition of $h(u, v)$ to as large a class of functions as possible. One method, suggested by the discussion in Section 1.6, is to define

$$h(u, v) = \lim_{n\to\infty} h(u_n, v_n) \tag{4.2.7}$$

whenever there are sequences $\{u_n\}$, $\{v_n\}$ of functions in C_0^∞ such that

$$u_n \to u \quad \text{in} \quad L^2, \qquad h(u_n - u_m) \to 0, \tag{4.2.8}$$

$$v_m \to v \quad \text{in} \quad L^2, \qquad h(v_n - v_m) \to 0, \tag{4.2.9}$$

provided limit (4.2.7) exists. The only problem is that one must make sure that this definition does not depend on the particular sequences $\{u_n\}$, $\{v_n\}$. We give a sufficient condition for this to happen.

Lemma 4.2.1. *If there exist constants $k, c > 0$ such that*

$$c(\|u'\|^2 + \|u\|^2) \leq h(u) + k\|u\|^2, \qquad u \in C_0^\infty, \tag{4.2.10}$$

then the limit (4.2.7) *exists whenever* (4.2.8) *and* (4.2.9) *hold, and it does not depend on the particular sequences chosen.*

We shall prove this lemma in the next section. Assuming it for the moment, we continue our search for a self-adjoint operator. Suppose (4.2.10) is satisfied. Then the limit (4.2.7) exists when (4.2.8) and (4.2.9) hold. Let $D(h)$ be the set of all $u \in L^2$ for which there is a sequence $\{u_n\} \subset C_0^\infty$ such that (4.2.8) holds. We define $h(u, v)$ for $u, v \in D(h)$ by (4.2.7), knowing that the limit is independent of the sequences chosen.

Next we define an operator H corresponding to h as follows. We say that $u \in D(H)$ and $Hu = f$ if $u \in D(h)$, $f \in L^2$, and

$$h(u, v) = (f, v), \qquad v \in D(h). \tag{4.2.11}$$

We must check to see whether this actually defines an operator, that is, we make sure that only one $f \in L^2$ can satisfy (4.2.11) for the same u. Suppose

$$h(u, v) = (f', v), \qquad v \in D(h),$$

holds as well as (4.2.11) with u the same. Then we have

$$(f - f', v) = 0, \qquad v \in D(h).$$

Since $C_0^\infty \subset D(h)$ and C_0^∞ is dense in L^2 (Theorem 7.8.2), we see that $f' = f$ (Lemma 1.5.2). Thus (4.2.11) defines an operator H. Clearly, it is linear. We call it the operator *associated* with the bilinear form $h(u, v)$. What we need is

Theorem 4.2.1. *If* (4.2.10) *holds, then the operator H associated with $h(u, v)$ is self-adjoint.*

PROOF. Set

$$b(u, v) = h(u, v) + k(u, v),$$

where k is the constant in (4.2.10). Then $b(u, v)$ is a bilinear form on $D(h)$, and

$$c\|u\|^2 \leq b(u), \qquad u \in D(h), \tag{4.2.12}$$

by (4.2.10). Note that $b(u, v)$ has all of the properties of a scalar product (see Section 1.5). Moreover, it can be claimed that $D(h)$ is a Hilbert space with this scalar product. For, suppose $b(u_n - u_m) \to 0$ as $m, n \to \infty$. Since $u_n \in D(h)$, there is a $v_n \in C_0^\infty$ such that $b(u_n - v_n) \le 1/n$ by (4.2.7) and (4.2.8). In fact, there is a sequence $\{\varphi_j\} \subset C_0^\infty$ such that $\varphi_j \to u_n$ and $h(\varphi_j - \varphi_l) \to 0$. Take j so large that

$$b(\varphi_j - \varphi_l) < 1/n, \qquad l > j.$$

Then

$$b(u_n - \varphi_j) = \lim_{l \to \infty} b(\varphi_l - \varphi_j) \le 1/n.$$

Now,

$$b(v_n - v_m)^{1/2} \le b(v_n - u_n)^{1/2} + b(u_n - u_m)^{1/2} + b(u_m - v_m)^{1/2} \to 0. \tag{4.2.13}$$

In particular, $\{v_n\}$ is a Cauchy sequence in L^2 by (4.2.12). Let u be its limit. Then $u \in D(h)$ by (4.2.13) and for any $\varepsilon > 0$

$$b(u - v_n) = \lim_{m \to \infty} b(v_m - v_n) < \varepsilon$$

for n sufficiently large. Thus, $b(u - v_n) \to 0$. Consequently,

$$b(u - u_n)^{1/2} \le b(u - v_n)^{1/2} + b(v_n - u_n)^{1/2} \to 0,$$

showing that $D(h)$ is complete with respect to the norm $b(u)^{1/2}$.

Next let f be any function in L^2. Then (v, f) is a bounded linear functional on $D(h)$ since

$$|(v, f)| \le \|v\| \, \|f\| \le b(v)^{1/2}\|f\|/c^{1/2}$$

by (4.2.12). Hence, by the Riesz representation theorem (Theorem 1.9.1), there is a $u \in D(h)$ such that

$$(v, f) = b(v, u), \qquad v \in D(h).$$

This is the same as

$$h(u, v) = (f - ku, v), \qquad v \in D(h).$$

Hence $u \in D(H)$ and $Hu = f - ku$ or $(H + k)u = f$. Since f was arbitrary, we see that $R(H + k) = L^2$. Next it can be claimed that $D(H)$ is dense in L^2. To see this let w be any function in $D(h)$ such that

$$b(w, v) = 0, \qquad v \in D(H).$$

This implies that

$$(w, (H + k)v) = 0, \qquad v \in D(H).$$

Since $R(H + k) = L^2$, w must vanish. This shows that $D(H)$ is dense in $D(h)$. Thus if $v \in D(h)$, there is a sequence $\{v_j\} \subset D(H)$ which converges to v in $D(h)$. By (4.2.12), v_j converges to v in L^2. Since $D(h)$ is dense in L^2, we see that $D(H)$ is dense as well. Since $H + k$ is Hermitian, densely defined and onto, it is self-adjoint by Theorem 2.6.3. Thus H is self-adjoint, and the theorem is proved. □

We have shown that if $h(u, v)$ is defined by (4.2.6) and (4.2.10) holds, then the operator H associated with $h(u, v)$ is self-adjoint. If $u \in D(H_1)$, where H_1 is given by (4.2.1), then

$$(H_1 u, v) = h(u, v), \qquad v \in D(h).$$

Thus $u \in D(H)$, and $Hu = H_1 u$. This shows that H is an extension of H_1. It is called the *forms extension* because it is obtained by means of bilinear forms. Moreover, it is uniquely determined by H_0 and V since they determine $h(u, v)$. We shall take H to be the Hamiltonian operator corresponding to the total energy since (1) it is self-adjoint, (2) it is determined uniquely by the kinetic and potential energies, (3) it is an extension of H_1, and (4) it coincides with H_1 when H_1 is self-adjoint. To prove this last statement, suppose H_1 is self-adjoint and $u \in D(H)$. Then for $v \in D(H_1)$ we have

$$(u, H_1 v) = (u, Hv) = (Hu, v).$$

Consequently, we see that $u \in D(H_1)$ and $H_1 u = Hu$.

4.3. The Remaining Proofs

In this section we shall give the proof of Lemma 4.2.1. First we prove two other lemmas.

Lemma 4.3.1. *Suppose $h(u, v)$ satisfies* (4.2.10) *and $\{u_n\}$ is a sequence of functions in C_0^∞ such that $u_n \to 0$ in L^2 and $b(u_n) \le C$ for some constant C. Then there is a subsequence $\{\tilde{u}_j\}$ of $\{u_n\}$ such that*

$$b(v, \tilde{u}_j) \to 0, \qquad v \in C_0^\infty. \tag{4.3.1}$$

PROOF. By (2.7.6),

$$|u_n(x)| \le C_1$$

for some constant C_1. Since $u_n \to 0$ in L^2, there is a subsequence $\{\tilde{u}_j\}$ of $\{u_n\}$ which converges to 0 a.e. Hence

$$b(v, \tilde{u}_j) = \left(\left[\frac{1}{2m} L^2 + k\right] v, \tilde{u}_j\right) + \int V(x) v(x) \tilde{u}_j(x)^* \, dx.$$

The first term on the right tends to 0 as $j \to \infty$. The integrand of the second term is bounded by $C_1 |V(x) v(x)|$, which is in L^1. Moreover, the integrand converges to 0 a.e. Hence the integral converges to zero by the Lebesgue dominated convergence theorem. □

Lemma 4.3.2. *In addition to the hypotheses of Lemma* 4.3.1, *assume that*

$$h(u_n - u_m) \to 0 \qquad \text{as} \quad m, n \to \infty. \tag{4.3.2}$$

Then

$$h(u_n) \to 0 \qquad \text{as} \quad n \to \infty. \tag{4.3.3}$$

PROOF. Let $\{\tilde{u}_j\}$ be the subsequence of Lemma 4.3.1.

$$b(u_n) = b(u_n, u_n - \tilde{u}_j) + b(u_n, \tilde{u}_j).$$

Let $\varepsilon > 0$ be given, and take N so large that $b(u_n - \tilde{u}_j) < \varepsilon^2/C$ for $n, j > N$. Then

$$b(u_n) < \varepsilon + b(u_n, \tilde{u}_j), \qquad n, j > N.$$

Let $j \to \infty$. By Lemma 4.3.1, $b(u_n, \tilde{u}_j) \to 0$. Hence

$$b(u_n) \leq \varepsilon, \qquad n > N.$$

This implies (4.3.3). □

Now we can give the

PROOF OF LEMMA 4.2.1. Suppose $\{u'_n\}$ and $\{v'_n\}$ are sequences in C_0^∞ such that $u'_n \to u$, $v'_n \to v$, $h(u'_n - u'_m) \to 0$ and $h(v'_n - v'_m) \to 0$. Set $u''_n = u_n - u'_n$ and $v''_n = v_n - v'_n$. Then $u''_n \to 0$, $v''_m \to 0$, and

$$b(u''_n - u''_m)^{1/2} \leq b(u_n - u_m)^{1/2} + b(u'_n - u'_m)^{1/2} \to 0,$$

with a similar statement for $b(v''_n - v''_m)$. By Lemma 4.3.2, $b(u''_n) \to 0$ and $b(v''_n) \to 0$. Thus

$$b(u_n, v_n) - b(u'_n, v'_n) = b(u''_n, v_n) + b(u'_n, v''_n),$$

and this converges to 0. This gives the desired result. □

Now that we have Theorem 4.2.1 we should give a criterion for (4.2.10) to hold. This is easily done by the methods of Chapter 2. Set

$$V_+(x) = \max(V(x), 0), \qquad V_-(x) = V_+(x) - V(x). \tag{4.3.4}$$

We have

Theorem 4.3.1. *If*

$$C_0 = \sup_x \int_x^{x+1} V_-(y)\,dy < \infty, \tag{4.3.5}$$

then there are constants $k, c > 0$ *such that* (4.2.10) *holds.*

PROOF. Set $w(x) = V_-(x)^{1/2}$. Then

$$C_0 = \sup_x \int_x^{x+1} w(y)^2\,dy.$$

By (2.7.7),

$$\|wu\|^2 \le \varepsilon C_0\|u'\|^2 + C_0(1+\varepsilon^{-1})\|u\|^2$$

for any $\varepsilon > 0$. Thus we have

$$h(u) \ge \frac{\hbar^2}{2m}\|u'\|^2 - \|wu\|^2 \ge \left(\frac{\hbar^2}{2m} - \varepsilon C_0\right)\|u'\|^2 - C_0(1+\varepsilon^{-1})\|u\|^2.$$

We need only take $\varepsilon < \hbar^2/2mC_0$. □

Summarizing, we have

Theorem 4.3.2. *If* (4.3.5) *holds, then* H_1 *has a self-adjoint forms extension.*

4.4. Negative Eigenvalues

In discussing the number of negative eigenvalues of the Hamiltonian operator, we shall need a few results for self-adjoint operators.

Theorem 4.4.1. *If* A *is self-adjoint and* $\sigma(A) \subset [0, \infty)$, *then*

$$(Au, u) \ge 0, \qquad u \in D(A). \tag{4.4.1}$$

PROOF. By the spectral theorem (Theorem 1.10.1),

$$(Au, u) = \int_{-\infty}^{\infty} \lambda \, d(E(\lambda)u, u).$$

Since $(-\infty, 0) \subset \rho(A)$, $E(\lambda)u$ is constant for $\lambda < 0$ (Theorem 2.1.3). Thus

$$(Au, u) = \int_0^{\infty} \lambda \, d(E(\lambda)u, u) \ge 0.$$

□

Corollary 4.4.1. *If* A *is self-adjoint,* $\sigma_e(A) \subset [0, \infty)$ *and there is a* $u \in D(A)$ *such that* $(Au, u) < 0$, *then* A *has a negative eigenvalue.*

Lemma 4.4.1. *If* M, N *are subspaces of a Hilbert space* $\mathcal{H}$ *and* $\dim N > \dim M$, *then there is a* $u \in N$ *such that* $u \ne 0$ *and* $u \perp M$.

PROOF. Let $u_1, \ldots, u_n$ be linearly independent vectors in N, where $n > \dim M$. By the projection theorem (Theorem 3.5.1), each u_j can be written as $u_j = u_j' + u_j''$, where $u_j' \in M$ and $u_j'' \perp M$. Since $\dim M < n$, the u_j' must be linearly dependent. Thus there are α_j not all zero such that $\Sigma \alpha_j u_j' = 0$. Hence $\Sigma \alpha_j u_j = \Sigma \alpha_j u_j''$ is in N and orthogonal to M. It cannot vanish because the α_j are not all zero and the u_j are linearly independent. □

If λ is an eigenvalue of A, then $n = \dim N(A - \lambda)$ is positive. It is called the *multiplicity of* λ. If n is finite, one can find an orthonormal set $\varphi_1, \ldots, \varphi_n$ of eigenelements corresponding to λ which span $N(A - \lambda)$. We can show this inductively as follows. If $0 \le k < n$ and $\varphi_1, \ldots, \varphi_k$ form an orthonormal set in $N(A - \lambda)$, the dimension of the subspace spanned by them is k. By Lemma 4.4.1 there is a unit vector φ_{k+1} in $N(A - \lambda)$ orthogonal to this subspace. Thus $\varphi_1, \ldots, \varphi_{k+1}$ form an orthonormal set in $N(A - \lambda)$. The induction will continue until $k = n$. But then the $\varphi_1, \ldots, \varphi_n$ span a subspace of $N(A - \lambda)$ of dimension n. Thus this subspace must be the whole of $N(A - \lambda)$. In discussing the number of eigenvalues of an operator we shall always consider each eigenvalue repeated as many times as its multiplicity.

Theorem 4.4.2. *Suppose A is self-adjoint and $\sigma_e(A) \subset [0, \infty)$. Then A has a finite number of negative eigenvalues iff there is a finite-dimensional subspace M such that*

$$(Au, u) \ge 0, \qquad u \in D(A), \qquad u \perp M. \tag{4.4.2}$$

PROOF. Suppose $\lambda_1, \ldots, \lambda_n$ are the only negative eigenvalues of A. Let $u_1, \ldots, u_n$ be corresponding eigenelements, and let M be the subspace spanned by them. Then $M \subset D(A)$. Now A maps M into itself. It also maps

$$M^\perp = \{u \mid u \perp M\} \tag{4.4.3}$$

into itself. For if $u \in D(A) \cap M^\perp$, then

$$(Au, u_k) = (u, Au_k) = \lambda_k(u, u_k) = 0.$$

Moreover, A is self-adjoint on $M^\perp$. For, suppose u, f are in $M^\perp$ and

$$(u, Av) = (f, v), \qquad v \in D(A) \cap M^\perp. \tag{4.4.4}$$

For any $w \in D(A)$, we have $w = w' + w''$, where $w' \in M$ and $w'' \perp M$. Since $M \subset D(A)$, we have $w'' \in D(A)$. Hence

$$(u, Aw) = (u, Aw'') = (f, w'') = (f, w).$$

Since A is self-adjoint, this implies that $u \in D(A)$ and $Au = f$. Thus A is self-adjoint in $M^\perp$. But the restriction of A to $M^\perp$ has no negative eigenvalues (all of the eigenelements corresponding to the negative eigenvalues of A are in M). Hence (4.4.2) holds. Conversely, assume that there is finite-dimensional subspace M such that (4.4.2) holds and that A has an infinite number of negative eigenvalues. If $\dim M < n$, then we can find n linearly independent eigenelements corresponding to negative eigenvalues. Let N be the subspace spanned by them. Then

$$(Au, u) < 0, \qquad u \in N, \qquad u \ne 0. \tag{4.4.5}$$

But by Lemma 4.4.1 there is a nonzero vector in $N \cap M^\perp$. For such a u, (4.4.2) and (4.4.5) contradict each other. This proves the theorem. □

Corollary 4.4.2. *If (4.4.2) holds, then the number of negative eigenvalues is $\le \dim M$.*

4.5. Existence of Bound States

Now we apply the theory of Section 4.4 to the Hamiltonian operator H. We shall assume throughout that

$$\sigma_e(H) \subset [0, \infty) \tag{4.5.1}$$

(sufficient conditions for (4.5.1) to hold were given in Section 3.8). The first question we ask is, when does H have at least one negative eigenvalue, or equivalently a *bound state* (eigenfunctions corresponding to eigenvalues not in $\sigma_e(H)$ are called bound states because a particle in such a state cannot leave the system without additional energy)? We give a sufficient condition in

Theorem 4.5.1. *If there is a number a such that*

$$\inf_{\alpha>0} \alpha^{-1} \int_{-\infty}^{\infty} V(x) e^{-2\alpha^2(x-a)^2}\, dx < -\frac{\hbar^2\sqrt{2\pi}}{4m}, \tag{4.5.2}$$

then H has a least one negative eigenvalue.

PROOF. Put $u(x) = e^{-\alpha^2(x-a)^2}$. Then

$$u' = -2\alpha^2(x-a)u, \qquad \|u'\|^2 = 4\alpha^4\|(x-a)u\|^2.$$

But

$$\|(x-a)u\|^2 = \int (x-a)^2 e^{-2\alpha^2(x-a)^2}\, dx$$

$$= (2\alpha^2)^{-3/2} \int y^2 e^{-y^2}\, dy = \sqrt{2\pi}\,/8\alpha^3.$$

Thus

$$(H_0u, u) = \frac{\hbar^2}{2m}\|u'\|^2 = \alpha\hbar^2\sqrt{2\pi}\,/4m.$$

Hence

$$(H_1u, u) = \alpha\left(\frac{\hbar^2\sqrt{2\pi}}{4m} + \alpha^{-1}\int V(x) e^{-2\alpha^2(x-a)^2}\, dx\right).$$

By (4.5.2) this is negative for some $\alpha > 0$. The theorem follows now from Corollary 4.4.1. □

Corollary 4.5.1. *If there is a set I such that*

$$\int_I V(x)\, dx < 0$$

and $V(x) \le 0$ outside I, then H has a negative eigenvalue.

PROOF. We have

$$\int_{-\infty}^{\infty} V(x)e^{-2\alpha^2x^2}\,dx \le \int_I \to \int_I V(x)\,dx < 0.$$

Apply Theorem 4.5.1. □

Corollary 4.5.2. *If*

$$\int_{-\infty}^{\infty} V(x)\,dx < 0,$$

then H has a negative eigenvalue.

4.6. Existence of Infinitely Many Bound States

In this section we shall give a sufficient condition that H have infinitely many negative eigenvalues. In searching for such a criterion, the basic idea is to make use of Theorem 4.4.2. We have

Lemma 4.6.1. *Suppose A is a self-adjoint operator such that $\sigma_e(A) \subset [0, \infty)$. Then A has an infinite number of negative eigenvalues iff there exists an infinite-dimensional subspace $N \subset D(A)$ such that*

$$(Au, u) < 0, \qquad u \in N, \qquad u \neq 0. \tag{4.6.1}$$

PROOF. If A has an infinite number of negative eigenvalues, we can take as N the subspace spanned by the corresponding eigenelements. On the other hand, if (4.6.1) holds, then (4.4.2) cannot be true. For if it were, Lemma 4.4.1 assures us of a nonvanishing $u \in N$ which is orthogonal to M. This u would satisfy both (4.4.2) and (4.6.1), an obvious impossibility. □

By using Lemma 4.6.1 we obtain

Theorem 4.6.1. *Suppose $V(x) \le 0$ for $x \ge a$ and that* (4.5.1) *holds. If there is a $b > 1$ such that*

$$\liminf_{R\to\infty} R\int_R^{bR} V(x)\,dx < -\frac{\hbar^2}{2m}, \tag{4.6.2}$$

then H has an infinite number of negative eigenvalues.

PROOF. Let $\delta > 0$ be such that the left-hand side of (4.6.2) equals $-\hbar^2(1 + 4\delta)/2m$. Then there is a sequence $\{R_k\}$ such that $R_k \to \infty$ and

$$R_k\int_{R_k}^{bR_k} V(x)\,dx \le -\frac{\hbar^2}{2m}(1 + 3\delta). \tag{4.6.3}$$

Next we let $\varphi(x)$ be a test function such that (cf. Section 4.2)

1. $0 \leq \varphi(x) \leq 1$ for all x,
2. there is a $\rho > 0$ such that $\varphi(x) = 0$ for $x < \rho$,
3. $\varphi(x) = 1$ for $1 \leq x \leq b$,
4. $\varphi(x) = 0$ for $x > b + c$ for some $c > 2\delta^{-1}$,
5. $|\varphi'(x)|^2 \leq 1 + \delta$ for $\rho \leq x \leq 1$,
6. $|\varphi'(x)|^2 \leq 2/c^2$ for $b \leq x \leq b + c$.

To construct such a function we first construct a piecewise linear function having these properties and then "round off the corners." Now,

$$\|\varphi'\|^2 = \int_0^1 + \int_b^{b+c} |\varphi'(x)|^2\, dx < 1 + 2\delta.$$

Set $\psi_R(x) = R^{-1/2}\varphi(x/R)$, $R > 0$. Then $\psi_R'(x) = R^{-3/2}\varphi'(x/R)$, $\|\psi_R\| = \|\varphi\|$, and

$$\|\psi_R'\|^2 = R^{-3}\int |\varphi'(x/R)|^2\, dx = R^{-2}\|\varphi'\|^2.$$

Thus,

$$(H_0\psi_R, \psi_R) = \frac{\hbar^2}{2m}\|\psi_R'\|^2 < \hbar^2(1 + 2\delta)/2mR^2. \tag{4.6.4}$$

Also,

$$\begin{aligned}(V\psi_R, \psi_R) &= R^{-1}\int_{-\infty}^{\infty} V(x)|\varphi(x/R)|^2\, dx \\ &\leq R^{-1}\int_R^{bR} V(x)\, dx.\end{aligned} \tag{4.6.5}$$

Here we have used the fact that $V(x) \leq 0$ for $x > \rho R$ when R is sufficiently large (note that $\varphi(x/R)$ vanishes for $x < \rho R$). Thus we have

$$(H\psi_R, \psi_R) \leq R^{-2}\left[\frac{\hbar^2}{2m}(1 + 2\delta) + R\int_R^{bR} V(x)\, dx\right]. \tag{4.6.6}$$

Now by taking a subsequence of $\{R_k\}$ we can arrange that

$$R_k > (b + c)R_{k-1}/\rho. \tag{4.6.7}$$

Set $\psi_k = \psi_{R_k}$. Then (4.6.7) ensures that the supports of the functions ψ_k do not intersect (the *support* of a function is the closure of the set of points where it does not vanish). Thus

$$(\psi_j, \psi_k) = (H\psi_j, \psi_k) = 0, \qquad j \neq k. \tag{4.6.8}$$

In particular, the ψ_j span an infinite-dimensional space. By (4.6.3) and (4.6.6),

$$(H\psi_k, \psi_k) \le R_k^{-2}\left[\frac{\hbar^2}{2m}(1+2\delta) + R_k\int_{R_k}^{bR_k} V(x)\,dx\right]$$

$$\le R_k^{-2}\left[\frac{\hbar^2}{2m}(1+2\delta) - \frac{\hbar^2}{2m}(1+3\delta)\right] = -\frac{\hbar^2\delta}{2mR_k^2} < 0.$$

Thus, if $\psi = \Sigma\alpha_k\psi_k$,

$$(H\psi, \psi) = \sum |\alpha_k|^2 (H\psi_k, \psi_k) < 0.$$

Thus we have $(H\psi, \psi) < 0$ on an infinite-dimensional subspace. The result follows from Lemma 4.6.1. □

If $V \le 0$ near $-\infty$, we have the following variation of Theorem 4.6.1.

Theorem 4.6.2. *Suppose* (4.5.1) *holds and* $V(x) \le 0$ *for* $x \le a$. *If there is a* $b > 1$ *such that*

$$\liminf_{R\to\infty} R\int_{-bR}^{-R} V(x)\,dx < -\frac{\hbar^2}{2m}, \tag{4.6.9}$$

then H *has an infinite number of negative eigenvalues.*

The changes required to prove Theorem 4.6.2 are obvious. The following is a simple corollary of Theorems 4.6.1 and 4.6.2.

Corollary 4.6.1. *Suppose* (4.5.1) *holds and* $V(x) \le 0$ *for* $x \ge a$. *If*

$$\int_a^\infty V(x)\,dx = -\infty, \tag{4.6.10}$$

then H *has an infinite number of negative eigenvalues. The same conclusion holds if* $V(x) \le 0$ *for* $x \le a$ *and* (4.6.10) *is replaced by*

$$\int_{-\infty}^a V(x)\,dx = -\infty. \tag{4.6.11}$$

PROOF. Suppose (4.6.10) holds. Then for j sufficiently large,

$$\sum_{k=j}^\infty \int_{2^k}^{2^{k+1}} V(x)\,dx = -\infty.$$

Set

$$\alpha_k = \int_{2^k}^{2^{k+1}} V(x)\,dx.$$

Then $\Sigma\alpha_k$ is a divergent series. In particular, this implies that $\{2^k\alpha_k\}$ is not a bounded sequence. For if it were bounded by a constant C_0, we would have

$$|\alpha_k| \le C_0/2^k,$$

which implies that $\Sigma|\alpha_k|$ converges. Thus there is a sequence $\{l_k\}$ such that

$$2^{l_k}\alpha_{l_k} \to -\infty.$$

This implies (4.6.2) with $R_k = 2^{l_k}$. Apply Theorem 4.6.1. In the case of (4.6.11) we apply Theorem 4.6.2. □

As an illustration, suppose

$$V(x) \le -\frac{\beta}{x^s}, \qquad x > a, \tag{4.6.12}$$

where $0 \le s \le 2$ and $\beta > 0$. Then

$$R\int_R^{bR} V(x)\,dx \le -\beta R^{2-s}(1 - b^{1-s})/(s-1). \tag{4.6.13}$$

Case 1: $0 \le s < 1$. The right-hand side of (4.6.13) converges to $-\infty$ for any $\beta > 0$ as $R \to \infty$.

Case 2: $s = 1$. The left-hand side of (4.6.13) equals

$$-\beta R \log b \to -\infty \qquad \text{as} \quad R \to \infty.$$

Case 3: $1 < s < 2$. The right-hand side of (4.6.13) converges to $-\infty$ for any β in this case as well.

Case 4: $s = 2$. The right-hand side of (4.6.13) equals $\beta(b^{-1} - 1)$. Taking b large, we see that (4.6.2) holds iff $\beta > \hbar^2/2m$. Thus we have proved

Corollary 4.6.2. *Suppose $V(x)$ satisfies* (4.6.12) *for $\beta > 0$ and $0 \le s \le 2$, and* (4.5.1) *holds. If $s < 2$, then H has an infinite number of negative eigenvalues. The same holds true if $s = 2$ and $\beta > \hbar^2/2m$.*

We can improve the last statement in Corollary 4.6.2. In fact we have

Theorem 4.6.3. *Suppose* (4.5.1) *holds and*

$$V(x) + \frac{\hbar^2}{8mx^2} \le 0, \qquad x > a. \tag{4.6.14}$$

If

$$\liminf_{R\to\infty} R\int_{e^R}^{e^{bR}} \left[\frac{1}{4x} + \frac{2mx}{\hbar^2} V(x)\right] dx < -1, \tag{4.6.15}$$

then H has an infinite number of negative eigenvalues. In particular, if

$$V(x) \le -\beta x^{-2}, \qquad x > a, \tag{4.6.16}$$

and $\beta > \hbar^2/8m$, the conclusion holds.

PROOF. We follow the proof of Theorem 4.6.1 and use a little trick. Let $v \in C_0^\infty$ be real valued, and set $v = x^{1/2}u$, $x = e^y$. Then

$$\|v'\|^2 = \int_{-\infty}^{\infty} \left| \frac{du}{dy} + \frac{1}{2}u \right|^2 dy = \int_{-\infty}^{\infty} \left[\left(\frac{du}{dy} \right)^2 + \frac{u^2}{4} \right] dy \qquad (4.6.17)$$

since

$$\int_{-\infty}^{\infty} u \frac{du}{dy}\, dy = 0. \qquad (4.6.18)$$

Hence

$$(Hv, v) = \frac{\hbar^2}{2m}\|u'\|^2 + \int \left[\frac{\hbar^2}{8m} + \tilde{V}(y) \right] u^2\, dy, \qquad (4.6.19)$$

where $\tilde{V}(y) = e^{2y}V(e^y)$. Take $u = \psi_R$, the function defined in the proof of Theorem 4.6.1. Then we have

$$(Hv, v) \le \frac{\hbar^2(1 + 2\delta)}{2mR^2} + \frac{1}{R}\int_R^{bR} \left[\frac{\hbar^2}{8m} + \tilde{V}(y) \right] dy$$

$$\le \frac{\hbar^2}{2mR^2}\left[1 + 2\delta + R\int_{e^R}^{e^{bR}} \left(\frac{1}{4x} + \frac{2mx}{\hbar^2} V(x) \right) dx \right].$$

The result now follows as in the proof of Theorem 4.6.1. □

4.7. Existence of Only a Finite Number of Bound States

In this section we shall give a criterion for H to have a finite number of negative eigenvalues. We shall prove

Theorem 4.7.1. *Assume that* (4.3.5) *and* (4.5.1) *hold. If there is an N such that*

$$R\int_R^{\infty} V_-(\pm x)\, dx \le \frac{\hbar^2}{8m}, \qquad R > N, \qquad (4.7.1)$$

then H has at most a finite number of negative eigenvalues. In particular, this is true if

$$V(x) \ge -\frac{\hbar^2}{8mx^2}, \qquad |x| > N. \qquad (4.7.2)$$

The proof of Theorem 4.7.1 will follow from a sequence of lemmas. For each $N > 0$ we set

$$b_N(u) = \int_{|x|<N} \left[\frac{\hbar^2}{2m}|u'(x)|^2 + V(x)|u(x)|^2 \right] dx, \qquad (4.7.3)$$

and we let $b^N(u)$ denote the integral of the same integrand over the set $|x| > N$. Note that

$$(Hu, u) = b_N(u) + b^N(u). \tag{4.7.4}$$

Also, we set

$$r_N(u) = |u(N)|^2 + |u(-N)|^2, \tag{4.7.5}$$

$$\|u\|_N^2 = \int_{|x|<N} |u(x)|^2\,dx, \qquad |u|_{1,N}^2 = \|u'\|_N^2 + \|u\|_N^2. \tag{4.7.6}$$

Lemma 4.7.1. *For each $\varepsilon > 0$ we have*

$$r_N(u) \le 2\varepsilon\|u'\|_N^2 + 2(1 + \varepsilon^{-1})\|u\|_N^2 \tag{4.7.7}$$

and

$$b_N(u) \ge \left(\frac{\hbar^2}{2m} - \varepsilon C_0\right)\|u'\|_N^2 - C_0(1 + \varepsilon^{-1})\|u\|_N^2, \tag{4.7.8}$$

where C_0 is defined by (4.3.5).

PROOF. Inequality (4.7.7) follows from (2.7.6). By (2.7.7),

$$\int_{|x|<N} V_-(x)|u(x)|^2\,dx \le \varepsilon C_0\|u'\|_N^2 + C_0(1 + \varepsilon^{-1})\|u\|_N^2. \tag{4.7.9}$$

This implies (4.7.8). □

Lemma 4.7.2. *If $|u_n|_{1,N} \le C$, then the sequence $\{u_n\}$ has a subsequence which converges uniformly in $I = [-N, N]$.*

PROOF. If $u(x)$ is real valued, then

$$u(x) - u(x') = \int_{x'}^{x} u'(y)\,dy,$$

and consequently,

$$|u(x) - u(x')|^2 \le |x - x'|\left|\int_{x'}^{x} |u'(y)|^2\,dy\right|$$

by the Schwarz inequality. Thus,

$$|u(x) - u(x')|^2 \le |x - x'|\,\|u'\|_N^2, \qquad x, x' \in I. \tag{4.7.10}$$

This inequality also holds for complex-valued $u(x)$. It shows that the functions satisfying the hypothesis of the lemma are equicontinuous. Moreover, they are uniformly bounded by (2.7.6). The result now follows by the Arzela–Ascoli theorem. □

Lemma 4.7.3. *If $\{u_n\}$ is an infinite orthonormal sequence in $L^2(I)$, then*

$$\sup_n \|u_n'\|_N = \infty. \tag{4.7.11}$$

PROOF. If $\|u_n'\|_N \le C$, we would have by Lemma 4.7.2 that a subsequence of the u_n must converge uniformly in I, and *a fortiori* in $L^2(I)$. But an orthonormal sequence cannot converge in norm (see (3.6.3)). □

Lemma 4.7.4. *For each pair of constants C, N there is a finite-dimensional subspace M of $L^2(I)$ that*

$$b_N(u) \ge Cr_N(u), \qquad u \perp M. \tag{4.7.12}$$

PROOF. If this were not true, there would exist an orthonormal sequence $\{u_n\}$ in $L^2(I)$ such that

$$b_N(u_n) < Cr_N(u_n). \tag{4.7.13}$$

In view of Lemma 4.7.1, this implies

$$\|u_n'\|_N \le K\|u_n\|_N = K \tag{4.7.14}$$

for some constant K. This is impossible by Lemma 4.7.3. □

Lemma 4.7.5. *Set*

$$A^2 = \int_N^\infty x^{-2}|u(x)|^2\,dx, \quad B^2 = \int_N^\infty |u'(x)|^2\,dx, \quad C = |u(N)|^2/N. \tag{4.7.15}$$

Then $A^2 \le 2AB + C$ and $2AB \le 4B^2 + C$.

PROOF. First assume that $u(x)$ is real valued. We have

$$\begin{aligned} A^2 &= \int_N^\infty \left[\int_N^x 2u(y)u'(y)\,dy + u(N)^2\right] x^{-2}\,dx \\ &= 2\int_N^\infty u(y)u'(y)\left[\int_y^\infty x^{-2}\,dx\right] dy + u(N)^2\int_N^\infty x^{-2}\,dx \\ &= 2\int_N^\infty y^{-1}u(y)u'(y)\,dy + N^{-1}u(N)^2. \end{aligned}$$

If we apply the Schwarz inequality, we obtain the first desired inequality. Clearly, this will hold for complex-valued $u(x)$ as well. To prove the second inequality, note that

$$2AB \le \tfrac{1}{2}A^2 + 2B^2 \le AB + \tfrac{1}{2}C + 2B^2$$

by the first inequality. Thus

$$AB \le \tfrac{1}{2}C + 2B^2,$$

which is merely the second inequality. □

Lemma 4.7.6. *If V satisfies* (4.7.1), *then*

$$\int_N^\infty V_-(x)|u(x)|^2\,dx \le \frac{\hbar^2}{4m}(2B^2 + C). \tag{4.7.16}$$

PROOF. Again it suffices to prove the lemma for $u(x)$ real valued. The left-hand side of (4.7.16) equals

$$\int_N^\infty V_-(x)\left[2\int_N^x u(y)u'(y)\,dy + u(N)^2\right]dx$$

$$= 2\int_N^\infty u(y)u'(y)\left[\int_y^\infty V_-(x)\,dx\right]dy + u(N)^2\int_N^\infty V_-(x)\,dx$$

$$\leq \frac{\hbar^2}{4m}\int_N^\infty y^{-1}|u(y)u'(y)|\,dy + \frac{\hbar^2}{8mN}u(N)^2$$

$$\leq \frac{\hbar^2}{8m}(2AB + C) \leq \frac{\hbar^2}{4m}(2B^2 + C). \qquad \square$$

Corollary 4.7.1. *If* (4.7.1) *holds, then*

$$b^N(u) + \frac{\hbar^2}{4mN}r_N(u) \geq 0. \tag{4.7.17}$$

PROOF. This follows from (4.7.16) and the comparable inequality for the interval $(-\infty, -N)$. $\square$

Lemma 4.7.7. *If there exist constants* C, N *such that*

$$b^N(u) + Cr_N(u) \geq 0, \qquad u \in D(H), \tag{4.7.18}$$

then H *has at most a finite number of negative eigenvalues.*

PROOF. By Lemma 4.7.4 there is a finite-dimensional subspace M of $L^2(I)$ such that (4.7.12) holds. Define the functions in M to vanish outside I. Then M becomes a finite-dimensional subspace of L^2. In view of (4.7.4) we have

$$(Hu, u) \geq 0, \qquad u \perp M. \tag{4.7.19}$$

This implies the conclusion by Theorem 4.4.4. $\square$

Theorem 4.7.1 is an immediate consequence of Corollary 4.7.1 and Lemma 4.7.7. It is interesting to apply the results of this section and those of Section 4.6 to a potential satisfying

$$V(x) = -\beta/|x|^s, \qquad |x| > a > 0, \tag{4.7.20}$$

for some $\beta > 0$. We have

Corollary 4.7.2. *H has an infinite number of negative eigenvalues if either*

a. $s < 2$ *or*
b. $s = 2$ *and* $\beta > \hbar^2/8m$.

It has at most a finite number of negative eigenvalues if either

c. $s = 2$ *and* $\beta \leq \hbar^2/8m$ *or*
d. $s > 2$.

4.8. Another Criterion

In this section we shall show how the hypothesis $V(x) \leq 0$ for $x \geq a$ can be removed in Corollary 4.6.1. We shall prove

Theorem 4.8.1. *If* (4.5.1) *and either* (4.6.10) *or* (4.6.11) *holds for some a, then H has an infinite number of negative eigenvalues.*

PROOF. Assume (4.6.10) holds. Let $\rho > a$ be given and pick b so large that

$$\int_{\rho}^{\rho+1} |V(x)|\, dx + \int_{\rho+1}^{b} V(x)\, dx < -\frac{2\hbar^2}{m} \tag{4.8.1}$$

and

$$\int_b^x V(y)\, dy \leq 0, \qquad b < x \leq b+1. \tag{4.8.2}$$

Let $\varphi(x)$ be a test function such that

1. $0 \leq \varphi(x) \leq 1$, all x,
2. $\varphi(x) = 0$, $x < \rho$,
3. $\varphi(x) = 1$, $\rho + 1 \leq x \leq b$,
4. $\varphi(x) = 0$, $x > b + 1$,
5. $|\varphi'(x)|^2 \leq 3/2$, $\rho < x < \rho + 1$,
6. $|\varphi'(x)|^2 \leq 3/2$, $b < x < b + 1$.

Then $\|\varphi'\|^2 \leq 3$ and thus

$$(H_0\varphi, \varphi) \leq \frac{3\hbar^2}{2m}. \tag{4.8.3}$$

On the other hand,

$$(V\varphi, \varphi) = \int_{\rho}^{\rho+1} V(x)\varphi(x)^2\, dx + \int_{\rho+1}^{b} V(x)\, dx + \int_b^{b+1} V(x)\varphi(x)^2\, dx.$$

Set

$$W(x) = \int_b^x V(y)\, dy.$$

Then

$$\begin{aligned}\int_b^{b+1} V(x)\varphi(x)^2\, dx &= \int_b^{b+1} W'(x)\varphi(x)^2\, dx \\ &= W(b+1)\varphi(b+1)^2 - W(b)\varphi(b)^2 \\ &\quad -2\int_b^{b+1} \varphi(x)\varphi'(x) W(x)\, dx \leq 0\end{aligned}$$

since $W(x) \le 0$, $\varphi'(x) \le 0$, $\varphi(x) \ge 0$ in the interval $b < x < b + 1$, and $W(b) = \varphi(b + 1) = 0$. Thus,

$$(V\varphi, \varphi) \le \int_{\rho}^{\rho+1} |V(x)|\, dx + \int_{\rho+1}^{b} V(x)\, dx. \tag{4.8.4}$$

By (4.8.1), (4.8.3), and (4.8.4), we have

$$(H\varphi, \varphi) < 0. \tag{4.8.5}$$

By varying ρ we can find a sequence of functions φ whose supports do not intersect satisfying (4.8.5). The result follows from Lemma 4.6.1. □

Exercises

1. If A is self-adjoint and B is Hermitian and bounded, show that $A + B$ is self-adjoint.

2. Show that the operator H defined by (4.2.11) is linear.

3. If V satisfies (4.3.5), give a value of c such that (4.2.10) holds.

4. Show that there is a test function φ satisfying (1)–(6) of Section 4.6.

5. Prove (4.6.17), (4.6.18), and (4.6.19).

6. Prove the two statements in the proof of Lemma 4.7.3.

7. Construct a function φ satisfying (1)–(6) of Section 4.8.

5
Estimating the Spectrum

5.1. Introduction

If the potential $V(x)$ is not a simple function, it may be very difficult to calculate the spectrum of H. In this case there are methods of determining parts of the spectrum or estimating lower bounds. In this chapter we shall describe some of these methods. In particular, we shall be concerned with obtaining lower bounds for the spectrum and essential spectrum of H. Also, we shall seek sufficient conditions on $V(x)$ to ensure that $\sigma(H)$ contains an interval of the form $[\mu, \infty)$. We assume throughout that $V(x)$ is locally integrable.

5.2. Some Crucial Lemmas

In this section we give some important lemmas which will be needed for our methods. The first is fairly simple. Let I be an interval of length $|I|$. We use the notation

$$(u, v)_I = \int_I u(x)v(x)^* \, dx, \qquad \|u\|_I^2 = (u, u)_I. \tag{5.2.1}$$

We have

Lemma 5.2.1. *If $u(x)$ is continuous and vanishes somewhere in $\bar{I}$, then*

$$\pi\|u\|_I \le 2|I| \, \|u'\|_I. \tag{5.2.2}$$

The proof of Lemma 5.2.1 is not difficult. In Section 5.5 we shall give a proof in the spirit of the methods of Chapter 4. The next lemma is the foundation upon which most of the results of this chapter are built.

Lemma 5.2.2. *Suppose*

$$\int_I V(x)\,dx \geq \nu|I| \tag{5.2.3}$$

and

$$\int_I V_-(x)\,dx \leq \tau|I|. \tag{5.2.4}$$

Then for every $b > \max(2\tau, -\nu)$, *we have*

$$\left(\nu - \frac{(\nu + 2\tau)^2}{b + \nu}\right)\|u\|_I^2 \leq bk_0^2|I|^2\|u'\|_I^2 + (Vu, u)_I, \tag{5.2.5}$$

where

$$k_0 = 2/\pi. \tag{5.2.6}$$

PROOF. Clearly, it suffices to prove (5.2.5) for u real. Set

$$\rho^2 = V_+, \qquad \sigma^2 = V_-, \tag{5.2.7}$$

and

$$u_1^2 = \min_{\bar{I}} u^2, \qquad u_2^2 = \max_{\bar{I}} u^2.$$

Then by Lemma 5.2.1,

$$\|u - u_1\|_I \leq k_0|I|\,\|u'\|_I \tag{5.2.8}$$

and

$$\|u_2 - u\|_I \leq k_0|I|\,\|u'\|_I \tag{5.2.9}$$

(we take u_1 and u_2 to have signs such that $u - u_1$ and $u - u_2$ vanish in $\bar{I}$). Moreover,

$$\|\rho u\|^2 \geq \|\rho u_1\|^2 = \|\rho\|^2 u_1^2 = \|\rho\|^2\|u_1\|^2/|I| \tag{5.2.10}$$

and

$$\|\sigma u\|^2 \leq \|\sigma u_2\|^2 = \|\sigma\|^2 u_2^2 = \|\sigma\|^2\|u_2\|^2/|I|, \tag{5.2.11}$$

where we have suppressed the subscript I on the norms (this will be done throughout the remainder of the proof). By (5.2.8) and (5.2.10) we have

$$\|u\| \leq k_0|I|\,\|u'\| + |I|^{1/2}\|\rho u\|/\|\rho\|.$$

Thus for any $\alpha > 0$,

$$\|u\|^2 \leq k_0^2|I|^2(1 + \alpha^{-1})\|u'\|^2 + |I|(1 + \alpha)\|\rho u\|^2/\|\rho\|^2,$$

and consequently,

$$(1 + \alpha)^{-1}|I|^{-1}\|\rho\|^2\|u\|^2 \leq \alpha^{-1}k_0^2|I|\,\|\rho\|^2\|u'\|^2 + \|\rho u\|^2. \tag{5.2.12}$$

By (5.2.9) and (5.2.11),

$$|I|^{1/2}\|\sigma u\| \leq k_0|I|\,\|\sigma\|\,\|u'\| + \|\sigma\|\,\|u\|.$$

Thus for any $\beta > 0$,

$$\|\sigma u\|^2 \leq (1 + \beta)k_0^2|I|\ \|\sigma\|^2\|u'\|^2 + (1 + \beta^{-1})\|\sigma\|^2\|u\|^2/|I|. \tag{5.2.13}$$

Set

$$A = \|\rho\|^2/|I|, \qquad B = \|\sigma\|^2/|I|,$$

and

$$\alpha = \frac{A + B}{b - 2B}, \qquad \beta = \frac{b + A - B}{A + B}.$$

If $A + B \neq 0$, then α and β are positive. Assuming this for the moment, we obtain from (5.2.12) and (5.2.13)

$$\frac{(A - B)b - 4AB}{b + A - B}\|u\|^2 \leq bk_0^2|I|^2\|u'\|^2 + (Vu, u). \tag{5.2.14}$$

On the other hand, (5.2.14) is trivially true if $A = B = 0$. By (5.2.3), $A - B \geq \nu$. Moreover, if

$$f(A) = [(A - B)b - 4AB]/(b + A - B),$$

then

$$f'(A) = (b - 2B)^2/(b + A - B)^2.$$

Thus,

$$f(A) \geq f(B + \nu) = \nu - \frac{(\nu + 2B)^2}{\nu + b}. \tag{5.2.15}$$

This implies (5.2.5) except when

$$-b < \nu < -\tau - B. \tag{5.2.16}$$

To take care of this case note that $A \geq 0$ always. Thus when (5.2.16) holds, the statement $A \geq B + \nu$ does not tell us anything. Therefore it will be better to replace (5.2.15) by

$$f(A) \geq f(0) = -bB/(b - B). \tag{5.2.17}$$

Let $g(B)$ denote the right-hand side of (5.2.17). Then

$$g'(B) = -b^2/(b - B)^2.$$

This shows that

$$g(B) \geq g(\tau) = -\tau b/(b - \tau).$$

Next set

$$h(\nu) = \nu - \frac{(\nu + 2\tau)^2}{b + \nu}. \tag{5.2.18}$$

Then

$$h'(\nu) = (b - 2\tau)^2/(b + \nu)^2.$$

Hence in the interval

$$-b < \nu < -\tau \tag{5.2.19}$$

we have

$$h(\nu) \leq h(-\tau) = g(\tau) \leq g(B) \leq f(A).$$

Again this implies (5.2.5). □

5.3. A Lower Bound for the Spectrum

In this section we show how Lemma 5.2.2 can be used to obtain an estimate for the lowest point of $\sigma(H)$. For any $t > 0$ we define

$$\nu_t = \inf_y t^{-1}\int_y^{y+t} V(x)\,dx \tag{5.3.1}$$

$$\tau_t = \sup_y t^{-1}\int_y^{y+t} V_-(x)\,dx. \tag{5.3.2}$$

We have

Theorem 5.3.1. *If*

$$t^2\tau_t < \pi^2\hbar^2/16m, \tag{5.3.3}$$

then

$$\sigma(H) \subset \left[\nu_t - \frac{8mt^2(\nu_t + 2\tau_t)^2}{\hbar^2\pi^2 + 8mt^2\nu_t}, \infty\right). \tag{5.3.4}$$

PROOF. Let I be any interval of length t. Now for each y,

$$t^{-1}\int_y^{y+t} V(x)\,dx \geq -t^{-1}\int_y^{y+t} V_-(x)\,dx \geq -\tau_t.$$

Thus we have

$$\nu_t + \tau_t \geq 0. \tag{5.3.5}$$

Set $b = \pi^2\hbar^2/8mt^2$ in Lemma 5.2.2. Inequality (5.3.3) says that $b > 2\tau_t$. Thus by (5.2.5) we have

$$\left(\nu_t - \frac{(\nu_t + 2\tau_t)^2}{b + \nu_t}\right)\|u\|_I^2 \leq \frac{\hbar^2}{2m}\|u'\|_I^2 + (Vu, u)_I. \tag{5.3.6}$$

If we consider the real line as the union of disjoint intervals of length t and sum (5.3.6) over these intervals, we obtain

$$\left(\nu_t - \frac{(\nu_t + 2\tau_t)^2}{b + \nu_t}\right)\|u\|^2 \leq (Hu, u). \tag{5.3.7}$$

Thus if λ is any point below the interval in (5.3.4), we have

$$a\|u\|^2 \leq ([H-\lambda]u, u)$$

for some positive a. Thus λ cannot be in the spectrum of H (Theorem 2.2.1). □

Next, we let

$$\nu_0 = \limsup_{t\to 0} \nu_t, \tag{5.3.8}$$

$$\tau_0 = \limsup_{t\to 0} t\tau_t \tag{5.3.9}$$

(note the factor of t appearing in (5.3.9), which does not appear in (5.3.8)). The following is a useful consequence of Theorem 5.3.1.

Theorem 5.3.2. *If $\nu_0 < \infty$, then*

$$\sigma(H) \subset \left[\nu_0 - \frac{32m\tau_0^2}{\hbar^2\pi^2}, \infty\right). \tag{5.3.10}$$

PROOF. There is a sequence $\{t_n\}$ such that $t_n \to 0$ and $\nu_{t_n} \to \nu_0$. We may assume $\tau_0 < \infty$. Let $\varepsilon > 0$ be given. Then for n sufficiently large we have

$$t_n\tau_{t_n} < \tau_0 + \varepsilon, \qquad t_n^2\tau_{t_n} < \pi^2\hbar^2/16m. \tag{5.3.11}$$

By Theorem 5.3.1,

$$\sigma(H) \subset [\mu_n, \infty),$$

where

$$\mu_n = \nu_{t_n} - \frac{8m(t_n\nu_{t_n} + 2t_n\tau_{t_n})^2}{\hbar^2\pi^2 + 8mt_n^2\nu_{t_n}}.$$

Moreover, by (5.3.11)

$$\liminf_{n\to\infty} \mu_n \geq \nu_0 - \frac{32m(\tau_0+\varepsilon)^2}{\hbar^2\pi^2}. \tag{5.3.12}$$

Since this is true for each $\varepsilon > 0$, (5.3.10) follows. □

5.4. Lower Bounds for the Essential Spectrum

In this section we shall show how Lemma 5.2.2 can be used to obtain information concerning the essential spectrum of H. First we shall need

Theorem 5.4.1. *Suppose* (4.3.5) *of Chapter* 4 *holds and there is an* N *such that*

$$\int_{|x|>N} (Hu)u^*\,dx \geq \lambda_0 \int_{|x|>N} |u(x)|^2\,dx, \qquad u \in D(H). \tag{5.4.1}$$

Then

$$\sigma_e(H) \subset [\lambda_0, \infty). \tag{5.4.2}$$

PROOF. Suppose $\lambda \in \sigma_e(H)$. Then there is a sequence $\{u_n\} \subset D(H)$ such that $\|u_n\| = 1$, $(H - \lambda)u_n \to 0$, and u_n converges weakly to 0 (Theorem 3.6.4). Thus

$$b_N(u_n) + b^N(u_n) = (Hu_n, u_n) \to \lambda \tag{5.4.3}$$

(see Section 4.7). By (5.4.1),

$$\limsup_{n\to\infty} b_N(u_n) \le \lambda + |\lambda_0|$$

(note that $b^N(u)$ is the left-hand side of (5.4.1)). From inequality (4.7.8) we see that the $\|u_n'\|_N$ are uniformly bounded. Consequently, in view of Lemma 4.7.2, there is a subsequence (also denoted by $\{u_n\}$) which converges in $L^2(I)$ to a function u (here I is the interval $[-N, N]$). Since u_n converges weakly to 0 in L^2, the limit u must vanish. Thus

$$\|u_n\|_N \to 0, \qquad \int_{|x|>N} |u_n(x)|^2\, dx \to 1. \tag{5.4.4}$$

But by inequality (4.7.8), (5.4.1), and (5.4.3),

$$(Hu_n, u_n) \ge \lambda_0 \int_{|x|>N} |u_n(x)|^2\, dx - K\|u_n\|_N^2$$

for some constant K. Hence $\lambda \ge \lambda_0$ by (5.4.4). □

Now we give an application. Let

$$\lambda_t = \liminf_{|y|\to\infty} t^{-1} \int_y^{y+t} V(x)\, dx, \tag{5.4.5}$$

$$\omega_t = \limsup_{|y|\to\infty} t^{-1} \int_y^{y+t} V_-(x)\, dx. \tag{5.4.6}$$

We have

Theorem 5.4.2. *If*

$$t^2\omega_t < \pi^2\hbar^2/16m, \tag{5.4.7}$$

then

$$\sigma_e(H) \subset \left[\lambda_t - \frac{8mt^2(\lambda_t + 2\omega_t)^2}{\hbar^2\pi^2 + 8mt^2\lambda_t}, \infty\right). \tag{5.4.8}$$

PROOF. Let $\varepsilon > 0$ be given. Take N so large that

$$t^{-1}\int_y^{y+t} V(x)\, dx > \lambda_t - \varepsilon,\ t^{-1}\int_y^{y+t} V_-(x)\, dx < \omega_t + \tfrac{1}{2}\varepsilon \tag{5.4.9}$$

whenever $|y| > N$. Let I be any interval of length t and a distance greater than N from the origin. Set $b = \pi^2\hbar^2/8mt^2$. As in the case of (5.3.5), we have

$$\lambda_t + \omega_t \geq 0. \tag{5.4.10}$$

Now (5.4.7) implies $b > 2\omega_t + \varepsilon$ for ε sufficiently small, and (5.4.10) implies $b > \varepsilon - \lambda_t$ as well. Thus we may apply Lemma 5.2.2 to obtain

$$\left(\lambda_t - \varepsilon - \frac{(\lambda_t + 2\omega_t)^2}{b + \lambda_t - \varepsilon}\right)\|u\|_I^2 \leq (Hu, u)_I. \tag{5.4.11}$$

Consider the set $|x| > N$ as the union of a set of nonoverlapping intervals of length t and sum (5.4.11) over this set. This gives

$$\left(\lambda_t - \varepsilon - \frac{(\lambda_t + 2\omega_t)^2}{b + \lambda_t - \varepsilon}\right)\int_{|x|>N} |u(x)|^2\,dx \leq b^N(u). \tag{5.4.12}$$

Theorem 5.4.1 now tells us that any point λ in the essential spectrum of H satisfies

$$\lambda \geq \lambda_t - \varepsilon - \frac{(\lambda_t + 2\omega_t)^2}{b + \lambda_t - \varepsilon}.$$

Since this is true for every $\varepsilon > 0$ sufficiently small, we obtain (5.4.8). □

Next we set

$$\lambda_0 = \limsup_{t\to 0} \lambda_t, \tag{5.4.13}$$

$$\omega_0 = \limsup_{t\to 0} t\omega_t. \tag{5.4.14}$$

(Again, note the factor t, which appears in (5.4.14) but not in (5.4.13)). We have

Theorem 5.4.3. *If $\lambda_0 < \infty$, then*

$$\sigma_e(H) \subset \left[\lambda_0 - \frac{32m\omega_0^2}{\hbar^2\pi^2}, \infty\right) \tag{5.4.15}$$

PROOF. Again, there is nothing to prove if $\omega = \infty$. We follow the proof of Theorem 5.3.2 replacing ν_t by $\lambda_t - \varepsilon$ and τ_t by $\omega_t + \frac{1}{2}\varepsilon$. We apply Theorem 5.4.2 instead of Theorem 5.3.1. □

We note some consequences of Theorems 5.4.2 and 5.4.3.

Corollary 5.4.1. *If $\omega_0 = 0$, then*

$$\sigma_e(H) \subset [\lambda_0, \infty). \tag{5.4.16}$$

In this case if 0 is in $\sigma_e(H)$, then we must have $\lambda_0 \leq 0$.

Corollary 5.4.2. *If $\omega_0 < \infty$ and $\lambda_0 = \infty$, then H has no essential spectrum. Thus the only points in $\sigma(H)$ are isolated eigenvalues having no finite accumulation point.*

PROOF. Let M be any positive number. Take t_0 so small that

$$tM + 2t\omega_t < 2(\omega_0 + 1)$$

and (5.4.7) holds for $t < t_0$. Then

$$\frac{t^2(M + 2\omega_t)^2}{\pi^2\hbar^2 + 8mt^2M} < \frac{4(\omega_0 + 1)^2}{\pi^2\hbar^2}.$$

Take $t < t_0$ such that $\lambda_t > M$. Let $\varepsilon > 0$ be given and take N so large

$$t^{-1}\int_y^{y+t} V(x)\,dx > M, \qquad t^{-1}\int_y^{y+t} V_-(x)\,dx < \omega_t + \varepsilon$$

for $|y| > N$. Following the proof of Theorem 5.4.2 we see that any point $\lambda \in \sigma_e(H)$ must satisfy

$$\lambda \geq M - \frac{8mt^2(M + 2\omega_t)^2}{\hbar^2\pi^2 + 8mt^2M} > M - \frac{32m(\omega_0 + 1)^2}{\pi^2\hbar^2}.$$

Since M was arbitrary, there can be no points in $\sigma_e(H)$. □

5.5. An Inequality

In this section we give a proof of Lemma 5.2.1. Our method makes use of the basic ideas of the last few chapters.

First we note that it suffices to prove the inequality

$$\int_0^1 |u(x)|^2\,dx \leq \frac{4}{\pi^2}\int_0^1 |u'(x)|^2\,dx \tag{5.5.1}$$

for continuously differentiable functions $u(x)$ such that

$$u(0) = 0. \tag{5.5.2}$$

For it follows from this that

$$\int_0^a |u(x)|^2\,dx \leq \frac{4a^2}{\pi^2}\int_0^a |u'(x)|^2\,dx \tag{5.5.3}$$

holds for $u(x)$ satisfying (5.5.2). In fact, we have

$$\begin{aligned}\int_0^a |u(x)|^2\,dx &= a\int_0^1 |u(ay)|^2\,dy \\ &\leq \frac{4a}{\pi^2}\int_0^1 |du(ay)/dy|^2\,dy = \frac{4a^2}{\pi^2}\int_0^a |u'(x)|^2\,dx.\end{aligned}$$

Next we note that (5.5.3) implies

$$\int_a^b |u(x)|^2\,dx \leq \frac{4(b-a)^2}{\pi^2}\int_a^b |u'(x)|^2\,dx \tag{5.5.4}$$

when

$$u(a) = 0. \tag{5.5.5}$$

This follows if we replace x by $y + a$ and a by $b - a$ in (5.5.3). Moreover, if we replace x by $-y$ in (5.5.4), we see that it holds when (5.5.5) is replaced by

$$u(b) = 0. \tag{5.5.6}$$

If $a < c < b$ and $u(c) = 0$, we have by two applications of (5.5.4)

$$\int_a^c |u(x)|^2\,dx \le \frac{4(c-a)^2}{\pi^2}\int_a^c |u'(x)|^2\,dx$$

and

$$\int_c^b |u(x)|^2\,dx \le \frac{4(b-c)^2}{\pi^2}\int_c^b |u'(x)|^2\,dx.$$

These combine to give (5.5.4) for $u(x)$ vanishing anywhere in $I = [a, b]$. Finally we note that (5.5.4) is inequality (5.2.2).

Thus it remains only to prove (5.5.1) under the condition (5.5.2). To this end we use the scalar product

$$(u, v)_I = \int_0^1 u(x)v(x)^*\,dx$$

and suppress the subscript I. Let D be the set of continuously differentiable functions on I satisfying (5.5.2). We define an operator A on $L^2(I)$ as follows. We say that $u \in D(A)$ and $Au = f$ iff $u \in D$ and

$$(u', v') = (f, v), \qquad v \in D.$$

Note that A is well defined since D is dense in $L^2(I)$. Clearly, A is Hermitian. Moreover, $R(A)$ is the whole of $L^2(I)$. For if f is any function in $L^2(I)$, then

$$u(x) = \int_0^x \int_y^1 f(t)\,dt\,dy \tag{5.5.7}$$

is in D. Moreover,

$$\begin{aligned}(u', v') &= \int_0^1 \int_x^1 f(y)\,dy\,v'(x)^*\,dx \\ &= \int_0^1 f(y)\int_0^y v'(x)^*\,dx\,dy = (f, v), \qquad v \in D.\end{aligned}$$

Thus $u \in D(A)$ and $Au = f$. Thus A is self-adjoint by Theorem 2.6.3. Next we note that A is one-to-one. In fact, if $u \in D$, then

$$u(x) = \int_0^x u'(y)\,dy,$$

$$|u(x)|^2 \le x\int_0^x |u'(y)|^2\,dy,$$

and consequently,

$$\int_0^1 |u(x)|^2\,dx \le \frac{1}{2}\int_0^1 |u'(x)|^2\,dx \tag{5.5.8}$$

(if we had been willing to accept $\frac{1}{2}$ in place of $4/\pi^2$ in (5.5.1), we would have been finished long ago). Thus,

$$(Au, u) \ge 2\|u\|^2, \qquad u \in D(A). \tag{5.5.9}$$

This shows that all $\lambda < 2$ are in the resolvent set of A (Theorem 2.2.1).

Next we note that A has no essential spectrum. For if $\lambda \in \sigma_e(A)$, there would be a sequence $\{u_n\} \subset D(A)$ such that $(A - \lambda)u_n \to 0$, $\|u_n\| = 1$ and the u_n converge weakly to 0 (Theorem 3.6.4). This would imply

$$\|u_n'\|^2 = (Au_n, u_n) \to \lambda,$$

and consequently there would be a constant C such that $\|u_n'\| \le C$. This would imply that there is a subsequence (also denoted by $\{u_n\}$) such that u_n converges in $L^2(I)$ to some function u (Lemma 4.7.2). Since u_n converges weakly to 0, the function u must vanish. But this is impossible since $\|u\| = \lim\|u_n\| = 1$. Hence A has no essential spectrum. This implies that the smallest point μ of $\sigma(A)$ must be an eigenvalue. Thus

$$(A - \mu)u = 0 \tag{5.5.10}$$

for some nonvanishing $u \in D(A)$. This implies

$$u(x) = \mu A^{-1}u = \mu\int_0^x \int_y^1 u(t)\,dt\,dy. \tag{5.5.11}$$

Hence

$$u'(x) = \mu\int_x^1 u(y)\,dy \tag{5.5.12}$$

and

$$u''(x) = -\mu u(x). \tag{5.5.13}$$

The most general solution of (5.5.13) is

$$u(x) = a\sin\mu^{1/2}x + b\cos\mu^{1/2}x. \tag{5.5.14}$$

In order for u to be in D, we must have $b = 0$. Moreover, (5.5.12) tells us that $u'(1) = 0$. Since

$$u'(x) = \mu^{1/2}a\cos\mu^{1/2}x,$$

we must have

$$\mu^{1/2}a\cos\mu^{1/2} = 0.$$

We know that $\mu \ne 0$. Moreover, $a = 0$ would make $u \equiv 0$, preventing u from being an eigenfunction. Thus we must have

$$\cos\mu^{1/2} = 0. \tag{5.5.15}$$

Thus μ is the smallest positive solution of (5.5.15), namely, $\mu = \pi^2/4$. Since this is the smallest eigenvalue of A, we have

$$(Au, u) \geq \mu\|u\|^2, \qquad u \in D(A) \tag{5.5.16}$$

(Theorem 4.4.1). This implies that (5.5.1) holds for $u \in D$. For suppose $u \in D$. Then there is a sequence $\{v_n\}$ of continuously differentiable functions such that

$$v_n(1) = 0 \tag{5.5.17}$$

and $v_n \to u'$ in $L^2(I)$ (Theorem 7.8.2). Set

$$u_n(x) = \int_0^x v_n(y)\, dy. \tag{5.5.18}$$

Now $u_n \in D(A)$ and

$$\|u_n - u\|^2 \leq \tfrac{1}{2}\|v_n - u'\|^2 \to 0$$

by (5.5.8). Now, (5.5.16) says

$$\|v_n\|^2 \geq \mu\|u_n\|^2.$$

Taking the limit we obtain (5.5.1).

5.6. Bilinear Forms

In this section we shall look for sufficient conditions such that

$$\sigma_e(H) \supset [\mu, \infty) \tag{5.6.1}$$

for some μ. Our first result is

Theorem 5.6.1. *Assume that* (4.3.5) *holds and that there is a sequence* $\{I_n\}$ *of intervals such that* $|I_n| \to \infty$ *and*

$$|I_n|^{-1} \int_{I_n} |V(x) - \mu|\, dx \to 0, \tag{5.6.2}$$

where $|I|$ *denotes the length of* I. *Then* (5.6.1) *holds.*

In proving Theorem 5.6.1 we shall make use of two theorems in Hilbert space. They are related to the ideas of Section 4.2. Let $\mathcal{H}$ be a Hilbert space. A *bilinear form* $a(u, v)$ is a scalar function linear in u and conjugate linear in v defined for u, v in some subspace of $\mathcal{H}$. The subspace is called the *domain* of a and is denoted by $D(a)$. We shall write $a(u)$ for $a(u, u)$. A bilinear form a is called *closed* if

$$\{u_n\} \subset D(a), \qquad u_n \to u, \qquad a(u_n - u_m) \to 0 \tag{5.6.3}$$

implies that $u \in D(a)$ and $a(u_n - u) \to 0$ (compare this with the definition of a closed operator given in Section 1.6). A bilinear form is called *Hermitian* if

$$a(v, u) = a(u, v)^*, \qquad u, v \in D(a). \tag{5.6.4}$$

If the domain of a bilinear form a is dense, we can define an operator A on $\mathcal{H}$ as follows. We say that $u \in D(A)$ and $Au = f$ iff $u \in D(a)$ and

$$a(u, v) = (f, v), \qquad v \in D(a). \tag{5.6.5}$$

Note that f is unique. For if f' satisfies (5.6.5) as well, then

$$(f - f', v) = 0, \qquad v \in D(a).$$

Since $D(a)$ is dense, this implies $f' = f$ (Lemma 1.5.2). Clearly, A is a linear operator. It is called the operator *associated with a*. First we have

Theorem 5.6.2. *Let a be a closed Hermitian bilinear form with dense domain. If*

$$a(u) + N\|u\|^2 \geq 0, \qquad u \in D(a), \tag{5.6.6}$$

for some N, then the operator A associated with a is self-adjoint and $\sigma(A) \subset [-N, \infty)$.

PROOF. Let λ be any number greater than N, and set

$$a_\lambda(u) = a(u) + \lambda\|u\|^2. \tag{5.6.7}$$

It is easily checked that $D(a)$ is a Hilbert space with scalar product $a_\lambda(u, v)$. The fact that it is complete follows from the closedness of a and (5.6.6). Let f be any element of $\mathcal{H}$. Then (v, f) is a bounded linear functional on $D(a)$. Hence, by the Riesz representation theorem (Theorem 1.9.1), there is a $u \in D(a)$ such that

$$(v, f) = a_\lambda(v, u), \qquad v \in D(a).$$

This is the same as

$$a(u, v) = (f - \lambda u, v), \qquad v \in D(a).$$

Thus $u \in D(A)$ and $Au = f - \lambda u$. Since f was arbitrary, we see that $R(A + \lambda) = \mathcal{H}$. Next we note that A is Hermitian. For we have

$$(Au, v) = a(u, v) = a(v, u)^* = (Av, u)^* = (u, Av), \qquad u, v \in D(A).$$

Finally, we note that $D(A)$ is dense in $\mathcal{H}$. For, suppose $v \in D(a)$ and

$$a_\lambda(u, v) = 0, \qquad u \in D(A).$$

This means that

$$([A + \lambda]u, v) = 0, \qquad u \in D(A).$$

Since $R(A + \lambda) = \mathcal{H}$, we see that v must vanish. Thus the closure of $D(A)$ in $D(a)$ is the whole of $D(a)$. This implies that for each $w \in D(a)$ and each $\varepsilon > 0$ there is a $u \in D(A)$ such that $\|v - u\| < \varepsilon$. Since $D(a)$ is dense in $\mathcal{H}$, we see that the same is true of $D(A)$. Now we apply Theorem 2.6.3 to conclude that A is self-adjoint. By (5.6.6), $A + \lambda$ is one-to-one for any $\lambda > N$. We have shown that $R(A + \lambda) = \mathcal{H}$ for such λ. Hence their negatives must be in $\rho(A)$. □

Next we have

Theorem 5.6.3. *Suppose a is a bilinear form satisfying the hypotheses of Theorem 5.6.2. Let $c(u)$ be a Hermitian bilinear form such that $D(a) \subset D(c)$ and*

$$|c(u)| \le Ka(u), \qquad u \in D(a). \tag{5.6.8}$$

Assume that every sequence $\{u_n\} \subset D(a)$ such that

$$\|u_n\|^2 + a(u_n) \le C \tag{5.6.9}$$

has a subsequence $\{v_j\}$ such that

$$c(v_j - v_k) \to 0. \tag{5.6.10}$$

Assume also that (5.6.9), (5.6.10), *and $v_j \to 0$ imply that $c(v_k) \to 0$. Set*

$$b(u) = a(u) + c(u), \tag{5.6.11}$$

and let A, B be the operators associated with a, b, respectively. Then

$$\sigma_e(A) = \sigma_e(B). \tag{5.6.12}$$

PROOF. First we show that there is a constant C_0 such that

$$|c(u)| \le \tfrac{1}{2}a(u) + C_0\|u\|^2, \qquad u \in D(a). \tag{5.6.13}$$

If this were not the case, there would be a sequence $\{u_n\} \subset D(a)$ such that

$$a(u_n) + \|u_n\|^2 = 1, \qquad |c(u_n)| > \tfrac{1}{2} + n\|u_n\|^2. \tag{5.6.14}$$

By hypothesis, there is a subsequence $\{v_j\}$ such that $c(v_j - v_k) \to 0$. Moreover, (5.6.8) and (5.6.14) imply that $v_j \to 0$. Thus, $c(v_j) \to 0$ by hypothesis. This contradicts the fact that $|c(v_j)| > \frac{1}{2}$ in (5.6.14). This establishes (5.6.13). Hence

$$a(u) \le 2b(u) + 2C_0\|u\|^2. \tag{5.6.15}$$

Thus, there is an M such that

$$\|u\|^2 \le a(u) + M\|u\|^2, \qquad \|u\|^2 \le b(u) + M\|u\|^2, \qquad u \in D(a).$$

Now (5.6.12) is equivalent to

$$\sigma_e(A + M) = \sigma_e(B + M).$$

Thus it suffices to prove the theorem with $a(u, v)$ replaced by $a(u, v) + M(u, v)$. Note that none of the hypotheses are altered. Consequently, we may assume that

$$\|u\|^2 \le a(u), \qquad \|u\|^2 \le b(u), \qquad u \in D(a). \tag{5.6.16}$$

In particular, $D(a)$ is a Hilbert space with scalar product $a(u, v)$. By the Riesz representation theorem (Theorem 1.9.1), for each $u \in D(a)$ there is an element $Gu \in D(a)$ such that

$$c(u, v) = a(Gu, v), \qquad v \in D(a). \tag{5.6.17}$$

Clearly, G is a linear operator defined everywhere on $D(a)$. It is bounded since

$$a(Gu) = c(u, Gu) \leq Ka(u)^{1/2}a(Gu)^{1/2},$$

and consequently,

$$a(Gu) \leq K^2 a(u). \tag{5.6.18}$$

Moreover, it can be claimed that G is a compact operator on $D(a)$. Let us assume this for the moment and see what consequences follow. By (5.6.17)

$$b(u, v) = a(u + Gu, v), \qquad v \in D(a).$$

Hence $u \in D(B)$ and $Bu = f$ iff $u \in D(a)$, $u + Gu \in D(A)$, and $A(u + Gu) = f$. In other words,

$$A(1 + G) = B. \tag{5.6.19}$$

Now suppose that $\lambda \in \sigma_e(A)$. Then there is a sequence $\{u_n\} \subset D(A)$ such that

$$\|u_n\| = 1, \qquad \|(A - \lambda)u_n\| \to 0, \qquad \text{and} \qquad u_n \to 0 \quad \text{weakly} \tag{5.6.20}$$

(Theorem 3.6.4). By (5.6.16) and Theorem 5.6.2, 0 is in $\rho(A) \cap \rho(B)$. Set $v_n = B^{-1}Au_n$. Then $\|v_n\| \leq C$ for some constant C. Moreover,

$$b(v_n) = (Bv_n, v_n) = (Au_n, v_n)$$

is likewise uniformly bounded. Also, by (5.6.15) the same is true of $a(v_n)$. Since G is a compact operator on $D(a)$, there is a subsequence (also denoted by $\{v_n\}$) such that

$$a(Gv_n - Gv_m) \to 0. \tag{5.6.21}$$

It is easy to show that

$$a(Gu, v) = a(u, Gv), \qquad c(Gu, v) = c(u, Gv). \tag{5.6.22}$$

Thus, if $v \in D(a)$,

$$b(Gv_n, v) = b(v_n, Gv) = (Au_n, Gv) = (Au_n - \lambda u_n, Gv) + \lambda(u_n, Gv) \to 0. \tag{5.6.23}$$

But by (5.6.21) there is a $g \in D(a)$ such that

$$b(Gv_n - g) \to 0.$$

This implies

$$b(Gv_n, v) \to (g, v), \qquad v \in D(a).$$

Consequently, we must have $g = 0$ and

$$a(Gv_n) \to 0. \tag{5.6.24}$$

Now by (5.6.19),

$$Au_n = Bv_n = A(1 + G)v_n.$$

Since A is one-to-one, we find

$$u_n = (1 + G)v_n. \tag{5.6.25}$$

Thus,

$$(B - \lambda)v_n = Au_n - \lambda(u_n - Gv_n) \to 0. \tag{5.6.26}$$

Moreover, $v_n \to 0$ weakly because

$$(v_n, v) = (Au_n, B^{-1}v) = (Au_n - \lambda u_n, B^{-1}v) + \lambda(u_n, B^{-1}v) \to 0. \tag{5.6.27}$$

But it is not true that $\|v_n\| = 1$. However, we have $\|v_n\| = \|u_n - Gv_n\| \to 1$. This is good enough for we can divide v_n by its norm without spoiling (5.6.26) or (5.6.27). Thus $\lambda \in \sigma_e(B)$. Conversely, suppose λ is in $\sigma_e(B)$. Then there is a sequence $\{v_n\} \subset D(B)$ such that

$$\|v_n\| = 1, \qquad \|(B - \lambda)v_n\| \to 0, \qquad \text{and} \qquad v_n \to 0 \quad \text{weakly}. \tag{5.6.28}$$

This time we set $u_n = A^{-1}Bv_n$. Then (5.6.24) and (5.6.25) hold. Thus,

$$(A - \lambda)u_n = Bv_n - \lambda(v_n + Gv_n) \to 0,$$

and $\lambda \in \sigma_e(A)$.

It remains only to prove that G is a compact operator. Since G is Hermitian and bounded, it is self-adjoint on $D(a)$. Set

$$G_1 = \int_0^\infty \lambda \, dE(\lambda), \qquad G_2 = G - G_1,$$

where $\{E(\lambda)\}$ is the corresponding spectral family. First we show that G_1 is a compact operator. To see this, suppose $a(u_n) \le C$. If $E_1 = E(0, \infty)$, then we have $a(E_1 u_n) \le C$. By hypothesis, there is a subsequence (also denoted by $\{u_n\}$) such that $c(E_1 u_n - E_1 u_m) \to 0$. By (5.6.17),

$$a(GE_1(u_n - u_m), u_n - u_m) \to 0.$$

Since $G_1 = GE_1$, we have

$$a[G_1^{1/2}(u_n - u_m)] \to 0.$$

This shows that $G_1^{1/2}$ and consequently G_1 is compact. The same reasoning applies to G_2. Thus $G = G_1 + G_2$ is compact, and the proof is complete. □

We give the proof of Theorem 5.6.1 in the next section.

5.7. Intervals Containing the Essential Spectrum

Before proving Theorem 5.6.1 we note the following corollary.

Theorem 5.7.1. *If* (4.3.5) *holds and there is a sequence* $\{I_n\}$ *of intervals such that* $|I_n| \to \infty$ *and*

$$|I_n|^{-1} \int_{I_n} |V(x) - \mu|^2 \, dx \to 0, \tag{5.7.1}$$

then (5.6.1) *holds.*

The proof of Theorem 5.7.1 consists of showing that (5.7.1) implies (5.6.2) via Schwarz's inequality.

In proving Theorem 5.6.1 we shall use

Lemma 5.7.1. *If* (5.6.2) *holds, then there is a sequence* $\{I_n\}$ *of intervals such that* $|I_n| \to \infty$ *and*

$$\int_{I_n} |V(x) - \mu| \, dx \to 0. \tag{5.7.2}$$

PROOF. If (5.7.2) did not hold, there would be positive constants c_0, N such that

$$\int_I |V(x) - \mu| \, dx > c_0 \tag{5.7.3}$$

whenever $|I| > N$. Now suppose $|I_n| \to \infty$. For $|I_n| > N$ break I_n into equal intervals $I_{n,k}$ of length between N and $2N$. There will be at least $|I_n|/2N$ such intervals in the interval I_n. Thus,

$$|I_n|^{-1} \int_{I_n} \geq |I_n|^{-1} \sum_k \int_{I_{n,k}} \geq c_0/2N,$$

contradicting (5.6.2). Thus (5.6.2) implies (5.7.2). □

Lemma 5.7.2. *Assume that* $V(x)$ *is locally square integrable and that*

$$\int_x^{x+1} |V(y)|^2 \, dy \to 0 \qquad \text{as} \quad |x| \to \infty. \tag{5.7.4}$$

Then every sequence $\{u_n\}$ *satisfying*

$$\|u_n'\|^2 + \|u_n\|^2 \leq C \tag{5.7.5}$$

has a subsequence $\{w_k\}$ *such that* Vw_k *converges in* L^2.

PROOF. By Lemma 4.7.2 there is a subsequence $\{w_k\}$ which converges uniformly on any bounded interval. Let $\varepsilon > 0$ be given. By (2.7.7) we can pick R so large that

$$\int_{|x|>R} |V(x)u(x)|^2 \, dx \leq \varepsilon(\|u'\|^2 + \|u\|^2). \tag{5.7.6}$$

On the other hand, we have

$$\int_{|x|<R} |V(x)[w_j(x) - w_k(x)]|^2\,dx \le \int_{|x|<R} |V(x)|^2\,dx$$
$$\max_{|x|\le R} |w_j(x) - w_k(x)|^2 \to 0 \tag{5.7.7}$$

as $j, k \to \infty$. Take N so large that (5.7.7) is $< \varepsilon$ for $j, k > N$. Hence we have by (5.7.5)–(5.7.7)

$$\int |V(x)[w_j(x) - w_k(x)]|^2\,dx < \varepsilon(C + 1), \qquad j, k > N.$$

Since ε was arbitrary, the result follows. □

One should compare Lemma 5.7.2 with Theorem 3.7.5. We are now ready for the

PROOF OF THEOREM 5.6.1. By Lemma 5.7.1 there is a sequence $\{I_n\}$ of intervals such that $|I_n| \to \infty$ and (5.7.2) holds. Set

$$V_1(x) = \begin{cases} \mu & \text{in } \bigcup I_n \\ V(x) & \text{elsewhere,} \end{cases}$$

and let N be so large that

$$h_1(u) = \frac{\hbar^2}{2m}\|u'\|^2 + (V_1 u, u) \tag{5.7.8}$$

satisfies

$$c_0(\|u'\|^2 + \|u\|^2) \le h_1(u) + N\|u\|^2, \qquad u \in C_0^\infty, \tag{5.7.9}$$

for some $c_0 > 0$. This can be accomplished by Theorem 4.3.1. Let λ be any number $\ge \mu$, and set $\tau = [2m(\lambda - \mu)]^{1/2}/\hbar$. Let $\varphi \in C_0^\infty$ be such that

$$\varphi(x) = \begin{cases} 1 & \text{for } |x| < \frac{1}{2} \\ 0 & \text{for } |x| > 1. \end{cases}$$

Set

$$u_n(x) = 2^{1/2}|I_n|^{-1/2}\varphi\left(2\frac{x - b_n}{|I_n|}\right)e^{i\tau x},$$

where b_n is the center of I_n. We have

$$D_x u_n = \tau u_n - i2^{3/2}|I_n|^{-3/2}\varphi'\left(2\frac{x - b_n}{|I_n|}\right)e^{i\tau x},$$

$$D_x^2 u_n = \tau^2 u_n - i\tau 2^{5/2}|I_n|^{-3/2}\varphi'\left(2\frac{x - b_n}{|I_n|}\right)e^{i\tau x}$$
$$-2^{5/2}|I_n|^{-5/2}\varphi''\left(2\frac{x - b_n}{|I_n|}\right)e^{i\tau x},$$

and

$$V_1 u_n = \mu u_n.$$

Then one checks easily that $\|u_n\| = \|\varphi\|$ and

$$(H_0 + V_1 - \lambda)u_n \to 0. \tag{5.7.10}$$

Thus, if H_1 is the operator associated with h_1, we have $\lambda \in \sigma(H_1)$. Since λ was any number $\geq \mu$, we have

$$\sigma_e(H_1) \supset [\mu, \infty). \tag{5.7.11}$$

Next, set $V_2 = V - V_1$. Then $V_2 = V - \mu$ in I_n and vanishes outside I_n. Hence

$$\int_x^{x+1} |V_2(y)|\, dy \to 0 \qquad \text{as} \quad |x| \to \infty \tag{5.7.12}$$

by (5.7.2). Set

$$a(u) = h_1(u) + N\|u\|^2, \qquad c(u) = (V_2 u, u).$$

All of the hypotheses of Theorem 5.6.3 are satisfied. In fact, suppose (5.6.9) holds. Then (5.7.5) holds in view of (5.7.9). An application of Lemma 5.7.2 shows us that (5.6.10) holds. Moreover, (5.6.9), (5.6.10), and $v_j \to 0$ imply via (5.7.9) and Lemma 5.7.2 that $|V_2|^{1/2} v_k \to w$ in L^2. If $\varphi \in C_0^\infty$,

$$(w, \varphi) \leftarrow \left(|V_2|^{1/2} v_k, \varphi\right) = \left(v_k, |V_2|^{1/2} \varphi\right) \to 0.$$

Thus $w = 0$. Consequently,

$$|c(v_k)| \leq \left\||V_2|^{1/2} v_k\right\|^2 \to 0.$$

Now we can apply Theorem 5.6.3 to conclude that

$$\sigma_e(H_1 + N) = \sigma_e(H + N),$$

which is equivalent to

$$\sigma_e(H_1) = \sigma_e(H).$$

This in conjunction with (5.7.11) gives the desired result. □

5.8. Coincidence of the Essential Spectrum with an Interval

Now we give another application of Theorem 5.6.3 which gives a criterion for

$$\sigma_e(H) = [\mu, \infty). \tag{5.8.1}$$

Theorem 5.8.1. *Assume that*

$$\limsup_{t \to 0} \limsup_{|x| \to \infty} \int_x^{x+t} |V(y)|\, dy = 0 \tag{5.8.2}$$

and

$$\lim_{t\to 0}\lim_{|x|\to\infty} t^{-1}\int_x^{x+t} V(y)\,dy = \mu. \tag{5.8.3}$$

Then (5.8.1) *holds.*

PROOF. First note that we can take $\mu = 0$. For we can replace V by $V - \mu$ and show that $\sigma_e(H - \mu) = [0, \infty)$. This implies (5.8.1).

Let $\varepsilon > 0$ be given. Take δ so small that

$$\limsup_{|x|\to\infty} \int_x^{x+\delta} |V(y)|\,dy < \tfrac{1}{2}\varepsilon,$$

$$\lim_{|x|\to\infty} \left|\delta^{-1}\int_x^{x+\delta} V(y)\,dy\right| < \tfrac{1}{2}\varepsilon.$$

Then take R so large that

$$\int_x^{x+\delta} |V(y)|\,dy < \varepsilon, \qquad |x| > R,$$

$$\left|\int_x^{x+\delta} V(y)\,dy\right| < \varepsilon\delta, \qquad |x| > R.$$

Set $\nu = -\varepsilon$, $\tau = \varepsilon/\delta$, $b = \varepsilon(2\delta^{-2} + 1)$, and let I be any interval a distance $> R$ from the origin. An application of Lemma 5.2.2 to V gives

$$\left(-\varepsilon - \frac{\delta^2}{2\varepsilon}\left(\frac{2\varepsilon}{\delta} - \varepsilon\right)^2\right)\|u\|_I^2 \le \varepsilon(2 + \delta^2)k_0^2\|u'\|_I^2 + (Vu, u)_I.$$

Thus,

$$-(Vu, u)_I \le \varepsilon(2 + \delta^2)k_0^2\|u'\|_I^2 + \varepsilon\left[1 + \tfrac{1}{2}(2 - \delta)^2\right]\|u\|_I^2.$$

If we apply Lemma 5.2.2 to $-V$, we get the same bound for $(Vu, u)_I$. Set

$$V_N(x) = \begin{cases} V(x), & |x| \le N \\ 0, & |x| > N, \end{cases}$$

$$V^N(x) = V(x) - V_N(x).$$

We have shown that for any $\varepsilon > 0$ there is an N so large that

$$|(V^N u, u)| \le \varepsilon(\|u'\|^2 + \|u\|^2). \tag{5.8.4}$$

On the other hand, V_N satisfies (5.7.4) for each N. Consequently, if $\{u_n\}$ is a sequence satisfying (5.7.5), then it has a subsequence $\{v_k\}$ such that $|V_N|^{1/2}v_k$ converges for each N. If we pick N so that (5.8.4) holds, we have

$$\left|\left(V[v_j - v_k], v_j - v_k\right)\right| \le 2\varepsilon C + \left\||V_N|^{1/2}(v_j - v_k)\right\|^2,$$

which can be made $< \varepsilon(2C + 1)$ by taking j, k sufficiently large. Since ε was arbitrary, we see that

$$\left(V[v_j - v_k], v_j - v_k\right) \to 0 \qquad \text{as} \quad j, k \to \infty.$$

As in the proof of Theorem 5.6.1 given in Section 5.7, the additional assumption $v_j \to 0$ implies that

$$(Vv_j, v_j) \to 0.$$

If we put $a(u) = ([H_0 + 1]u, u)$ and $c(u) = (Vu, u)$, we see that the hypotheses of Theorem 5.6.3 are satisfied. This gives the desired conclusion. □

The following criterion is also useful.

Theorem 5.8.2. *If V is locally integrable and*

$$\int_{|x-y|<1} |V(y) - \mu| \, dy \to 0 \qquad as \quad |x| \to \infty, \tag{5.8.5}$$

then (5.8.1) *holds.*

PROOF. Set

$$a(u) = ([H_0 + 1]u, u), \qquad c(u) = ([V - \mu]u, u).$$

As before it is not difficult to show that the hypotheses of Theorem 5.6.3 are satisfied. This gives $\sigma_e(H + 1 - \mu) = [1, \infty)$, which is equivalent to (5.8.1). □

5.9. The Harmonic Oscillator

For a particle bound to the origin by the potential

$$V(x) = \tfrac{1}{2} m\omega^2 x^2, \tag{5.9.1}$$

ω a constant, Corollary 5.4.2 tells us that H has no essential spectrum. We want to study its spectrum in detail. We write H_0 in the form

$$H_0 = \frac{1}{2m} p^2, \tag{5.9.2}$$

where

$$p = -i\hbar \frac{\partial}{\partial x}$$

is the momentum operator (see Section 1.3). Thus

$$H = \frac{1}{2m}(p^2 + m^2\omega^2 x^2). \tag{5.9.4}$$

Note that

$$px - xp = -i\hbar. \tag{5.9.5}$$

Thus, we may write H in the form

$$H = \frac{1}{2m}(p + im\omega x)(p - im\omega x) + \tfrac{1}{2}\omega\hbar$$

$$= \frac{1}{2m}(p - im\omega x)(p + im\omega x) - \tfrac{1}{2}\omega\hbar.$$

If we set

$$a = p + im\omega x, \tag{5.96}$$

this becomes

$$H = \frac{1}{2m} a^* a - \tfrac{1}{2}\omega\hbar = \frac{1}{2m} aa^* + \tfrac{1}{2}\omega\hbar. \tag{5.9.7}$$

Set

$$N = \frac{1}{2m\omega\hbar} a^* a. \tag{5.9.8}$$

Then

$$H = \omega\hbar\left(N - \tfrac{1}{2}\right). \tag{5.9.9}$$

Now suppose

$$(N - \mu)\psi = 0. \tag{5.9.10}$$

This is the same as

$$(a^* a - 2m\omega\hbar\mu)\psi = 0.$$

Applying a, we obtain

$$(aa^* - 2m\omega\hbar\mu)a\psi = 0,$$

which is equivalent to

$$\left(a^* a - 2m\omega\hbar\left[\,\mu + 1\right]\right)a\psi = 0$$

or

$$(N - \mu - 1)a\psi = 0. \tag{5.9.11}$$

Moreover, (5.9.11) is the same as

$$\left(aa^* - 2m\omega\hbar\left[\,\mu - 1\right]\right)\psi = 0. \tag{5.9.12}$$

If we apply a^*, we obtain

$$\left(a^* a - 2m\omega\hbar\left[\,\mu - 1\right]\right)a^*\psi = 0,$$

which is equivalent to

$$(N - \mu + 1)a^*\psi = 0. \tag{5.9.13}$$

In particular, we see that if μ is an eigenvalue with eigenfunction ψ, then $\mu - 1$ is also an eigenvalue if $a^*\psi$ is in L^2 and does not vanish. By (5.9.8), $(N\psi, \psi) \geq 0$, and consequently it has no negative eigenvalues. Since $\sigma_e(H)$ is empty, the same is true of $\sigma_e(N)$. Hence N has only isolated eigenvalues. Since none of them can be negative, there must be a smallest one. Let us check to see whether it is 0. If $N\psi = 0$, then $a\psi = 0$. This means

$$-i\hbar\psi' + im\omega x\psi = 0.$$

Thus ψ must be of the form

$$\psi(x) = c\exp\left\{m\omega x^2/2\hbar\right\},$$

which is decidedly not in L^2 unless it vanishes. Thus 0 is not an eigenvalue of N. Let us try 1. If

$$(N-1)\psi = 0, \tag{5.9.14}$$

then (5.9.13) says $Na^*\psi = 0$. Since 0 is not an eigenvalue of N, this implies that $a^*\psi = 0$. Thus

$$-i\hbar\psi' - im\omega x\psi = 0,$$

which has the solution

$$\psi_0(x) = c\exp\{-m\omega x^2/2\hbar\}. \tag{5.9.15}$$

Since ψ_0 is in L^2, $\mu = 1$ is an eigenvalue of N. Moreover, (5.9.14) implies

$$(N-2)a\psi_0 = 0$$

in view of (5.9.11). Since $a\psi_0$ is in L^2, we see that $\mu = 2$ is an eigenvalue. In fact, repeated applications of (5.9.11) show that

$$(N-n)a^{n-1}\psi_0 = 0, \tag{5.9.16}$$

and since $a^n\psi_0 \in L^2$ for any n, we see that N has eigenvalues $\mu_n = n$ with corresponding eigenfunctions

$$\psi_n = c_n a^{n-1}\psi_0, \tag{5.9.17}$$

where the constants c_n are chosen to make $\|\psi_n\| = 1$. Next we note that N has no other eigenvalues. For suppose $0 < \mu < 1$ and (5.9.10) holds. Then (5.9.13) holds as well. Since $N \geq 0$, this implies $a^*\psi = 0$. But (5.9.12) then says $(\mu - 1)\psi = 0$. Thus $\psi = 0$ and μ cannot be an eigenvalue. Similarly, if $1 < \mu < 2$ and (5.9.10) holds, then (5.9.13) holds. Since $\mu - 1$ cannot be an eigenvalue, we must have $a^*\psi = 0$. This in turn implies $(\mu - 1)\psi = 0$ by (5.9.12). Consequently, μ cannot be an eigenvalue. We see by induction that the positive integers are the only eigenvalues (and points of the spectrum) of N. It is up to the reader to give a name to the operator N. At any rate we see from (5.9.9) that the spectrum of H consists of eigenvalues

$$\lambda_n = \omega\hbar\left(n - \tfrac{1}{2}\right), \qquad n = 1, 2, \ldots, \tag{5.9.18}$$

with corresponding eigenfunctions given by (5.9.17).

5.10. The Morse Potential

A very interesting potential is given by

$$V(x) = c\left(e^{-2a(x-\mu)} - 2e^{-a(x-\mu)}\right), \tag{5.10.1}$$

named for P. M. Morse. Note that $V(x) \to 0$ as $x \to \infty$, $V(x) \to \infty$ as $x \to -\infty$, and it has a minimum $-c$ at $x = \mu$. Let us analyze the spectrum of the corresponding Hamiltonian H. By Theorem 3.1.1 we see that

$$\sigma(H) \subset [-c, \infty), \tag{5.10.2}$$

but this is all we can conclude from the results of that chapter. On the other hand, Corollary 5.4.1 tells us that

$$\sigma_e(H) \subset [0, \infty).$$

Moreover, Theorem 5.6.1 tells us that

$$\sigma_e(H) \supset [0, \infty).$$

Hence

$$\sigma_e(H) = [0, \infty), \tag{5.10.3}$$

and all negative points of $\sigma(H)$ are isolated eigenvalues. Moreover, an application of Theorem 4.7.1 shows that H has at most a finite number of negative eigenvalues. To complete the picture, we will show that for a sufficiently small, H has at least one negative eigenvalue. In fact, we have

$$\int_{-\infty}^{\infty} e^{-ax-2\alpha^2x^2}\,dx = \frac{\sqrt{2\pi}}{2\alpha}\,e^{a^2/8\alpha^2}. \tag{5.10.4}$$

Hence

$$\alpha^{-1}\int_{-\infty}^{\infty} V(x)e^{-2\alpha^2(x-\mu)^2}\,dx = \frac{c\sqrt{2\pi}}{2\alpha^2}\left[e^{a^2/2\alpha^2} - 2e^{a^2/8\alpha^2}\right]. \tag{5.10.5}$$

Set $\beta = a^2/8\alpha^2$. Then the right-hand side of (5.10.5) becomes

$$\frac{4\beta\sqrt{2\pi}}{a^2}\,e^{\beta}(e^{3\beta} - 2). \tag{5.10.6}$$

Take β so small that $e^{3\beta} < 2$. Then take a so small that the expression (5.10.6) is $< -\hbar^2\sqrt{2\pi}/4m$. It follows from Theorem 4.5.1 that H has at least one negative eigenvalue for such a. □

Exercises

1. Prove (5.2.12)–(5.2.14).
2. Show that (5.2.14) and (5.2.15) imply (5.2.5) when (5.2.16) does not hold.
3. Prove (5.3.11) and (5.3.12).
4. Show that $u_n \to 0$ weakly in L^2 and $u_n \to u$ in $L^2(I)$ imply that $u = 0$ in I.
5. Prove (5.4.10).
6. Fill in the details of the proof of Theorem 5.4.3.
7. What determined the choice of α and β in (5.2.12) and (5.2.13)?

8. If $\mu < 0$, is (5.5.14) a solution of (5.5.10)? Could μ be an eigenvalue of A?

9. Show that u_n defined by (5.5.18) is in $D(A)$.

10. Show that $D(a)$ is complete when a is closed and (5.6.6) holds.

11. Show that $R(A + \lambda) = \mathcal{H}$ implies that $D(A)$ is dense in $D(a)$.

12. Show that (5.6.14) is true if (5.6.13) is false.

13. Prove (5.6.22) and (5.6.24).

14. Prove the last two statements in the proof of Theorem 5.6.3.

15. Show that the operator G_2 defined in Section 5.6 is compact.

16. Show that K compact implies K^2 compact.

17. Prove Theorem 5.8.2.

18. Prove (5.9.5).

6

Scattering Theory

6.1. Time Dependence

We have not yet considered the effect of time on a particle. As we mentioned in the very beginning, every state function $\psi(x, t)$ depends on time as well as position, but so far we have studied everything for a fixed time. Actually, some of the quantities and operators we have discussed can depend on time. For instance, the potential can be a function $V(x, t)$ depending on time as well as position. Of course, this would cause the Hamiltonian operator to depend on time as well.

To discuss the dependence of ψ on time we shall need another axiom:

Postulate 5. If $H(t)$ is the Hamiltonian (total energy) operator and $\psi(t) = \psi(x, t)$ is the state of a particle at the time t, then $\psi(t)$ is a solution of

$$i\hbar\psi'(t) = H\psi(t). \tag{6.1.1}$$

We should explain this postulate a bit. First of all, the derivative $\psi'(t)$ appearing in (6.1.1) is defined as

$$\psi'(t) = \lim_{h\to 0}\left[\psi(t+h) - \psi(t)\right]/h,$$

where the limit is taken in $\mathcal{H} = L^2$. The postulate states that this derivative exists in the sense described for all t and satisfies (6.1.1).

In this chapter we shall assume that $H = H(t)$ does not depend on t, and we shall study the dependence of $\psi(t)$ on t. Let $\psi_0 = \psi_0(x)$ be a function in $\mathcal{H}$, and consider the solution $\psi(t) = \psi(x, t)$ of

$$i\hbar\psi'(t) = H\psi(t), \qquad -\infty < t < \infty, \qquad \psi(0) = \psi_0. \tag{6.1.2}$$

By Postulate 5, $\psi(x, t)$ is the state function at time t of a particle whose state function at time 0 is $\psi_0(x)$. At this point you might feel a bit apprehensive. It is all very good and well to postulate that $\psi(t)$ is a solution of (6.1.2), but the existence of solutions of (6.1.2) is a mathematical question; we should check with mathematicians to see whether solutions exist before making such a postulate. Moreover, the postulate would also be in trouble if a solution of (6.1.2) were not unique. Well, we cannot turn back the course of history, but we shall see presently that (6.1.2) indeed has a unique solution for every $\psi_0 \subset D(H)$ whenever H is a self-adjoint operator on a Hilbert space $\mathcal{H}$.

To get an idea how to solve (6.1.2), let us imagine that H is not an operator but a constant. In this case one easily obtains the solution

$$\psi(t) = e^{-itH/\hbar}\psi_0. \tag{6.1.3}$$

Returning to reality, we realize that (6.1.3) does not usually make sense when H is an operator. But it does have meaning when H is self-adjoint. In fact, we can define it as

$$e^{-itH/\hbar}\psi_0 = \int_{-\infty}^{\infty} e^{-it\lambda/\hbar}\, dE(\lambda)\, \psi_0, \tag{6.1.4}$$

where $\{E(\lambda)\}$ is the spectral family of H. The expression (6.1.4) is defined for all $\psi_0 \in \mathcal{H}$ (cf. Theorem 1.10.1).

Moreover, it can be claimed that (6.1.4) is a solution of (6.1.2). To see this, note that

$$\frac{1}{\sigma}\left(e^{-i(t+\sigma)H/\hbar} - e^{-itH/\hbar}\right)\psi_0 = \int \frac{1}{\sigma} e^{-it\lambda/\hbar}\left(e^{-i\sigma\lambda/\hbar} - 1\right) dE(\lambda)\, \psi_0.$$

The integrand on the right-hand side is bounded in absolute value by

$$\frac{1}{|\sigma|}\left|e^{-i\sigma\lambda/\hbar} - 1\right| \le |\lambda|/\hbar$$

and converges for each λ to

$$-\, i\lambda\hbar^{-1}e^{-it\lambda/\hbar}$$

as $\sigma \to 0$. Thus the integral will converge to

$$-\frac{i}{\hbar}\int \lambda e^{-it\lambda/\hbar}\, dE(\lambda)\, \psi_0 = -\frac{i}{\hbar} He^{-itH/\hbar}\psi_0$$

provided

$$\int_{-\infty}^{\infty} \lambda^2\, d(E(\lambda)\psi_0, \psi_0) < \infty,$$

that is, provided $\psi_0 \in D(H)$ (cf. Theorem 1.10.1). Next we note that (6.1.4) is the only solution of (6.1.2). For, if there were another, their difference $\varphi(t)$ would be a solution of

$$i\hbar\varphi'(t) = H\varphi(t), \qquad \varphi(0) = 0.$$

But

$$(\varphi(t), \varphi(t))' = (\varphi'(t), \varphi(t)) + (\varphi(t), \varphi'(t))$$
$$= \frac{1}{i\hbar}\left[H\varphi, \varphi) - (\varphi, H\varphi)\right] = 0,$$

since H is Hermitian. Hence $\|\varphi(t)\|$ is a constant for all t. But $\varphi(0) = 0$. Hence $\|\varphi(t)\| = 0$ for all t, and $\varphi(t)$ vanishes identically.

Summing up, we have

Theorem 6.1.1. *If H is a self-adjoint operator on a Hilbert space $\mathcal{H}$, then* (6.1.2) *has a unique solution for each $\psi_0 \in D(H)$. It is given by* (6.1.4) *and satisfies*

$$\|\psi(t)\| = \|\psi_0\|, \qquad -\infty < t < \infty. \tag{6.1.5}$$

The identity (6.1.5) says that the operator $e^{-itH/\hbar}$ is an *isometry*. This fact is important in quantum mechanics because we always insist that our state functions have norm 1 (see Section 1.1).

You may have noticed that the constant $\hbar$ has started to become a nuisance in some of our formulas. To keep it from getting in our way we shall choose units that will make $\hbar = 1$.

6.2. Scattering States

Suppose a particle is shot into a region where it is affected by potential forces. Assume that these forces are of the *short range* type, that is, they affect the particle essentially only in the region. Outside the region, the effect of the forces is assumed negligible. Let H_0 be the Hamiltonian governing the particle without the forces and let H denote the Hamiltonian with the forces included. Long before the particle comes near the region (i.e., when t is very negative) the particle is essentially governed by the Hamiltonian H_0, and hence its state function ψ should be close to a solution of $i\psi' = H_0\psi$ (recall that we have taken $\hbar = 1$). Thus there should be a state $\psi_-(t)$ satisfying

$$\psi_-(t) = e^{-itH_0}\psi_-(0) \tag{6.2.1}$$

such that

$$\|\psi(t) - \psi_-(t)\| \to 0 \qquad \text{as} \quad t \to -\infty. \tag{6.2.2}$$

When it exists, $\psi_-(t)$ is called the *incoming asymptotic state* for $\psi(t)$. Similarly, long after the particle has left the region (assuming that it does not get trapped there), it should approach a state governed by H_0. Thus there should be a state $\psi_+(t)$ satisfying

$$\psi_+(t) = e^{-itH_0}\psi_+(0) \tag{6.2.3}$$

such that

$$\|\psi(t) - \psi_+(t)\| \to 0 \quad \text{as} \quad t \to \infty. \tag{6.2.4}$$

The state $\psi_+(t)$ is called the *outgoing asymptotic state* for $\psi(t)$. If $\psi(t)$ possesses both incoming and outgoing asymptotic states, it is called a scattering state.

Now $\psi(t)$ is a solution of (6.1.2). Hence it is of the form

$$\psi(t) = e^{-itH}\psi(0). \tag{6.2.5}$$

By (6.2.1)–(6.2.5) we have

$$\|e^{-itH}\psi(0) - e^{-itH_0}\psi_\pm(0)\| \to 0 \quad \text{as} \quad t \to \pm\infty. \tag{6.2.6}$$

Since

$$e^{itH}e^{-itH} = 1 \tag{6.2.7}$$

and

$$\|e^{itH}u\| = \|u\|, \tag{6.2.8}$$

this implies

$$\|\psi(0) - e^{itH}e^{-itH_0}\psi_\pm(0)\| \to 0, \qquad t \to \pm\infty. \tag{6.2.9}$$

In other words, the limits

$$\psi(0) = \lim_{t\to\pm\infty} e^{itH}e^{-itH_0}\psi_\pm(0) \tag{6.2.10}$$

exist. Thus we have shown that if the potential forces are of short range, then there are states $\psi_\pm(0)$ such that (6.2.10) holds. Let us examine this situation in a bit more detail.

Let $M_\pm$ denote the set of all $u \in \mathcal{H}$ such that the limits

$$W_\pm u = \lim_{t\to\pm\infty} W(t)u \tag{6.2.11}$$

exist, where

$$W(t) = e^{itH}e^{-itH_0}. \tag{6.2.12}$$

First we note the following:

Theorem 6.2.1. *The $M_\pm$ are closed subspaces of $\mathcal{H}$.*

PROOF. Suppose $\{u_n\}$ is a sequence of elements in M_+ such that $u_n \to u$ in $\mathcal{H}$. Then

$$\|W(t_1)u - W(t_2)u\| \le \|W(t_1)(u - u_k)\| + \|[W(t_1) - W(t_2)]u_k\|$$
$$+ \|W(t_2)(u_k - u)\| \le 2\|u - u_k\| + \|[W(t_1) - W(t_2)]u_k\|$$

by (6.2.8). Let $\varepsilon > 0$ be given. Pick k so large that the first term on the right is $< \frac{1}{2}\varepsilon$. Then fix k and take t_1, t_2 so large that the second term is $< \frac{1}{2}\varepsilon$. Thus $W(t)u$ converges as $t \to \infty$. Consequently, $u \in M_+$. The same argument applies in the case of M_-. □

We see from Theorem 6.2.1 that the operators $W_\pm$ are defined on the subspaces $M_\pm$. They are isometries since

$$\|W_\pm u\| = \lim_{t\to\pm\infty} \|W(t)u\| = \|u\| \tag{6.2.13}$$

by (6.2.8). Let $R_\pm$ denote the range of $W_\pm$. We have

Theorem 6.2.2. *$\psi(t)$ has an incoming (outgoing) asymptotic state iff $\psi(0) \in R_-$ (R_+).*

PROOF. Suppose $\psi(t)$ has an incoming asymptotic state ψ_-. By (6.2.9), $W(t)\psi_-(0) \to \psi(0)$ as $t \to -\infty$. Thus $\psi(0) \in R_-$. Conversely, if $\psi(0) \in R_-$, then there is an $f \in M_-$ such that $W_- f = \psi(0)$. Thus $W(t)f \to \psi(0)$ as $t \to -\infty$. Set $\psi(t) = e^{-itH_0}f$. Then

$$\|\psi(t) - \psi_-(t)\| = \|\psi(0) - W(t)f\| \to 0 \quad \text{as} \quad t \to -\infty.$$

Hence ψ_- is an incoming asymptotic state for ψ. □

Corollary 6.2.1. *$\psi(t)$ is a scattering state iff $\psi(0) \in R_- \cap R_+$.*

Next we note

Theorem 6.2.3. *$R_\pm$ are closed in $\mathcal{H}$.*

PROOF. Suppose $f_k \in R_+$ and $f_k \to f$ in $\mathcal{H}$. Then there are $u_k \in M_+$ such that $f_k = W_+u_k$. By (6.2.13),

$$\|u_j - u_k\| = \|W_+(u_j - u_k)\| = \|f_j - f_k\| \to 0.$$

Hence the u_j converge in $\mathcal{H}$ to a $u \in \mathcal{H}$. Since M_+ is closed (Theorem 6.2.1), $u \in M_+$, and

$$\|f_j - W_+u\| = \|W_+(u_j - u)\| = \|u_j - u\| \to 0.$$

Thus $f = W_+u \in R_+$. □

If $\psi(t)$ is a scattering state, then

$$\psi(0) = W_\pm\psi_\pm(0),$$

where the $\psi_\pm(t)$ are its asymptotic states. Since W_+ is one-to-one, we have

$$\psi_+(0) = W_+^{-1}W_-\psi_-(0) = S\psi_-(0).$$

S is called the *scattering operator*. Its importance stems from the fact that it compares the outgoing asymptotic state with the incoming one. Clearly, S maps M_- into M_+, but its domain need not be the whole of M_-. We have

Theorem 6.2.4. *$D(S) = M_-$ iff $R_- \subset R_+$. $R(S) = M_+$ iff $R_+ \subset R_-$.*

PROOF. $D(S)$ consists of those $u \in M_-$ such that $W_-u \in R_+$. If $R_- \subset R_+$, then $W_-u \in R_+$ for all $u \in M_-$, and consequently $D(S)$ consists of the whole of M_-. If $R_- \not\subset R_+$, there is a $u \in M_-$ such that $W_- u$ is not in R_+ and consequently $u \notin D(S)$. This proves the first statement. To see the second, note that $R(S)$ consists of those $f \in M_+$ such that $W_+ f \in R_-$. If $R_+ \subset R_-$, $R(S)$ contains the whole of M_+. Otherwise, there is an $f \in M_+$ such that $W_+ f$ is not in R_-. Consequently, $f \notin R(S)$. □

An operator is called *unitary* from $\mathcal{H}_1$ to $\mathcal{H}_2$ if it is an isometry of $\mathcal{H}_1$ onto the whole of $\mathcal{H}_2$. In particular, $W_\pm$ is unitary from $M_\pm$ to $R_\pm$. As a consequence of Theorem 6.2.4 we have

Corollary 6.2.2. *S is unitary from M_- to M_+ iff $R_- = R_+$.*

The operators $W_\pm$ defined by (6.2.11) are called *wave operators*. We shall study their properties in the next few sections.

6.3. Properties of the Wave Operators

In this section we shall discuss some of the properties of spaces $M_\pm$ and $R_\pm$. First we shall need

Lemma 6.3.1. *Suppose H is self-adjoint and $u \in D(H)$. Then $u \in N(H)$ iff $u = e^{itH}u$ for all real t.*

PROOF. Suppose $u \in N(H)$, and set $\psi(t) \equiv u$. Then $\psi(t)$ is a solution of

$$i\psi'(t) = H\psi(t), \qquad \psi(0) = u. \tag{6.3.1}$$

By Theorem 6.1.1, the only solution of (6.3.1) is

$$\psi(t) = e^{-itH}u. \tag{6.3.2}$$

Thus this must equal u. Conversely, suppose $u = e^{itH}u$ for all t. Then the function $\psi(t)$ given by (6.3.2) does not depend on t, and consequently $\psi'(t) = 0$. But it is a solution of (6.3.1). Hence $H\psi(t) = 0$ for all t. In particular, $Hu = H\psi(0) = 0$. □

Corollary 6.3.1. *Under the same hypotheses,*

$$e^{itH}u = e^{it\lambda}u \qquad \text{for all} \quad t \tag{6.3.3}$$

iff

$$(H - \lambda)u = 0. \tag{6.3.4}$$

Thus, if $u \neq 0$, then (6.3.3) holds iff λ is an eigenvalue of H and u is a corresponding eigenelement.

Theorem 6.3.1. *If $\psi(0)$ is an eigenelement of H, then $\psi(t)$ has no asymptotic states (incoming or outgoing) unless $\psi(0)$ is also an eigenelement of H_0 corresponding to the same eigenvalue.* [In this case, $\psi(0) = \psi_\pm(0)$.]

PROOF. Suppose $W(t)v \to u = \psi(0)$ as $t \to -\infty$. Then

$$e^{-itH_0}v - e^{-itH}u \to 0. \tag{6.3.5}$$

Suppose $(H - \lambda)u = 0$. Then by Corollary 6.3.1,

$$e^{-itH_0}v - e^{-it\lambda}u \to 0$$

and consequently,

$$v - e^{it(H_0-\lambda)}u \to 0. \tag{6.3.6}$$

Thus, for any s,

$$e^{i(t+s)(H_0-\lambda)}u - e^{it(H_0-\lambda)}u \to 0 \qquad \text{as} \quad t \to -\infty$$

and hence

$$e^{is(H_0-\lambda)}u \to u \qquad \text{as} \quad t \to -\infty.$$

Since there is no dependence on t now, this implies

$$e^{is(H_0-\lambda)}u = u \qquad \text{for all} \quad s, \tag{6.3.7}$$

and consequently $(H_0 - \lambda)u = 0$ (Corollary 6.3.1). Moreover, we see that $v = u$ by (6.3.6) and (6.3.7). □

The same method of proof gives

Theorem 6.3.2. *If $\psi(t)$ has an asymptotic state which for $t = 0$ is an eigenstate (eigenelement) of H_0, then $\psi(0)$ is an eigenstate of H corresponding to the same eigenvalue.*

Theorem 6.3.3. *For each s, e^{isH_0} maps $M_\pm$ into itself, e^{isH} maps $R_\pm$ into itself, and*

$$W_\pm e^{isH_0} = e^{isH}W_\pm. \tag{6.3.8}$$

PROOF. Suppose $W_- u = f$. Then $\| W(t)e^{isH_0}u - e^{isH}f\| = \| W(t-s)u - f\| \to 0$ as $t \to -\infty$. Thus $e^{isH_0}u \in M_-$ and $W_- e^{isH_0}u = e^{isH}f$. From this we see that $e^{isH}f \in R_-$ and (6.3.8) holds. □

For any subset N of a Hilbert space $\mathcal{H}$, we let $N^\perp$ denote the set of all elements of $\mathcal{H}$ which are orthogonal to all the elements of N. It is easily verified that $N^\perp$ is always a closed subspace. We have

Theorem 6.3.4. *For every s, e^{isH_0} maps $M_\pm^\perp$ into itself and e^{isH} maps $R_\pm^\perp$ into itself.*

PROOF. If $u \in M_-$, so is $e^{-isH_0}u$ (Theorem 6.3.3). Thus if $f \perp M_-$, we have

$$\left(e^{isH_0}f, u\right) = \left(f, e^{-isH_0}u\right) = 0.$$

Hence $e^{isH_0}f \perp M_-$. Similarly, if $v \in R_-$, so is $e^{-isH}v$ and hence if $g \perp R_-$,

$$\left(e^{-isH}g, v\right) = \left(g, e^{isH}v\right) = 0. \qquad \square$$

A closed subspace N of $\mathcal{H}$ is said to *reduce* an operator A if the orthogonal projection P onto N maps $D(A)$ into itself and satisfies

$$PA \subset AP$$

(see Section 3.5). Note that (6.3.9) implies that A maps $D(A) \cap N$ into N and $D(A) \cap N^{\perp}$ into $N^{\perp}$. We see from Theorems 6.3.3 and 6.3.4 that the following holds.

Theorem 6.3.5. *For each s the spaces $M_{\pm}$ reduce e^{isH_0} and the spaces $R_{\pm}$ reduce e^{isH}.*

Next we note that we can even say more. We shall use

Theorem 6.3.6. *Suppose H is self-adjoint. Then $u \in D(H)$ iff $t^{-1}[e^{-itH}u - u]$ converges in $\mathcal{H}$ as $t \to 0$.*

PROOF. Suppose $u \in D(H)$, and define $\psi(t)$ by (6.3.2). Then the expression converges to $\psi'(0) = -iHu$. Conversely, suppose the expression converges to w, and let v be any element in $D(H)$. Set $\varphi(t) = e^{itH}v$. Since

$$\left(\left[e^{-itH} - 1\right]u, v\right) = \left(u, \left[e^{itH} - 1\right]v\right),$$

we have

$$(w, v) = (u, \varphi'(0)) = -i(u, Hv).$$

Since H is self-adjoint, we see that $u \in D(H)$ and $Hu = iw$. □

Theorem 6.3.7. *Let $P_{\pm}$ be the orthogonal projections onto $M_{\pm}$. Then the operators $P_{\pm}$ map $D(H_0)$ into itself.*

PROOF. If $u \in D(H_0)$, then

$$t^{-1}\left[e^{-itH_0} - 1\right]P_{\pm}u = P_{\pm}\left\{t^{-1}\left[e^{-itH_0} - 1\right]u\right\}$$

by Theorem 6.3.5. The right-hand side converges by Theorem 6.3.6. □

Theorem 6.3.8. *The $M_{\pm}$ reduce H_0 and the $R_{\pm}$ reduce H.*

PROOF. If $u \in D(H_0)$, then

$$H_0 P_{\pm}u = i \lim t^{-1}\left[e^{-itH_0}P_{\pm}u - P_{\pm}u\right]$$

$$= iP_{\pm} \lim t^{-1}\left[e^{-itH_0}u - u\right] = P_{\pm}H_0 u.$$

Thus the $M_{\pm}$ reduce H_0. The proof for the $R_{\pm}$ is similar. □

Theorem 6.3.9. *$HW_{\pm}P_0 = W_{\pm}H_0$, where P_0 is the orthogonal projection onto $N(H_0)^{\perp}$.*

PROOF. First we note that φ is in $N(H_0)$ iff $(\varphi, H_0 v) = 0$ for all $v \in D(H_0)$. Thus

$$N(H_0) = R(H_0)^{\perp}. \tag{6.3.10}$$

This implies that

$$N(H_0)^{\perp} = \overline{R(H_0)} \tag{6.3.11}$$

(see Appendix B). Thus P_0 is the orthogonal projection onto the closure of $R(H_0)$. Next we note that

$$P_0 P_{\pm} P_0 = P_{\pm} P_0. \tag{6.3.12}$$

In fact, we have by Theorem 6.3.8

$$P_0 P_{\pm} H_0 v = P_0 H_0 P_{\pm} v = H_0 P_{\pm} v = P_{\pm} H_0 v$$

for $v \in D(H_0)$. Thus $P_0 P_{\pm} = P_{\pm}$ on $R(H_0)$. Since they are bounded operators, this holds on the closure of $R(H_0)$ as well. This gives (6.3.12). Now suppose $u \in D(W_{\pm} H_0)$. Then $u \in D(H_0)$ and $H_0 u \in M_{\pm}$. Thus

$$H_0(1 - P_{\pm})P_0 u = (1 - P_{\pm})H_0 u = 0.$$

Hence $P_0(1 - P_{\pm})P_0 u = 0$. By (6.3.12) this yields

$$P_0 u = P_{\pm} P_0 u. \tag{6.3.13}$$

Hence $P_0 u \in M_{\pm}$ and

$$\begin{aligned} W_{\pm} H_0 u &= W_{\pm} H_0 P_0 u \\ &= i W_{\pm} \lim t^{-1}\left[e^{-itH_0} - 1\right] P_0 u \\ &= i \lim t^{-1}\left[e^{-itH} - 1\right] W_{\pm} P_0 u. \end{aligned} \tag{6.3.14}$$

The existence of this limit shows that $W_{\pm} P_0 u \in D(H)$ and

$$W_{\pm} H_0 u = H W_{\pm} P_0 u. \tag{6.3.15}$$

Thus $u \in D(H W_{\pm} P_0)$ and (6.3.15) holds.

Conversely, if $u \in D(H W_{\pm} P_0)$, then $P_0 u \in M_{\pm}$ and $W_{\pm} P_0 u \in D(H)$. Thus the last limit in (6.3.14) exists. Hence $W_{\pm} t^{-1}[e^{-itH_0} - 1] P_0 u$ converges in $\mathcal{H}$, and since $R_{\pm}$ is a closed subspace (Theorem 6.2.3) it converges to an element of the form $W_{\pm} f$. Thus

$$\|t^{-1}\left[e^{-itH_0} - 1\right] P_0 u - f\| = \|W_{\pm}\left\{t^{-1}\left[e^{-itH_0} - 1\right] P_0 u - f\right\}\| \to 0,$$

which shows that $u \in D(H_0)$ and (6.3.15) holds. □

If A_1, A_2 are operators on Hilbert spaces $\mathcal{H}_1, \mathcal{H}_2$, respectively, we say that they are *unitarily equivalent* if there is a unitary operator B from $\mathcal{H}_1$ to $\mathcal{H}_2$ such that

$$A_1 = B^{-1} A_2 B.$$

The importance of this concept is that two such operators are essentially identical in their behavior. In particular, all basic properties of one can be determined from the other. The wave operators we have defined give an excellent example of this.

Corollary 6.3.2. *If $N(H_0)$ is empty, then*

$$HW_{\pm} = W_{\pm}H_0. \tag{6.3.16}$$

Thus the restriction of H_0 to $M_{\pm}$ is unitarily equivalent to the restriction of H to $R_{\pm}$.

The proof of Corollary 6.3.2 is immediate from Theorem 6.3.9 if we note that $W_{\pm}$ is unitary from $M_{\pm}$ to $R_{\pm}$.

Since unitary equivalence is of interest in mathematics irrespective of applications to quantum theory, one sees from Corollary 6.3.2 why wave operators are of interest in pure mathematics.

6.4. The Domains of the Wave Operators

The theory just presented poses many problems. The first is to determine the subspaces $M_{\pm}$. For all we know they may consist of only the zero vector. In this case the whole theory is meaningless. Fortunately there is a fairly simple criterion which is very useful in many situations. We describe it in this section.

The idea is very simple. If we want to obtain

$$\lim_{t\to\infty} W(t)u, \tag{6.4.1}$$

it suffices to show that the integral

$$\int_a^{\infty} \|W'(t)u\|\,dt \tag{6.4.2}$$

exists for some a. In fact, we have

$$|(W(t_2)u - W(t_1)u, v)| = \left|\int_{t_1}^{t_2}(W'(t)u, v)\,dt\right| \le \|v\|\int_{t_1}^{t_2}\|W'(t)u\|\,dt.$$

Thus, $\|W(t_2)u - W(t_1)u\| \to 0$ as $t_1, t_2 \to \infty$ if the integral exists. Now

$$W'(t)u = ie^{itH}(H - H_0)e^{-itH_0}u, \tag{6.4.3}$$

provided

$$e^{-itH_0}u \in D(H_0) \cap D(H). \tag{6.4.4}$$

Thus we have

Theorem 6.4.1. *Suppose* (6.4.4) *holds for all $t \ge a$ and*

$$\int_a^{\infty} \|(H - H_0)e^{-itH_0}u\|\,dt < \infty \tag{6.4.5}$$

for some a. Then $u \in M_+$. If there is a b such that (6.4.4) *holds when*

$t \le b$ and

$$\int_{-\infty}^{b} \|(H - H_0)e^{-itH_0}u\| \, dt < \infty, \tag{6.4.6}$$

then $u \in M_-$.

Let us see what the theorem can yield when applied to our situation of a single particle with Hamiltonian $H = H_0 + V$. We have

Theorem 6.4.2. *Assume that for each real s there is an $a(s)$ such that*

$$\int_{a(s)}^{\infty} t^{-3/2}\left(\int_{-\infty}^{\infty} |V(x)(x-s)|^2 \exp\left\{-\frac{(x-s)^2}{2(1+t^2)}\right\} dx\right)^{1/2} dt < \infty \tag{6.4.7}$$

with the inner integral finite for each $t > a(s)$. Then M_+ is the whole of L^2. If this holds with (a, ∞) replaced by $(-\infty, -a)$, then $M_- = L^2$.

Before proving Theorem 6.4.2 let us note the following consequence.

Corollary 6.4.1. *If $(1 + |x|)^\alpha V(x)$ is in L^2 for some $\alpha > \frac{1}{2}$, then $M_+ = M_- = L^2$.*

To derive the corollary from the theorem note that

$$e^{-h} \le (1 + \beta^{-1}h)^{-\beta}, \qquad h, \beta \ge 0. \tag{6.4.8}$$

Thus

$$\exp\left\{-\frac{(x-s)^2}{2(1+t^2)}\right\} \le C(1+t^2)^{1-\alpha}|x-s|^{2\alpha-2}. \tag{6.4.9}$$

Since

$$|x - s| \le (1 + |s|)(1 + |x|),$$

the left-hand side of (6.4.7) is bounded by a constant times

$$\int_1^{\infty} t^{-3/2+1-\alpha} \, dt \left(\int |V(x)\rho(x)^\alpha|^2 \, dx\right)^{1/2},$$

where

$$\rho(x) = 1 + |x|. \tag{6.4.10}$$

Since $\alpha > \frac{1}{2}$, the integral in t is finite. Hence (6.4.7) holds for each real s. This proves Corollary 6.4.1.

In proving Theorem 6.4.2 we shall make use of

Lemma 6.4.1. *The set of (finite) linear combinations of functions whose Fourier transforms are of the form*

$$\psi_s(k) = ke^{-k^2-iks}, \qquad s \ \text{real}, \tag{6.4.11}$$

is dense in L^2.

PROOF. We shall show that the only function orthogonal to this set vanishes. The result then follows from Corollary 3.5.3. Suppose

$$(w, \check{\psi}_s) = 0 \qquad \text{for all real} \quad s.$$

Then

$$\int_{-\infty}^{\infty} ke^{-k^2+iks}\hat{w}(k)\, dk = (\hat{w}, \psi_s) = 0$$

for all s. This says that $\check{h}(s) \equiv 0$, where

$$h(k) = ke^{-k^2}\hat{w}(k).$$

Parseval's identity says that $h(k) \equiv 0$, which in turn means that $\hat{w}(k) \equiv 0$ and consequently that $w(x) \equiv 0$. □

Now we can give the

PROOF OF THEOREM 6.4.2. Multiplying t by a positive constant does not affect the validity of (6.4.7). Thus we lose nothing by taking $2m = \hbar = 1$. Set $u = \check{\psi}_s$. Then

$$\begin{aligned} e^{-itH_0}u &= \left[e^{-itk^2}\psi_s\right]^{\vee} = \frac{1}{\sqrt{2\pi}}\int_{-\infty}^{\infty} e^{ik(x-s)-(1+it)k^2}k\, dk \\ &= \frac{1}{\sqrt{2\pi}}\exp\left\{-\frac{(x-s)^2}{4(1+it)}\right\} \\ &\quad \times \int e^{-z^2}\left[\frac{z}{(1+it)^{1/2}} + \frac{i(x-s)}{2(1+it)}\right]\frac{dz}{(1+it)^{1/2}}, \end{aligned} \tag{6.4.12}$$

where we have made the substitution

$$z = (1+it)^{1/2}k - \tfrac{1}{2}i(x-s)(1+it)^{-1/2}$$

and completed the square. Now I claim that

$$\int e^{-z^2}\, dz = \sqrt{\pi}, \qquad \int e^{-z^2}z\, dz = 0. \tag{6.4.13}$$

This is well known if z is a real variable and the integration is taken from $-\infty$ to ∞. But an examination of (6.4.12) shows that this is not the case. However, in each case the integrand is an analytic function of z tending to 0 rapidly as $|z| \to \infty$ in the region between the real axis and the actual path of integration. Hence by Cauchy's theorem, the integrals in (6.4.13) are independent of the path in that region. Thus (6.4.13) holds, and (6.4.12) can be written as

$$e^{-itH_0}u = \frac{i(x-s)}{(2+2it)^{3/2}} \exp\left\{ -\frac{(x-s)^2}{4(1+it)} \right\}. \tag{6.4.14}$$

Thus,

$$|e^{-itH_0}u| = \frac{|x-s|}{(4+4t^2)^{3/4}} \exp\left\{ -\frac{(x-s)^2}{4(1+t^2)} \right\} \tag{6.4.15}$$

and consequently,

$$\| Ve^{-itH_0}u\|^2 = (4+4t^2)^{-3/2} \int_{-\infty}^{\infty} |V(x)(x-s)|^2 \exp\left\{ -\frac{(x-s)^2}{2(1+t^2)} \right\} dx. \tag{6.4.16}$$

By hypothesis, this is finite for all real s and $t > a$. Thus $e^{-itH_0}u \in D(V)$ for all such s and t. Moreover, (6.4.12) shows us that $e^{-itH_0}u \in D(H_0)$ for each s and t. Finally, note that (6.4.7) implies (6.4.5) in view of (6.4.16). Hence (6.4.5) holds for u in the set described in Lemma 6.4.1. Since this set is dense in L^2, we see by Theorem 6.4.1 that M_+ contains a dense set. Since M_+ is a closed subspace (Theorem 6.2.1), we see that M_+ must be the whole space L^2. The same reasoning applies to M_-. □

We note the following convenient consequence of Corollary 6.4.1.

Corollary 6.4.2. *If $V(x)$ is locally square integrable and there is an $\alpha > 1$ such that $|x|^\alpha V$ is bounded for $|x|$ large, then $M_+ = M_- = L^2$.*

6.5. Local Singularities

Theorem 6.4.2 does not apply unless $V(x)$ is locally in L^2. In this section we shall show that the behavior of V on a bounded interval does not effect the existence of the wave operators. We shall prove

Theorem 6.5.1. *Suppose that $V(x)$ is locally square integrable for $|x| > b$ and that for each real s there is an $a(s)$ such that*

$$\int_{a(s)}^{\infty} t^{-3/2} \left(\int_{|x|>b} |V(x)(x-s)|^2 \exp\left\{ -\frac{(x-s)^2}{2(1+t^2)} \right\} dx \right)^{1/2} dt < \infty \tag{6.5.1}$$

with the inner integral finite for each $t > a(s)$. Then $M_+ = L^2$. A similar statement holds for M_-.

PROOF. Let $\varphi(x)$ be a function in C_0^∞ which equals 1 for $|x| < b$ and such that $0 \le \varphi(x) \le 1$ (for the existence of such a function see Theorem 7.8.3). Thus,

$$W(t) = e^{itH}\varphi e^{-itH_0} + e^{itH}(1-\varphi)e^{-itH_0} = W_1(t) + W_2(t).$$

Now if $\hat{u} = \psi_s$, then by (6.4.15)

$$|e^{-itH_0}u| \le |x - s|(4 + 4t^2)^{-3/4}.$$

Hence

$$\|\varphi e^{-itH_0}u\| \le (4 + 4t^2)^{-3/4}\left(\int_{-\infty}^{\infty} |\varphi(x)(x-s)|^2\, dx\right)^{1/2}. \tag{6.5.2}$$

This shows that

$$W_1(t)u \to 0 \qquad \text{as} \quad t \to \pm\infty. \tag{6.5.3}$$

Moreover, we have

$$\begin{aligned} W_2'(t)u &= ie^{itH}\left[H(1-\varphi) - (1-\varphi)H_0\right]e^{-itH_0}u \\ &= ie^{itH}(1-\varphi)Ve^{-itH_0}u + ie^{itH}\varphi'' e^{-itH_0}u + 2ie^{itH}\varphi' e^{-itH_0}u' \\ &= W_3(t)u + W_4(t)u + W_5(t)u. \end{aligned} \tag{6.5.4}$$

Now (6.5.2) implies

$$W_4(t)u \to 0 \qquad \text{as} \quad t \to \pm\infty. \tag{6.5.5}$$

Moreover, (6.4.14) implies

$$|e^{-itH_0}u'| \le (4 + 4t^2)^{-5/4}\left[(4 + 4t^2)^{1/2} + |x - s|^2\right]. \tag{6.5.6}$$

Thus, we have

$$\|\varphi' e^{-itH_0}u'\| \to 0 \qquad \text{as} \quad t \to \pm\infty.$$

Finally, we note that (6.5.1) implies that $W_3(t)u$ converges to a limit as $t \to \infty$ as we showed in the proof of Theorem 6.4.2. □

In Section 8.4 we shall show how local square integrability can be removed completely.

Finally, we note that the transformation $\tau = (1/2)(1 + t^2)$ converts (6.5.1) into

$$\int_0^{c(s)} \tau^{-3/4}\left(\int_{|x|>b} |V(x)(x-s)|^2 \exp\{-\tau(x-s)^2\}\, dx\right)^{1/2} d\tau < \infty. \tag{6.5.7}$$

Thus we have

Corollary 6.5.1. *Suppose $V(x)$ is locally square integrable for $|x| > b$ and for each real s there is a $c(s) > 0$ such that* (6.5.7) *holds with the inner integral finite for each τ. Then $M_+ = L^2$ with a similar statement holding for M_-.*

Exercises

1. If A is an isometry, show that $(Au, Av) = (u, v)$ for all u, v.

2. If A is unitary, show that $AA^* = 1$ and $A^* = A^{-1}$.

3. Show that $d(E(\lambda)u, u)/d\lambda \geq 0$ when it exists.

4. Prove Theorem 6.3.2.

5. Show that the $R_\pm$ reduce H.

6. Prove (6.4.8) and (6.4.9).

7. Prove (6.5.4).

8. Prove (6.5.6).

9. Show that (6.5.1) is equivalent to (6.5.7).

7

Long-Range Potentials

7.1. The Coulomb Potential

We showed in Chapter 6 that $M_{\pm} = L^2$ when $V(x)$ satisfies

$$|V(x)| \le C\rho(x)^{-\alpha}, \qquad |x| > 1, \tag{7.1.1}$$

for some $\alpha > 1$ (Corollary 6.4.2). Here

$$\rho(x) = 1 + |x|. \tag{7.1.2}$$

A very important potential in physics is the Coulomb potential

$$V(x) = c_0|x|^{-1} \tag{7.1.3}$$

(it has more significance in higher dimensions). It would therefore be of interest to determine what happens when $V(x)$ satisfies (7.1.1) with $0 < \alpha \le 1$. The result is not what we would like. Among other things we shall show that $M_{\pm}$ consist of only the zero function in the case of (7.1.3) and other potentials having a weak decay at infinity. Such potentials are called *long-range potentials* because their effects are obviously felt further than the vicinity of a bounded set (otherwise the arguments of Section 6.2 would imply the existence of the wave operators).

We shall prove the negative result in the following way. For a function $f(k)$, define the operator $f(p)$ by

$$f(p)\psi = [f(k)\hat{\psi}]^{\vee}. \tag{7.1.4}$$

The domain of $f(p)$ consists of those $\psi \in L^2$ such that the right-hand side of (7.1.4) is in L^2. For the potential (7.1.3) we shall show that the limits

$$\tilde{W}_{\pm}\psi = \lim_{t \to \pm\infty} W(t)t^{-ic_0/|p|}\psi \tag{7.1.5}$$

exist for all $\psi \in L^2$. The operators (7.1.5) are called the *modified wave operators*. The irony of the matter is that the existence of the modified wave operators (7.1.5) precludes the existence of the ordinary wave operators. To see this, let f be any function in L^2 and let g be any function in M_+. If $f \in D(\tilde{W}_+)$, then

$$(t^{-ic_0/|p|}f, g) = (W(t)t^{-ic_0/|p|}f, W(t)g) \to (\tilde{W}_+ f, W_+ g)$$

as $t \to \infty$. But this equals

$$\int_{-\infty}^{\infty} t^{-ic_0/|p|}\hat{f}\hat{g}^* \, dp, \tag{7.1.6}$$

which certainly cannot approach a limit unless the integrand vanishes identically. Thus if $D(\tilde{W}_+) = L_2$, then $M_+ = \{0\}$.

This negative result is rather upsetting since the theory excludes such an important potential. However, the fact that the modified wave operators exist does constitute a sort of consolation prize since they do have many of the properties of the ordinary wave operators. In this chapter we shall explore the existence of the modified wave operators for certain classes of potentials. For any arbitrary potential $V(x)$ we shall define

$$X_t = X_t^V(p) = \int_0^t V(sp/m) \, ds. \tag{7.1.7}$$

We shall prove

Theorem 7.1.1. *Assume that* $V = V_1 + V_2$, *where* V_1 *satisfies* (7.1.1) *with* $\alpha > 1$ *and* V_2 *satisfies*

$$|V'(x)| \le C\rho(x)^{-\beta}, \tag{7.1.8}$$

with $\beta > 3/2$. *Then the limits*

$$\tilde{W}_\pm \psi = \lim_{t \to \pm\infty} W(t)e^{-iX_t}\psi \tag{7.1.9}$$

exist for every $\psi \in L^2$.

As in the case of the ordinary wave operators, we let $\tilde{M}_\pm$ be the subspaces consisting of those $\psi \in L^2$ for which the limits (7.1.9) exist. Theorem 7.1.1 states that $\tilde{M}_\pm = L^2$ if $V(x)$ satisfies (7.1.8) with $\beta > 3/2$. In the case of the Coulomb potential (7.1.3) we have (taking $m = 1$)

$$X_t(p) = c_0|p|^{-1}\log(tp) \tag{7.1.10}$$

if we take $V_1 = V(x)$ for $|x| \le 1$ and let it vanish otherwise. Thus

$$e^{-iX_t} = (tp)^{-ic_0/|p|}.$$

Since the operator $p^{-ic_0/|p|}$ is bounded and independent of t, the existence of the limits (7.1.5) follows from Theorem 7.1.1.

This chapter is devoted to the proof of Theorem 7.1.1 and some of its generalizations. The proofs are not difficult, but one must pay attention to details. We give some illustrations in the next section.

7.2. Some Examples

Let us consider potentials of the form

$$V(x) = c_0|x|^{-\alpha} \tag{7.2.1}$$

with $\frac{1}{2} < \alpha < 1$. If we take $V_1(x) = V(x)$ for $|x| \leq 1$ and have it vanish otherwise, we have

$$X_t = c_0(1-\alpha)^{-1}|p|^{-\alpha}(t^{1-\alpha} - p^{\alpha-1}). \tag{7.2.1}$$

The hypotheses of Theorem 7.1.1 are satisfied. On the other hand, consider potentials of the form

$$V(x) = c_0|x|^{-\alpha}\sin(|x|^{\sigma}), \qquad \alpha < \sigma + 1. \tag{7.2.3}$$

Take $V_1(x)$ as above. It is easily checked that the hypotheses of Theorem 7.1.1 are satisfied if

$$\alpha - \tfrac{1}{2} > \sigma \geq 0. \tag{7.2.4}$$

Moreover, we have

$$X_t = \pm c_0|p|^{-1}\int_1^{|tp|} r^{-\alpha}\sin(r^{\sigma})\,dr,$$

where the sign is that of tp. An interesting phenomenon occurs here. When $\alpha + \sigma > 1$, it is a simple matter to show that X_t converges to limits as $t \to \pm\infty$. In fact, a simple integration by parts shows that the integral

$$\int_1^{\infty} r^{-\alpha}\sin(r^{\sigma})\,dr$$

exists. In this case the limits

$$\begin{aligned} W_{\pm}\psi &= \lim W(t)e^{-iX_t}e^{iX_t}\psi \\ &= \tilde{W}_{\pm}\lim e^{iX_t}\psi \end{aligned}$$

exist. Hence the ordinary wave operators exist, but the method of Chapter 6 cannot be applied.

As another example, consider a potential satisfying

$$V(x) = \begin{cases} V_+, & x > N \\ V_-, & x < -N, \end{cases}$$

where $V_{\pm}$ are constants and $V(x)$ is bounded in the interval $[-N, N]$. It is easily checked that the hypotheses of Theorem 7.1.1 are satisfied. In fact, let $\varphi(x)$ be a function in C_0^{∞} vanishing for $|x| > N$ such that

$$\int \varphi(x)\,dx = 1 \tag{7.2.5}$$

(for the existence of such a function see Section 7.8). If we put

$$h(x) = V_- + (V_+ - V_-)\int_{-\infty}^{x}\varphi(y)\,dy, \tag{7.2.6}$$

we have $h \in C^\infty$ and $h(x) = V(x)$ for $|x| > N$. Thus h satisfies (7.1.8) for any β and $V - h$ satisfies (7.1.1) for any α. Hence the modified wave operators exist for this potential, but the regular wave operators do not.

7.3. The Estimates

In this section we discuss the estimates required to prove Theorem 7.1.1. They are not difficult, but they require a bit of care. First we assume that there is a function $X_t(p)$ [not necessarily the function given by (7.1.7)] which satisfies

$$|x^k X_t| \leq C_k(p)\rho(t)^{\sigma(k)}, \qquad k = 2, 3, \ldots, \tag{7.3.1}$$

where

$$x = i\frac{d}{dp}, \tag{7.3.2}$$

the $C_k(p)$ are functions of p, and $\sigma(t)$ is a nondecreasing function. We assume that the $C_k(p)$ are bounded on each interval of the form $0 < a \leq |p| \leq b < \infty$ (i.e., have no singularities other than 0 and $\pm\infty$). Also we assume that

$$\sigma(k) \leq k - 1, \qquad k = 2, 3, \ldots . \tag{7.3.3}$$

[The notation (7.3.2) is suggested by (f) of Section 1.3.] Next we let λ be a fixed parameter satisfying $0 \leq \lambda \leq 1$, and we define

$$Y = \frac{tp^2}{2\lambda m} + X_t(p). \tag{7.3.4}$$

Then we have

$$xe^{-iY}\psi = e^{-iY}(x + Q)\psi, \tag{7.3.5}$$

where

$$Q(p) = \frac{tp}{\lambda m} + xX_t. \tag{7.3.6}$$

We should pause to clarify the meaning of (7.3.5). The left-hand side can be interpreted to mean either

$$x\left[e^{-iY}\hat{\psi}\right]^{\vee} \tag{7.3.7}$$

[in this case e^{-iY} is the operator defined by (7.1.4)] or to mean

$$i\frac{d}{dp}\left(e^{-iY}\hat{\psi}\right) \tag{7.3.8}$$

(in this case e^{-iY} is the function). These interpretations are different, one being the Fourier transform of the other. However, as long as one is consistent no confusion will result since their norms are equal by Parseval's identity [see (1.3.6)]. Note that (7.3.1) and (7.3.6) imply

$$|xQ| \le \frac{|t|}{\lambda m} + C_2\rho(t)^{\sigma(2)} \tag{7.3.9}$$

and

$$|x^kQ| \le C_{k+1}\rho(t)^{\sigma(k+1)}, \qquad k = 2, 3, \ldots . \tag{7.3.10}$$

Next we assume that

$$|xX_t| \le \frac{|tp|}{2m}, \qquad |tp| > N. \tag{7.3.11}$$

This implies

$$|Q| \ge \frac{|tp|}{2\lambda m}, \qquad |tp| > N, \tag{7.3.12}$$

and

$$|xQ^{-1}| \le C(p)\lambda/|t|, \qquad |tp| > N, \tag{7.3.13}$$

where $C(p)$ and all subsequent constants have the same properties of the $C_k(p)$. Simple calculations also give

$$|x^kQ^{-1}| \le C(p)\lambda|t|^{\nu(k+1)-2}, \qquad |tp| > N, \qquad k = 2, 3, \tag{7.3.14}$$

where

$$\nu(k) = \max(\sigma(k), 1). \tag{7.3.15}$$

Next set

$$A_\gamma\psi = \rho(\lambda x)^{-\gamma}e^{-iY}\psi \tag{7.3.16}$$

By (7.3.5), we have

$$e^{-iY}\psi = xe^{-iY}Q^{-1}\psi - e^{-iY}xQ^{-1}\psi. \tag{7.3.17}$$

This implies

$$\|A_\gamma\psi\| \le \lambda^{-1}\|A_{\gamma-1}Q^{-1}\psi\| + \|A_\gamma xQ^{-1}\psi\|. \tag{7.3.18}$$

Assume that $\hat{\psi}$ is in C_0^∞ and it vanishes in a neighborhood of 0. Then by (7.3.12) and (7.3.13)

$$\begin{aligned}\|A_1\psi\| &\le \lambda^{-1}\|Q^{-1}\psi\| + \|xQ^{-1}\psi\| \\ &\le C(\|\psi\| + \|x\psi\|)/|t|, \qquad |t| > N'.\end{aligned} \tag{7.3.19}$$

This can be improved slightly by substituting (7.3.19) back into (7.3.18) to obtain

$$\begin{aligned}\|A_1\psi\| &\le \lambda^{-1}\|Q^{-1}\psi\| + \|A_1xQ^{-1}\psi\| \\ &\le C|t|^{-1}(\|\psi\| + \|xQ^{-1}\psi\| + \|x^2Q^{-1}\psi\|) \\ &\le C|t|^{-1}\|\psi\| + C|t|^{-2}(\|x\psi\| + \|x^2\psi\|)\end{aligned} \tag{7.3.20}$$

(here we used the fact that $\sigma(3) \le 2$). Using (7.3.18) and (7.3.19), we have

$$\begin{aligned}\|A_2\psi\| &\le \lambda^{-1}\|A_1Q^{-1}\psi\| + \|A_1xQ^{-1}\psi\| \\ &\le C(\|Q^{-1}\psi\| + \|xQ^{-1}\psi\| + \|x^2Q^{-1}\psi\|)/\lambda|t| \qquad (7.3.21)\\ &\le C|t|^{\nu(3)-3}\|\psi\| + C|t|^{-2}(\|x\psi\| + \|x^2\psi\|).\end{aligned}$$

Substituting (7.3.20) and (7.3.21) into (7.3.18) we get

$$\begin{aligned}\|A_2\psi\| &\le \lambda^{-1}\|A_1Q^{-1}\psi\| + \|A_2xQ^{-1}\psi\| \\ &\le C\lambda^{-1}(|t|^{-1}\|Q^{-1}\psi\| + |t|^{-2}(\|xQ^{-1}\psi\| + \|x^2Q^{-1}\psi\|)) \\ &\quad + C|t|^{\nu(3)-3}\|xQ^{-1}\psi\| + C|t|^{-2}(\|x^2Q^{-1}\psi\| + \|x^3Q^{-1}\psi\|) \\ &\le C(|t|^{-2}\|\psi\| + |t|^{\nu(3)-4}\|x\psi\| + |t|^{-3}(\|x^2\psi\| + \|x^3\psi\|)).\end{aligned} \qquad (7.3.22)$$

Next, we compute

$$\begin{aligned}\|A_1(xX_t)\psi\| &\le C|t|^{-1}\|(xX_t)\psi\| + C|t|^{-2}(\|x(xX_t)\psi\| \\ &\quad + \|x^2(xX_t)\psi\|) \le C(|t|^{\sigma(1)-1} + |t|^{\sigma(3)-2})\|\psi\| \qquad (7.3.23)\\ &\quad + C|t|^{\sigma(2)-2}\|x\psi\| + C|t|^{\sigma(1)-2}\|x^2\psi\|\end{aligned}$$

and

$$\begin{aligned}\|A_2(xX_t)\psi\| &\le C|t|^{-2}\|(xX_t)\psi\| + C|t|^{\nu(3)-4}\|x(xX_t)\psi\| \\ &\quad + C|t|^{-3}(\|x^2(xX_t)\psi\| + \|x^3(xX_t)\psi\|) \\ &\le (|t|^{\sigma(4)-3} + |t|^{\sigma(1)-2} + |t|^{\sigma(2)+\nu(3)-4})\|\psi\| \qquad (7.3.24)\\ &\quad + C(|t|^{\sigma(1)+\nu(3)-4} + |t|^{\sigma(3)-3})\|x\psi\| \\ &\quad + C|t|^{\sigma(2)-3}\|x^2\psi\| + C|t|^{\sigma(1)-3}\|x^3\psi\|.\end{aligned}$$

Finally, we compute

$$\begin{aligned}\|A_3\psi\| &\le \lambda^{-1}\|A_2Q^{-1}\psi\| + \|A_2xQ^{-1}\psi\| \\ &\le C|t|^{\nu(3)-3}\|Q^{-1}\psi\| + C|t|^{-2}(\|xQ^{-1}\psi\| + \|x^2Q^{-1}\psi\|) \\ &\quad + C|t|^{\nu(3)-3}\|xQ^{-1}\psi\| + C|t|^{-2}(\|x^2Q^{-1}\psi\| + \|x^3Q^{-1}\psi\|) \\ &\le C|t|^{\nu(3)-4}(\|\psi\| + \|x\psi\|) + |t|^{-3}(\|x^2\psi\| + \|x^3\psi\|). \qquad (7.3.25)\end{aligned}$$

It should be remembered that in inequalities (7.3.19)–(7.3.25) it is assumed that $\hat{\psi}$ is in C_0^∞ and vanishes near 0. This is the only way we could get the constants $C(p)$ to be bounded. Hence all of the constants in these inequalities depend on the set where ψ vanishes.

We end this section with the following observation.

Lemma 7.3.1. *If* $0 \le \theta \le 1$, *then*

$$\|A_{\gamma+\theta}\psi\| \le \|A_\gamma\psi\|^{1-\theta}\|A_{\gamma+1}\psi\|^\theta$$

The proof of Lemma 7.3.1 follows from Holder's inequality

$$\int_{-\infty}^{\infty}|f(x)g(x)|\,dx \le \left(\int_{-\infty}^{\infty}|f(x)|^p\,dx\right)^{1/p}\left(\int_{-\infty}^{\infty}|g(x)|^{p'}\,dx\right)^{1/p'}$$

for any $p > 1$ with

$$\frac{1}{p}+\frac{1}{p'}=1$$

(see Appendix C). If $\theta = 0$ or $\theta = 1$ there is nothing to prove. Otherwise we note that

$$\|\rho_\lambda^{-\theta}\varphi\|^2 = \int \rho(\lambda x)^{-2\theta}|\varphi(x)|^2\,dx = \int \rho_\lambda^{-2\theta}|\varphi(x)|^{2\theta}$$

$$\times|\varphi(x)|^{2-2\theta}\,dx \le \left(\int \rho_\lambda^{-2}|\varphi|^2\,dx\right)^{\theta}\left(\int|\varphi|^2\,dx\right)^{1-\theta},$$

where we have taken $\rho_\lambda(x) = \rho(x\lambda)$ and $p = 1/\theta$. Putting $\varphi = A_\gamma\psi$ and noting that $A_{\gamma+\theta} = \rho_\lambda^{-\theta}A_\gamma$ we obtain (7.3.26). □

7.4. The Derivatives of $V(x)$

In this section we shall prove some consequences of (7.1.8). First we have

Theorem 7.4.1. *If* (7.1.8) *holds with* $\beta > 1$, *then there exist constants* $V_\pm$ *such that*

$$|V(x) - V_\pm| \le C\rho(x)^{1-\beta}, \qquad \pm x > 0. \tag{7.4.1}$$

PROOF. We have for $y < x$

$$|V(x) - V(y)| \le \int_y^x |V'(s)|\,ds \le C\int_y^x \rho(s)^{-\beta}\,ds$$

$$\le C'|\rho(x)^{1-\beta} - \rho(y)^{1-\beta}|.$$

This shows that

$$V(x) \to V_\pm \qquad \text{as} \quad x \to \pm\infty \tag{7.4.2}$$

and that (7.4.1) holds. □

Theorem 7.4.2. *Suppose* (7.1.8) *holds with* $\beta - 1 > \delta > 0$, *where* $\delta < 1$. *Then* $V = V_1 + V_2$ *with* V_1 *satisfying* (7.1.1) *for some* $\alpha > 1$ *and* V_2 *is infinitely differentiable and satisfies*

$$|V^{(n)}(x)| \le C_n\rho(x)^{-\tau_n} \tag{7.4.3}$$

with

$$\tau_n \ge \beta + (n-1)\delta, \qquad n > 0. \tag{7.4.4}$$

PROOF. First assume that $V_{\pm} = 0$ in (7.4.1). Let $b \geq 1$ be given and let $\varphi \in C_0^\infty$ be such that

$$\varphi(x) = \begin{cases} 1, & |x| \leq 1 \\ 0, & |x| > 2, \end{cases} \tag{7.4.5}$$

and $0 \leq \varphi(x) \leq 1$ (for the existence of such a function see Section 7.8). Set

$$w(x) = [\varphi(x/b) - \varphi(2x/b)]V(x), \tag{7.4.6}$$

$$\varphi_b(x) = cb^{-\delta}\varphi(8x/b^\delta), \tag{7.4.7}$$

and

$$v(x) = \int_{-\infty}^{\infty} w(x+y)\varphi_b(y)\,dy, \tag{7.4.8}$$

where the constant c is chosen so that

$$\int \varphi_b(x)\,dx = 1.$$

Now,

$$w'(x) = b^{-1}[\varphi'(x/b) - 2\varphi'(2x/b)]V(x) + [\varphi(x/b) - \varphi(2x/b)]V'(x).$$

Hence by Theorem 7.4.1,

$$|w'(x)| \leq Cb^{-1}\rho(x)^{1-\beta} + C\rho(x)^{-\beta}, \tag{7.4.9}$$

where the constant C depends only on V and φ. Now $w(x)$ vanishes unless

$$\tfrac{1}{2}b < |x| < 2b. \tag{7.4.10}$$

By (7.4.9) and (7.4.10) we get

$$|w'(x)| \leq C\rho(x)^{-\beta} \tag{7.4.11}$$

or its equivalent

$$|w'(x)| \leq C\rho(b)^{-\beta}. \tag{7.4.12}$$

Since $v(x)$ can be written in the form

$$v(x) = \int w(z)\varphi_b(z-x)\,dz,$$

we see that it is infinitely differentiable and vanishes unless there is a z such that

$$\tfrac{1}{2}b \leq |z| \leq 2b, \qquad 4|z-x| \leq b^\delta,$$

which implies

$$(1/4)b \leq |x| \leq (5/4)b. \tag{7.4.13}$$

Now

$$\begin{aligned} v(x) - w(x) &= \int [w(x+y) - w(x)]\varphi_b(y)\,dy \\ &= \int w'(x+\theta y)y\varphi_b(y)\,dy. \end{aligned}$$

Hence

$$|v(x) - w(x)| \le C\rho(b)^{-\beta}\int |y|\varphi_b(y)\,dy$$
$$= C\rho(b)^{-\beta}b^{\delta}\int |z|\varphi(z)\,dz \le C\rho(x)^{\delta-\beta} \tag{7.4.14}$$

by (7.4.10) and (7.4.13). Note that $\beta - \delta > 1$. If we write

$$p = -i\frac{d}{dx}, \tag{7.4.15}$$

then

$$p^k v = \int w(z)p^k\varphi_b(x-z)\,dz$$
$$= i\int w'(z)p^{k-1}\varphi_b(x-z)\,dz$$

by integration by parts. Hence

$$|p^k v| \le C\rho(b)^{-\beta}b^{(k-1)\delta}\int |\varphi^{(k-1)}(y)|\,dy. \tag{7.4.16}$$

Thus we have shown that w is the sum of two functions, the first of which, $w(x) - v(x)$, satisfies (7.1.1) with $\alpha = \beta - \delta$ by (7.4.14), and the second, $v(x)$, is in C_0^∞ and satisfies (7.4.3) with the τ_n prescribed by (7.4.16).

Now set $b = 2^j$ and let $w_j(x)$ and $v_j(x)$ be the corresponding functions defined by (7.4.6) and (7.4.8), respectively. Then by (7.4.14) and (7.4.16),

$$|v_j(x) - w_j(x)| \le C\rho^{\delta-\beta} \tag{7.4.17}$$

and

$$|p^k v_j| \le C_k\rho^{(k-1)\delta-\beta}, \tag{7.4.18}$$

where the constants do not depend on j.

Set $w_0 = \varphi(x)V(x)$ and

$$V_2(x) = \sum_{j=1}^{\infty} v_j(x),$$

$$V_1(x) = V(x) - V_2(x) = \sum_{j=1}^{\infty}\left[w_j(x) - v_j(x)\right] + w_0.$$

The series converge since for each x at most a finite number of the $w_j(x)$ and $v_j(x)$ do not vanish. Since the constants in (7.4.17) and (7.4.18) do not depend on j, we have proved the theorem for the case $V_\pm = 0$. Now suppose they do not vanish. Let $\varphi(x)$ be a function in C_0^∞ satisfying (7.2.5) and define $h(x)$ by (7.2.6).

Since $h'(x) = (V_+ - V_-)\varphi(x)$, we see that h is infinitely differentiable. Moreover, $h(x) = V_+$ for x large and positive and $h(x) = V_-$ for x large and negative. Hence

$$|h^{(n)}(x)| \leq C_n \rho(x)^{-\sigma_n} \tag{7.4.19}$$

for any σ_n. Moreover, the function $\tilde{V}(x) = V(x) - h(x) \to 0$ as $|x| \to \infty$. If we apply the above proof to $\tilde{V}$ we obtain the theorem for V. □

7.5. The Relationship Between X_t and $V(x)$

Now we suppose that $V(x)$ satisfies (7.4.3) with arbitrary exponents τ_n. Define X_t by (7.1.7). Then X_t satisfies (7.3.1) for some function σ. We wish to determine σ in terms of the τ_n. We have

Theorem 7.5.1. *If $V(x)$ satisfies* (7.4.3) *and X_t is given by* (7.1.7), *then X_t satisfies* (7.3.1) *with*

$$\sigma(k) = \max_{n \leq k} (n - \tau_{n-1}), \qquad k \geq 1. \tag{7.5.1}$$

PROOF. We have

$$|X_t| \leq C\int_0^t \rho(sp)^{-\tau_0}\, ds = C' t^{1-\tau_0}.$$

Also,

$$\begin{aligned} xX_t &= i\int_0^t sV'(sp)\, ds = \frac{i}{p^2}\int_0^{tp} \mu\, dV(\mu) \\ &= \frac{i}{p^2}\left[tpV(tp) - X_t \right]. \end{aligned} \tag{7.5.2}$$

Hence

$$|xX_t| \leq C'' t^{1-\tau_0}.$$

Differentiating (7.5.2) repeatedly we have

$$x^k X_t = ip^{-2}\left(t^k p x^{k-1} V(tp) - x^{k-1} X_t\right).$$

The result now follows by induction. □

If $V(x)$ satisfies (7.4.3), we see from Theorem 7.4.1 that we may always assume

$$\tau_n \leq \tau_{n-1} + 1, \tag{7.5 3}$$

that is, the sequence $n - \tau_n$ is increasing. In this case we have by Theorem 7.5.1 that

$$\sigma(k) = k - \tau_{k-1}. \tag{7.5.4}$$

On the other hand, by Theorem 7.4.2 we can subtract off a potential satisfying (7.1.1) with $\alpha > 1$ so that

$$\tau_n \geq \tau_1 + (n-1)\delta, \tag{7.5.5}$$

provided $0 < \delta < \min(1, \tau_1 - 1)$. In particular, if $\tau_1 > 3/2$, we can take $\delta > 1/2$. In this case we have $\tau_2 > 2$ and $\tau_3 > 5/2$. This implies $\sigma(3) = 3 - \tau_2 < 1$ and $\sigma(4) = 4 - \tau_3 < 3/2$. We substitute these into (7.3.19)–(7.3.25) and find

Theorem 7.5.2. *If $\hat{\psi} \in C_0^\infty$ vanishes near the origin, then*

$$\|A_1\psi\| \le C|t|^{-1}, \qquad \|A_2\psi\| \le C|t|^{-2}, \qquad \|A_3\psi\| \le C|t|^{-3}, \quad (7.5.6)$$

and

$$\|A_1(xX_t)\psi\| \le C|t|^{1-\tau_1}, \qquad \|A_2(xX_t)\psi\| \le C|t|^{-\tau_1}, \qquad (7.5.7)$$

where the constant depends on ψ.

PROOF. The inequalities in (7.5.6) follow from (7.3.19), (7.3.22), (7.3.25), and the fact that $\sigma(3) < 1$. The first inequality in (7.5.7) follows from (7.3.23) and the facts that $-\tau_0 \le 1 - \tau_1$ and $\tau_1 < \tau_2$. The second inequality follows from (7.3.24) and the facts that $1 - \tau_3 < -\tau_1, \tau_1 \le \tau_1 < \tau_2$. □

If we apply Lemma 7.3.1 to (7.5.6) and (7.5.7) we obtain

Theorem 7.5.3. *Suppose $V(x)$ satisfies (7.4.3) with exponents τ_n such that (7.5.3) and (7.5.5) hold. Assume that $\beta = \tau_1 > 3/2$ and that $\hat{\psi} \in C_0^\infty$ vanishes near the origin. Let A_γ be defined by (7.3.16). Then there is a constant C not depending on t such that*

$$\begin{aligned} &\|A_{1+\theta}\psi\| \le C|t|^{-1-\theta}, \qquad \|A_{1+\theta}(xX_t)\psi\| \le C|t|^{1-\beta-\theta}, \\ &\|A_{2+\theta}\psi\| \le C|t|^{-2-\theta}, \qquad |t| > 1, \qquad 0 \le \theta \le 1. \end{aligned} \qquad (7.5.8)$$

7.6. An Identity

In this section we shall prove an important identity which will be used in the proof of Theorem 7.1.1. It is

$$e^{itH_0}F(mx)e^{-itH_0}\psi = e^{-imx^2/2t}F(tp)e^{imx^2/2t}\psi \qquad (7.6.1)$$

holding for certain classes of functions $F(x)$. We begin by noting that

$$xe^{-itH_0}\psi = i\frac{d}{dp}(e^{-itH_0}\psi) = e^{-itH_0}\left(x + \frac{tp}{m}\right)\psi. \qquad (7.6.2)$$

On the other hand,

$$pe^{imx^2/2t}\psi = -i\frac{d}{dx}(e^{imx^2/2t}\psi) = e^{imx^2/2t}\left(p + \frac{mx}{t}\right)\psi. \qquad (7.6.3)$$

It follows from (7.6.2) and (7.6.3) that

$$e^{itH_0}(mx)e^{-itH_0}\psi = e^{-imx^2/2t}tp\, e^{imx^2/2t}\psi = (mx + tp)\psi. \qquad (7.6.4)$$

From this it follows immediately that (7.6.1) holds for all polynomials

$F(x)$. Moreover, if z is not real, (7.6.2) implies

$$e^{itH_0}(mx - z)^{-1}e^{-itH_0}\psi = (mx + tp - z)^{-1},$$

while (7.6.3) implies

$$e^{-imx^2/2t}(tp - z)^{-1}e^{imx^2/2t}\psi = (mx + tp - z)^{-1}.$$

From this it follows that (7.6.1) is valid for all $F(x)$ which are ratios of polynomials and such that their denominators do not vanish for real x. Next we show that (7.6.1) holds if $F(x)$ is the uniform limit of such functions. For if $G(x)$ is of the type described and $F(x)$ satisfies $|F(x) - G(x)| < \varepsilon$, then

$$\|e^{itH_0}[F(mx) - G(mx)]e^{-itH_0}\psi\| \le \varepsilon\|\psi\|$$

and

$$\|e^{-imx^2/2t}[F(tp) - G(tp)]e^{imx^2/2t}\| \le \varepsilon\|\psi\|.$$

Hence the norm of the difference between both sides of (7.6.1) is $\le 2\varepsilon\|\psi\|$. Thus (7.6.1) holds.

7.7. The Reduction

In this section we show how the results of the preceding sections can be used in the proof of Theorem 7.1.1. Let $V(x)$ satisfy the hypotheses of that theorem. Then $V = V_1 + V_2$, where V_1 satisfies (7.1.1) with $\alpha > 1$ and V_2 satisfies (7.1.8) with $\beta > 3/2$. By Theorem 7.4.2, we can write $V_2 = V_3 + V_4$, where V_4 also satisfies (7.1.1) with $\alpha > 1$ and V_3 satisfies (7.4.3) with $\tau_1 = \beta$, $\tau_2 > 2$, and $\tau_3 > 5/2$. Set

$$X_t(p) = \int_0^t V_3(sp/m)\,ds. \tag{7.7.1}$$

We are going to show that the limits (7.1.9) exist if $\hat{\psi} \in C_0^\infty$ vanishes in a neighborhood of 0 and X_t is given by (7.7.1). This is not exactly what we require since we want the limits (7.1.9) to hold for $X_t = X_t^V$ and ψ any function in L^2. However, we shall show in the next section that this implies what we need.

We proceed as in the case of short range potentials considered in Chapter 6. We set

$$\tilde{W}(t) = W(t)e^{-X_t} \tag{7.7.2}$$

and we try to prove

$$\int_a^\infty \|\tilde{W}'(\pm t)\psi\|\,dt < \infty \tag{7.7.3}$$

for some a. Now

$$\tilde{W}'(t) = e^{itH}(V - X_t')e^{-iY_t}, \tag{7.7.4}$$

where

$$Y_t = tH_0 + X_t, \qquad X_t' = \frac{d}{dt}X_t. \tag{7.7.5}$$

We shall split this up into several summands and show that each of them is integrable from some a to ∞. Now $V = V_1 + V_3 + V_4$, where V_1 and V_4 satisfy (7.1.1) for some $\alpha > 1$. If $V_5 = V_1 + V_4$, we have

$$\|e^{itH}V_5 e^{-iY_t}\psi\| = \|V_5 e^{-iY_t}\psi\| \le C\|A_\alpha \psi\| \le C|t|^{-\alpha}$$

by Theorem 7.5.3. This is integrable near $\pm\infty$ since $\alpha > 1$. Thus it remains to show that

$$\int_{-\infty}^{-a} + \int_a^\infty \|(V_3 - X_t')e^{-iY_t}\psi\|\, dt < \infty \tag{7.7.6}$$

holds for some a. Since X_t is given by (7.7.1), the integrand in (7.7.6) involves only V_3 and not V_5. Thus we can consider the case $V = V_3$, and we may drop the subscript from V_3. Now by (7.6.1)

$$e^{itH_0}V(x)e^{-itH_0} = e^{-iZ_t}V(tp/m)e^{iZ_t}, \tag{7.7.7}$$

where

$$Z_t = mx^2/2t. \tag{7.7.8}$$

Thus

$$\begin{aligned} e^{itH_0}V(x)e^{-itH_0} - X_t' &= \int_0^1 \frac{d}{d\lambda}\left(e^{-i\lambda Z_t}V(tp/m)e^{i\lambda Z_t}\right) d\lambda \\ &= i\int_0^1 e^{-i\lambda Z_t}\left[V(tp/m)Z_t - Z_t V(tp/m)\right]e^{i\lambda Z_t}\, d\lambda \\ &= i\int_0^1 e^{-i\lambda Z_t}\left[\frac{t}{2m}V''(tp/m) - iV'(tp/m)x\right]e^{i\lambda Z_t}\, d\lambda \\ &= \frac{it}{2m}\int_0^1 e^{itH_0/\lambda}V''(\lambda x)e^{-itH_0/\lambda}\, d\lambda \\ &\quad + \int_0^1 e^{itH_0/\lambda}V'(\lambda x)e^{-itH_0/\lambda}x\, d\lambda, \end{aligned} \tag{7.7.9}$$

where we used the identity

$$x^2 g(p) = -g''(p) + 2ig'(p)x + g(p)x^2 \tag{7.7.10}$$

and applied (7.6.1) again. Now,

$$xe^{-iX_t}\psi = e^{-iX_t}[x + (xX_t)]\psi.$$

Hence

$$\begin{aligned} \|[V - X_t']e^{-iY_t}\psi\| \le \frac{t}{2m}&\int_0^1 \|V''(\lambda x)e^{-iY_t}\psi\|\, d\lambda \\ &+ \int_0^1 \|V'(\lambda x)e^{-iY}[x + (xX_t)]\psi\|\, d\lambda, \end{aligned} \tag{7.7.11}$$

where Y is given by (7.3.4). Now we know that V satisfies (7.4.3) with $\tau_1 = \beta$, $\tau_2 > 2$, and $\tau_3 > 5/2$. Hence

$$\|V''(\lambda x)e^{-iY}\psi\| \leq C\|A_{\tau_2}\psi\| \leq C'|t|^{-\tau_2}$$

by Theorem 7.5.3. (Note that the constants are independent of λ.) Since $\tau_2 > 2$, we see that the first term on the right of (7.7.11) is integrable near $t = \pm\infty$. Next we note that

$$\|V'(\lambda x)e^{-iY}[x + (xX_t)]\psi\| \leq C(\|A_{\tau_1}x\psi\| + \|A_{\tau_1}(xX_t)\psi\|)$$
$$\leq C(|t|^{-\tau_1} + |t|^{2-2\tau_1}).$$

Since $\tau_1 > 3/2$, this is also integrable near $t = \pm\infty$. Thus (7.7.3) holds and consequently the limits (7.1.9) exist for each ψ such that $\hat{\psi} \in C_0^\infty$ and vanishes near the origin when X_t is given by (7.7.1). In the next section we shall show how this implies Theorem 7.7.1.

7.8. Mollifiers

In this section we shall discuss the density of C_0^∞ in L^2. We need such results not only for the present chapter but also for other chapters as well. First we give some important background material.

Set

$$j(x) = \begin{cases} c_0 \exp\{(x^2-1)^{-1}\}, & |x| < 1 \\ 0, & |x| \geq 1. \end{cases} \tag{7.8.1}$$

It is easily checked that $j(x)$ is in C_0^∞. We pick the constant c_0 in (7.8.1) so that

$$\int_{-\infty}^{\infty} j(x)\,dx = 1. \tag{7.8.2}$$

Set

$$j_n(x) = nj(nx), \qquad n = 1, 2, \ldots, \tag{7.8.3}$$

and

$$J_nf(x) = \int_{-\infty}^{\infty} j_n(x-y)f(y)\,dy. \tag{7.8.4}$$

Since $j_n \in C_0^\infty$, the integral (7.8.4) exists for each x as long as f is locally integrable. We shall use

Theorem 7.8.1. *If $f \in L^2$, then*

$$\|J_nf\| \leq \|f\| \tag{7.8.5}$$

and

$$\|f - J_nf\| \to 0 \quad \text{as} \quad n \to \infty. \tag{7.8.6}$$

We postpone the proof of Theorem 7.8.1 until the end of this section. Now we show how it can be applied.

Theorem 7.8.2. *The set of those $\varphi \in C_0^\infty$ which vanish near the origin is dense in L^2.*

PROOF. Let u be an arbitrary function in L^2. We shall need the fact that for any $\varepsilon > 0$ there is a continuous function $w \in L^2$ such that

$$\|u - w\| < \varepsilon. \tag{7.8.7}$$

If you are unfamiliar with this fact, you can consider functions in L^2 as the limits in norm of continuous functions. Now there is an N so large that

$$\int_{|x|>N-1} |w(x)|^2 \, dx < \varepsilon^2 \tag{7.8.8}$$

(otherwise, w would not be in L^2). Set

$$M = \max_{|x| \le N} |w(x)|$$

(we know that M is finite since w is continuous). Now I am sure you will have no trouble constructing a continuous function v such that

$$v(x) = \begin{cases} 1, & 2\varepsilon^2 < |x| < N \\ 0, & |x| > N + 1 \text{ or } |x| < \varepsilon^2, \end{cases}$$

$$0 \le v(x) \le 1 \qquad \text{for all} \quad x.$$

Then

$$\int |w - vw|^2 \, dx = \int_{|x|<2\varepsilon^2} + \int_{N<|x|} (1 - v)^2 |w|^2 \, dx \le 4\varepsilon^2 M^2 + \varepsilon^2.$$

Hence we have

$$\|u - vw\|^2 \le (8M^2 + 4)\varepsilon^2.$$

Now vw is continuous, vanishes for $|x|$ large, and vanishes near 0. Hence it suffices to show that any such function is close in norm to a function in C_0^∞ which vanishes near 0. So suppose f is continuous and vanishes for $|x|$ near 0 and ∞. Define $J_n f$ by (7.8.4). I claim that $J_n f$ is in C_0^∞. It is infinitely differentiable because we can differentiate under the integral sign as many times as we want and still have an absolutely convergent integral. Moreover, since $f(y)$ vanishes for $|y|$ large and $j_n(x - y)$ vanishes for $|x - y| > 1/n$, we see that the integrand in (7.8.4) vanishes for $|x|$ large. Thus $J_n f \in C_0^\infty$. Next, note that it will vanish near 0 for n large for the same reason. Thus $J_n f$ is of the desired type for n large. Now all we need do is appeal to (7.8.6) of Theorem 7.8.1 to complete the proof. □

Another theorem we have needed on occasion is

Theorem 7.8.3. *Let M, N be closed subsets of the real line such that M is bounded and does not intersect N. Then there is a $\varphi \in C_0^\infty$ such that*

$$\varphi(x) = \begin{cases} 1 & \textit{for} \quad x \in M \\ 0 & \textit{for} \quad x \in N, \end{cases}$$

$$0 \le \varphi(x) \le 1 \qquad \textit{for all} \quad x.$$

PROOF. There is a positive number δ such that every point of M is a distance $> 2\delta$ from every point of N (prove this). Let M_δ be the set of all those points which are within a distance δ of M (i.e., $x \in M_\delta$ if there is a $y \in M$ such that $|x - y| < \delta$. In particular we have $M \subset M_\delta$.) Set

$$\varphi(x) = J_n \chi_{M_\delta} = \int_{M_\delta} j_n(x - y)\, dy, \tag{7.8.9}$$

where, as usual, χ_I is the characteristic function of the set I [see (1.7.1)]. Since χ_{M_δ} is in L^2 and vanishes for $|x|$ large, we know that $\varphi \in C_0^\infty$. Moreover,

$$0 \le \varphi(x) \le \int_{-\infty}^{\infty} j_n(x - y)\, dy = 1$$

by (7.8.2) and (7.8.3). Also, if $n > 1/\delta$, then $j_n(z)$ vanishes for $|z| > \delta$. Thus if $x \in M$ and $y \notin M_\delta$, then $j_n(x - y) = 0$. Hence

$$\varphi(x) = \int_{M_\delta} j_n(x - y)\, dy = \int_{-\infty}^{\infty} j_n(x - y)\, dy = 1.$$

On the other hand, if $x \in N$ and $y \in M_\delta$, then $|x - y| > \delta$, and consequently $j_n(x - y) = 0$. Thus $\varphi(x) = 0$ for $x \in N$. Thus we have shown that for n large the function φ given by (7.8.9) fulfills the requirements of the theorem. □

Now we can give the

PROOF OF THEOREM 7.1.1. In the beginning of Section 7.7 we showed that $V = V_3 + V_5$, where $V_5 = V_1 + V_4$ satisfies (7.1.1) for some $\alpha > 1$. Also we showed in that section that the limits (7.1.9) exist provided $X_t = X_t^{V_3}$ and $\hat{\psi} \in C_0^\infty$ and vanishes near the origin. Now

$$X_t^V = X_t^{V_3} + X_t^{V_5}$$

and

$$X_t^{V_5} \to \int_0^\infty V_5(sp/m)\, ds,$$

the limit existing because V_5 satisfies (7.1.1) with $\alpha > 1$. Hence the limits (7.1.9) exist with $X_t = X_t^V$. Moreover, since the operator (7.7.2) satisfies

$$\| \tilde{W}(t) \| = 1,$$

the sets $\tilde{M}_{\pm}$ are closed subspaces (see Theorem 6.2.1). Thus it suffices to show that the limits (7.1.9) exist for all ψ in a dense set of L^2. But the set of those $\psi \in C_0^\infty$ which vanish near the origin is dense in L^2 (Theorem 7.8.2). Consequently, those ψ whose Fourier transforms are in C_0^∞ and vanish near 0 are also dense by Parseval's identity. Thus the theorem is proved. □

It remains to give the

PROOF OF THEOREM 7.8.1. Apply Schwarz's inequality to (7.8.4). This gives

$$|J_nf(x)|^2 \le \int_{-\infty}^{\infty} j_n(x-y)|f(y)|^2\,dy \tag{7.8.10}$$

since

$$\int_{-\infty}^{\infty} j_n(x)\,dx = 1. \tag{7.8.11}$$

Thus

$$\int_I |J_nf(x)|^2\,dx \le \int_I \int_{-\infty}^{\infty} j_n(x-y)|f(y)|^2\,dy\,dx$$

$$= \int_{-\infty}^{\infty} |f(y)|^2 \int_I j_n(x-y)\,dx\,dy \le \int_{-\infty}^{\infty} |f(y)|^2\,dy$$

for any bounded interval I. Letting $I \to R$, we obtain (7.8.5). Next let u be any function in L^2 and let $\varepsilon > 0$ be given. There is a continuous function $w \in L^2$ such that (7.8.7) holds. Take N so large that (7.8.8) holds. Since w is uniformly continuous on bounded intervals, there is a positive $\delta \le 1$ such that

$$|w(x+y) - w(x)|^2 < \varepsilon^2/N, \qquad |x| \le N, \qquad |y| < \delta.$$

Thus,

$$\int_{-N}^{N} |w(x+y) - w(x)|^2\,dx < 2\varepsilon^2, \qquad |y| < \delta.$$

Also,

$$\int_{|x|>N} |w(x+y) - w(x)|^2\,dx \le 2\int_{|x|>N} |w(x+y)|^2\,dx$$

$$+ 2\int_{|x|>N} |w(x)|^2\,dx \le 4\int_{|x|>N-1} |w(x)|^2\,dx < 4\varepsilon^2.$$

Hence

$$\int_{-\infty}^{\infty} |w(x+y) - w(x)|^2\,dx < 6\varepsilon^2, \qquad |y| < \delta. \tag{7.8.12}$$

Now by (7.8.4) and (7.8.11),

$$J_nw(x) - w(x) = \int_{-\infty}^{\infty} j_n(y)[w(x-y) - w(x)]\,dy \tag{7.8.13}$$

and by the Schwarz inequality

$$|J_n w(x) - w(x)|^2 \le \int_{-\infty}^{\infty} j_n(y)|w(x-y) - w(x)|^2\,dy.$$

Consequently,

$$\int_{-\infty}^{\infty} |J_n w(x) - w(x)|^2\,dx \le \int_{-\infty}^{\infty}\int_{-\infty}^{\infty} j_n(y)|w(x-y) - w(x)|^2\,dx\,dy. \tag{7.8.14}$$

Note that $j_n(y)$ vanishes for $n|y| > 1$. Thus the integration with respect to y in (7.8.14) is only over the interval $|y| < 1/n$. If we take $n > 1/\delta$, then (7.8.12) holds for such y. Hence

$$\|J_n w - w\|^2 \le 6\varepsilon^2 \int_{-\infty}^{\infty} j_n(y)\,dy = 6\varepsilon^2.$$

This proves the theorem. □

For future reference we note that

$$\int_{-\infty}^{\infty} |J_n f(x)|\,dx \le \int_{-\infty}^{\infty} |f(x)|\,dx \tag{7.8.15}$$

and

$$\int_{-\infty}^{\infty} |J_n f(x) - f(x)|\,dx \to 0 \quad \text{as} \quad n \to \infty \tag{7.8.16}$$

hold for functions in $L^1(-\infty, \infty)$. These are even simpler to prove than Theorem 7.8.1. In fact, we have by (7.8.4)

$$\int_{-\infty}^{\infty} |J_n(x)|\,dx \le \int_{-\infty}^{\infty} |f(y)| \int_{-\infty}^{\infty} j_n(x-y)\,dx\,dy,$$

which gives (7.8.15). To prove (7.8.16) we let $\varepsilon > 0$ be given and find a continuous function $g \in L^1$ such that

$$\int_{-\infty}^{\infty} |g(x) - f(x)|\,dx < \varepsilon.$$

Then we take N so large that

$$\int_{|x|>N-1} |g(x)|\,dx < \varepsilon$$

and $0 < \delta < 1$ so that

$$|g(x+y) - g(x)| < \varepsilon/N, \qquad |x| \le N, \qquad |y| < \delta.$$

Thus,

$$\int_{-\infty}^{\infty} |g(x+y) - g(x)|\,dx < 4\varepsilon.$$

Using (7.8.13), we have

$$\int_{-\infty}^{\infty} |J_n g(x) - g(x)|\,dx \le \int_{-\infty}^{\infty}\int_{-\infty}^{\infty} j_n(y)|g(x-y) - g(x)|\,dx\,dy < 4\varepsilon$$

and (7.8.16) is proved.

Exercises

1. Show that (7.1.6) cannot converge as $t \to \infty$ unless the integrand vanishes identically.

2. Prove (7.1.10) under the conditions stated.

3. Prove (7.3.14).

4. In the proof of Theorem 7.4.2 show that at each point x only a finite number of the v_j, w_j do not vanish.

5. Prove Theorems 7.5.2 and 7.5.3.

6. Prove (7.7.10).

7. Show that $j(x)$ given by (7.8.1) is in C_0^∞.

8. Prove (7.8.8).

9. Let M, N be closed sets of real numbers which do not intersect. If M is bounded, show that they are a positive distance apart.

10. Give an example of closed, nonintersecting sets of real numbers which are not a positive distance from each other.

8
Time-Independent Theory

8.1. The Resolvent Method

The approach to scattering in Chapter 6 is direct. We define the wave operators and then show that they exist under certain conditions by applying the definition. It was discovered quite early that different approaches can yield more information than the direct method. Such methods are usually called *time-independent* or *stationary-state* scattering theory because the original definitions of the wave operators are camouflaged and the theory is developed without explicit reference to time. In contrast to this the original direct approach of Chapter 6 is called the *time-dependent* theory.

One way of converting from time-dependent to time-independent theory is as follows. From the identity

$$iz^{-1} = \int_0^\infty e^{izt}\,dt, \qquad \operatorname{Im} z > 0, \tag{8.1.1}$$

we see that

$$R(z)f = -i\int_0^\infty e^{i(z-H)t}f\,dt, \tag{8.1.2}$$

where, as usual, $R(z)$ is the resolvent operator:

$$R(z) = (z - H)^{-1}. \tag{8.1.3}$$

Let F denote the Fourier transform with respect to time:

$$Fh(s) = \frac{1}{\sqrt{2\pi}}\int_{-\infty}^\infty e^{-its}h(t)\,dt. \tag{8.1.4}$$

The inverse Fourier transform is given by

$$\bar{F}w(t) = \frac{1}{\sqrt{2\pi}}\int_{-\infty}^\infty e^{its}w(s)\,ds \tag{8.1.5}$$

(see Section 1.3). By (8.1.2) we have

$$R(s+ia)f = -i\int_0^\infty e^{ist}e^{i(ia-H)t}f\,dt$$
$$= -i\sqrt{2\pi}\,\bar{F}\{\chi_{(0,\infty)}e^{i(ia-H)t}f\}, \qquad a>0, \tag{8.1.6}$$

where $\chi_I(t)$ is the characteristic function for the set I:

$$\chi_I(t) = \begin{cases} 1, & t\in I \\ 0, & t\notin I. \end{cases} \tag{8.1.7}$$

By Parseval's identity we have

$$\int_{-\infty}^\infty (R_0(z)f, R(z)g)\,ds = 2\pi\int_0^\infty e^{-2at}(e^{-itH_0}f, e^{-itH}g)\,dt$$
$$= 2\pi\int_0^\infty e^{-2at}(W(t)f,g)\,dt, \tag{8.1.8}$$

where $z = s+ia$, $a>0$. If $f\in M_+$, we have

$$2a\int_0^\infty e^{-2at}(W(t)f,g)\,dt \to (W_+ f, g) \qquad \text{as} \quad a\to 0. \tag{8.1.9}$$

Consequently, we have

$$(W_+ f, g) = \lim_{0<a\to 0}\frac{a}{\pi}\int_{-\infty}^\infty (R_0(s+ia)f, R(s+ia)g)\,ds. \tag{8.1.10}$$

Note that there is no mention of time in (8.1.10).

A formula for W_- is obtained in a similar way. If $\operatorname{Im} z<0$,

$$-iz^{-1} = \int_{-\infty}^0 e^{izt}\,dt, \tag{8.1.11}$$

and consequently,

$$R(s-ia)f = i\int_{-\infty}^0 e^{ist}e^{-i(ia+H)t}f\,dt$$
$$= i\sqrt{2\pi}\,\bar{F}\{\chi_{(-\infty,0)}e^{-i(ia+H)t}f\}. \tag{8.1.12}$$

Hence

$$\int_{-\infty}^\infty (R_0(z)f, R(z)g)\,ds = 2\pi\int_{-\infty}^0 e^{2at}(W(t)f,g)\,ds, \tag{8.1.13}$$

where $z = s-ia$, $a>0$. Thus if $f\in M_-$,

$$(W_- f, g) = \lim_{0<a\to 0}\frac{a}{\pi}\int_{-\infty}^\infty (R_0(s-ia)f, R(s-ia)g)\,ds. \tag{8.1.14}$$

Offhand, this approach appears cumbersome. First, one must deal with the limits (8.1.10) and (8.1.14), which appear formidable. Second, even if these limits do exist, they do not guarantee that f is in M_+ or M_- (why?). Then why bother with such a method? The answer is that it produces stronger results than the time-dependent method. At this point it is unclear why this should be so. Part of the answer is that instead of the limits

(8.1.10) and (8.1.14) we can consider the limits

$$(W_{\pm}^{I} f, g) = \lim \frac{a}{\pi} \int_I (R_0(s \pm ia)f, R(s \pm ia)g)\, ds \tag{8.1.15}$$

for bounded intervals I (the elimination of t from the formulas allows this). We carry out the details in the next section.

8.2. The Theory

In this section we present an abstract theory which guarantees the existence of the wave operators. Strictly speaking, it is not completely a time-independent approach, but rather a mixture of both methods.

Let H_0, H be self-adjoint operators on a Hilbert space $\mathcal{H}$. As usual, we denote the resolvents by

$$R_0(z) = (z - H_0)^{-1}, \qquad R(z) = (z - H)^{-1}. \tag{8.2.1}$$

We define

$$j(z, f, g) = \frac{a}{\pi}(R_0(z)f, [R(z) - R_0(z)]g). \tag{8.2.2}$$

Let $\{E_0(\lambda)\}$, $\{E(\lambda)\}$ be the spectral families associated with H_0 and H, respectively. We shall say that f is in the *continuous subspace* $\mathcal{H}_c(H)$ of H if for each λ, $E(\mu)f \to E(\lambda)f$ when $\mu \to \lambda$ in any way [i.e., $E(\lambda)f$ is continuous from both sides, not only from the right]. It is easy to show that $\mathcal{H}_c(H)$ is indeed a closed subspace. In fact, it consists of those elements of $\mathcal{H}$ which are orthogonal to the eigenelements of H. Note that $E(\bar{I})f = E(I)f$ for any interval I when $f \in \mathcal{H}_c(H)$, where $E(I) = \chi_I(H)$. Our interest in the continuous subspace of an operator stems from the following considerations. As we saw in Section 6.3 (Theorem 6.3.2), if H_0 has an eigenvalue λ with eigenelement f, then f cannot be in M_+ unless λ is an eigenvalue of H and f is a corresponding eigenelement. Since such coincidences are not to be expected, it usually suffices to look for candidates for M_+ among the members of $\mathcal{H}_c(H_0)$. For the applications we have in mind, H_0 has no eigenvalues. In this case $\mathcal{H}_c(H_0) = \mathcal{H}$, and no problem arises.

For an interval $I = (a, b)$ we set $\nu_I = \min(-a, b)$. When it exists, we define

$$J_I^{\pm}(f, g) = \lim_{0<a\to 0} \int_I j(s \pm ia, f, g)\, ds. \tag{8.2.3}$$

The main result of this section is

Theorem 8.2.1. *Suppose that* $f \in \mathcal{H}_c(H_0)$ *is such that* (a) *for each bounded interval* I *with* ν_I *sufficiently large there is a dense subset* S_I *of* $\mathcal{H}$ *such that* $J_I^+(f, g)$ *exists for each* g *in* S_I *and* (b)

$$\limsup_{t\to+\infty} |J_I^+(e^{-itH_0}f, e^{-itH_0}f)| \to 0 \qquad \textit{as} \quad \nu_I \to \infty. \tag{8.2.4}$$

Then $f \in M_+$. If the plus signs in (a) *and* (b) *are changed to minus and the* lim sup *in* (8.2.4) *is taken as $t \to -\infty$, then $f \in M_-$.*

On the surface it might appear that Theorem 8.2.1 does not say very much, but we shall show in Section 8.3 that it produces better results than the theorems of Chapter 6. Hypothesis (8.2.4) may seem puzzling since it does not state that $J_I^+(e^{-itH_0}f, e^{-itH_0})$ exists. As we shall see in proof, this follows from hypothesis (a). In proving the theorem we shall need a result from spectral theory.

Lemma 8.2.1. *For any $f, g \in \mathcal{H}$ and any open interval I,*

$$\frac{a}{\pi}\int_I (R(s \pm ia)f, R(s \pm ia)g)\, ds \to (\tilde{E}(I)f, g), \tag{8.2.5}$$

where

$$\tilde{E}(I) = \tfrac{1}{2}\left[E(I) + E(\bar{I})\right] \tag{8.2.6}$$

and $E(J) = \chi_J(H)$.

PROOF. By Theorem 1.10.1 the left-hand side of (8.2.5) is

$$\int_I\left[\int_{-\infty}^{\infty} \delta_a(s-\lambda)\, d(E(\lambda)f, g)\right] ds = \int_{-\infty}^{\infty} f_a(\lambda)\, d(E(\lambda)f, g), \tag{8.2.7}$$

where

$$\delta_a(\lambda) = a/\pi(\lambda^2 + a^2) \tag{8.2.8}$$

and

$$f_a(\lambda) = \int_I \delta_a(s-\lambda)\, ds. \tag{8.2.9}$$

Now,

$$f_a(\lambda) \to \begin{cases} 1, & \lambda \in I \\ \frac{1}{2}, & \lambda = a \text{ or } \lambda = b \\ 0, & \lambda \notin \bar{I} \end{cases} \tag{8.2.10}$$

as $a \to 0$ and

$$|f_a(\lambda)| \le 1. \tag{8.2.11}$$

Hence the right-hand side of (8.2.7) approaches the right-hand side of (8.2.5) (Theorem 1.10.1). □

Lemma 8.2.2. *If $z = s \pm ia$, $a > 0$, then*

$$a\int_{-\infty}^{\infty} (R(z)f, R(z)g)\, ds = \pi(f, g). \tag{8.2.12}$$

PROOF. We use (8.2.7). In this case

$$I = (-\infty, \infty) \qquad \text{and} \qquad f_a(\lambda) \equiv 1. \qquad \square$$

Now we are ready for the

PROOF OF THEOREM 8.2.1. Set $z = s + ia$ and

$$h(z, f, g) = \frac{a}{\pi}(R_0(z)f, R(z)g). \tag{8.2.13}$$

Thus

$$h(z, f, g) = \frac{a}{\pi}(R_0(z)f, R_0(z)g) + j(z, f, g). \tag{8.2.14}$$

By Hypothesis (a) and Lemma 8.2.1,

$$\int_I h(z, f, g)\, ds \to (\tilde{E}_0(I)f, g) + J_I^+(f, g) \qquad \text{as} \quad a \to 0 \tag{8.2.15}$$

for each bounded interval I if $g \in S_I$, where $\tilde{E}_0(I)$ is the operator (8.2.6) for the case $H = H_0$. Let $[f, g]_I$ denote the right-hand side of (8.2.15). Since

$$\begin{aligned}\left|\int_I h(z, f, g)\, ds\right|^2 &\leq \frac{a}{\pi}\int_I \|R_0(z)f\|^2\, ds\ \frac{a}{\pi}\int_I \|R(z)g\|^2\, ds \\ &\to \|\tilde{E}_0(I)f\|^2 \|\tilde{E}(I)g\|^2\end{aligned} \tag{8.2.16}$$

by Lemma 8.2.1, we have

$$|[f, g]_I| \leq \|\tilde{E}_0(I)f\|\ \|\tilde{E}(I)g\|. \tag{8.2.17}$$

This shows that the limits (8.2.3) and (8.2.15) exist for every $g \in \mathcal{H}$. In fact, let $g \in \mathcal{H}$ and $\varepsilon > 0$ be given. Then there is a $g_1 \in S_I$ such that $\|g - g_1\| < \varepsilon$. Thus

$$\begin{aligned}\left|\int_I h(z, f, g)\, ds - \int h(z_1, f, g)\, ds\right| &\leq \left|\int_I h(z, f, g - g_1)\, ds\right| + \left|\int_I h(z, f, g_1)\, ds\right. \\ &\quad \left. - \int_I h(z_1, f, g_1)\, ds\right| + \left|\int_I h(z_1, f, g_1 - g)\, ds\right| \\ &\leq 2\|f\|\ \|g - g_1\| + \left|\int_I h(z, f, g_1)\, ds - \int_I h(z_1, f, g_1)\, ds\right|.\end{aligned}$$

The first term on the right is $< 2\varepsilon\|f\|$ while the second tends to 0 as $|\text{Im } z|$ and $|\text{Im } z_1|$ go to 0. Hence

$$[f, g]_I = \lim_{a \to 0} \int_I h(z, f, g)\, ds \tag{8.2.18}$$

exists for all $g \in \mathcal{H}$, equals the right-hand side of (8.2.15) and satisfies (8.2.17).

Let $C\bar{I}$ denote the complement of the closure of I. Thus if $I = (c, d)$, then

$$C\bar{I} = (-\infty, c) \cup (d, \infty).$$

Then we have by (8.2.16)

$$\lim\left|\int_I h\big(z, E_0(C\bar{I})f, g\big)\, ds\right| \leq \|\tilde{E}_0(I)E_0(C\bar{I})f\|\ \|g\| = 0.$$

Hence $[E_0(\bar{I})f, g]_I$ exists and

$$\big[E_0(\bar{I})f, g\big]_I = \big[f, g\big]_I. \tag{8.2.19}$$

Similarly,

$$\lim\left|\int_{CI} h(z, E_0(I)f, g)\, ds\right| \leq \|\tilde{E}_0(CI)E_0(I)f\|\ \|g\| = 0.$$

Since $f \in \mathcal{H}_c(H_0)$, we have $E_0(\bar{I})f = E_0(I)f$, and consequently,

$$\begin{aligned}\big[f, g\big]_I &= \lim_{a\to 0}\int_{-\infty}^{\infty} h(z, E_0(I)f, g)\, ds \\ &= \lim a\int_0^{\infty} e^{-at}\big(e^{-itH_0}E_0(I)f, e^{-itH}g\big)\, dt\end{aligned} \tag{8.2.20}$$

by (8.1.8). Thus

$$\begin{aligned}\big[f, e^{-i\sigma H}g\big]_I &= \lim a\int_0^{\infty} e^{-at}\big(e^{-itH_0}E_0(I)f, e^{-i(t+\sigma)H}g\big)\, dt \\ &= \lim a\int_\sigma^{\infty} e^{-a(\tau-\sigma)}\big(e^{-i(\tau-\sigma)H_0}E_0(I)f, e^{-i\tau H}g\big)\, d\tau \\ &= \lim e^{a\sigma}a\int_0^{\infty} e^{-a\tau}\big(e^{-i\tau H_0}E_0(I)e^{i\sigma H_0}f, e^{-i\tau H}g\big)\, d\tau \\ &= \lim e^{a\sigma}a\int_0^{\sigma} e^{-a\tau}\big(e^{-i\tau H_0}E_0(I)f, e^{-i\tau H}g\big)\, d\tau \\ &= \big[e^{i\sigma H_0}f, g\big]_I\end{aligned}$$

by (8.2.20). Hence $[e^{i\sigma H_0}f, g]_I$ exists for each σ and

$$\big[e^{i\sigma H_0}f, g\big]_I = \big[f, e^{-i\sigma H}g\big]_I, \qquad g \in \mathcal{H}. \tag{8.2.21}$$

By (8.2.17) the complex conjugate of $[f, g]_I$ is a bounded linear functional on $\mathcal{H}$. Consequently, by the Riesz representation theorem (Theorem 1.9.1), there is an element $W_I f$ such that

$$\big[f, g\big]_I = (W_I f, g). \tag{8.2.22}$$

By (8.2.17),

$$\|W_I f\| \leq \|E_0(I)f\|. \tag{8.2.23}$$

Moreover (8.2.21) implies

$$W_I e^{i\sigma H_0}f = e^{i\sigma H}W_I f. \tag{8.2.24}$$

Now if $I \subset I'$, (8.2.22) and (8.2.23) tell us that

$$\|[W_{I'} - W_I]f\| = \|W_{I'\setminus I}f\| \le \|E_0(I' \setminus I)f\| \le \|E_0(CI)f\|, \tag{8.2.25}$$

where $I' \setminus I$ denotes the set of points in I' but not in I. This shows that $W_I f$ converges to a limit f_+ as $\nu_I \to \infty$. Let $\varepsilon > 0$ be given, and take ν_I so large that

$$\|f_+ - W_I f\| < \varepsilon, \qquad \|E_0(CI)f\| < \varepsilon^2 \tag{8.2.26}$$

and

$$\limsup_{t\to\infty} |J_I^+(e^{-itH_0}f, e^{-itH_0}f)| < \varepsilon^2. \tag{8.2.27}$$

Now by (8.2.15), (8.2.18), and (8.2.22),

$$\begin{aligned}\|[W_I - E_0(I)]e^{-itH_0}f\|^2 &= \|W_I f\|^2 + \|E_0(I)f\|^2 \\ &\quad -2\,\mathrm{Re}\big(W_I e^{-itH_0}f, E_0(I)e^{-itH_0}f\big) = \|W_I f\|^2 \\ &\quad -\|E_0(I)f\|^2 - 2\,\mathrm{Re}\, J_I^+\big(e^{-itH_0}f, E_0(I)e^{-itH_0}f\big)\end{aligned}$$

Moreover, by (8.2.16)

$$|J_I^+(f, g)| \le 2\|E_0(I)f\|\,\|g\|.$$

Hence by (8.2.26) and (8.2.27),

$$\begin{aligned}|J_I^+(e^{-itH_0}f, E_0(I)e^{-itH_0}f)| &\le |J_I^+(e^{-itH_0}f, e^{-itH_0}f)| \\ &\quad + |J_I^+(e^{-itH_0}f, E_0(CI)e^{-itH_0}f)| < 2\varepsilon^2 + 2\|f\|\varepsilon^2\end{aligned}$$

for t sufficiently large. Thus

$$\|[W_I - E_0(I)]e^{-itH_0}f\| < 2(1 + \|f\|)^{1/2}\varepsilon,$$

and consequently,

$$\begin{aligned}\|f_+ - e^{itH}e^{-itH_0}f\| &\le \|f_+ - W_I f\| \\ &\quad + \|e^{itH}[W_I - E_0(I)]e^{-itH_0}f\| + \|E_0(CI)f\| \\ &< \varepsilon + 2(1 + \|f\|)^{1/2}\varepsilon + \varepsilon^2\end{aligned}$$

for t large. This shows that $f \in M_+$. □

8.3. A Simple Criterion

The hypotheses of Theorem 8.2.1 are not in a form that can be readily applied in practice. In this section we give one criterion that is easy to use. In Section 8.4 we shall apply it to our Hamiltonian operator. Later on we shall give other criteria that can be used for various purposes.

The situation we consider now is as follows. Let H_0, H be self-adjoint operators on a Hilbert space $\mathcal{H}$. We assume that there are a Hilbert space

$\mathcal{K}$ and operators A, B from $\mathcal{H}$ to $\mathcal{K}$ such that A is closed,

$$\|Bu\| \le C(\|Hu\| + \|u\|), \qquad u \in D(H), \tag{8.3.1}$$

and

$$(u, Hv) = (H_0u, v) + (Au, Bv)_{\mathcal{K}}, u \in D(A) \cap D(H_0), \qquad v \in D(H). \tag{8.3.2}$$

We have

Theorem 8.3.1. *If $e^{-itH_0}f \in D(A) \cap D(H_0)$ for $t > b$ and*

$$\int_b^{\infty} \|Ae^{-itH_0}f\|\, dt < \infty, \tag{8.3.3}$$

then $f \in M_+$.

In proving Theorem 8.3.1 we shall use two simple lemmas.

Lemma 8.3.1. *If* (8.3.2) *holds, then*

$$R(z)g - R_0(z)g = [BR(\bar{z})]^* AR_0(z)g \tag{8.3.4}$$

whenever $R_0(z)g \in D(A)$.

PROOF. Set $u = R_0(z)g$, $v = R(\bar{z})h$ in (8.3.2). This gives

$$(R_0(z)g, h) = (g, R(\bar{z})h) - (AR_0(z)g, BR(\bar{z})h)_{\mathcal{K}} \tag{8.3.5}$$

holding for all $h \in \mathcal{H}$. □

Lemma 8.3.2. *If $z = s \pm ia$, $0 < a < 1$, then for any bounded interval I*

$$a\int_I \|R(z)h\|^2\, ds \le C_I\|R(i)h\|^2 \tag{8.3.6}$$

and

$$a\int_I \|[BR(z)]^*h\|^2\, ds \le C_I\|[BR(i)]^*h\|^2, \tag{8.3.7}$$

where the constant C_I does not depend on a.

PROOF. We prove (8.3.7) since (8.3.6) is only a special case. The first resolvent equation states that

$$R(z) - R(\zeta) = (\zeta - z)R(z)R(\zeta). \tag{8.3.8}$$

Thus the left-hand side of (8.3.7) is bounded by

$$2a\int_I \|[BR(i)]^*h\|^2\, ds$$
$$+2a\int_I |z - i|^2 \|R(z)[BR(i)]^*h\|^2\, ds \le C\|[BR(i)]^*h\|^2$$

by Lemma 8.2.2. □

Lemma 8.3.3. *Suppose $D(B) \supset D(H^\sigma)$ for some $\sigma > 0$ and*

$$\|Bu\| \le C(\|H^\sigma u\| + \|u\|), \qquad u \in D(H^\sigma) \tag{8.3.9}$$

Then $D[E(I)] \subset D(B)$ for each bounded interval I and $BE(I)$ is a bounded operator on $\mathcal{H}$.

PROOF. By (8.3.9), $BR(i)^\sigma$ is a bounded operator. Moreover,

$$\|(i - H)^\sigma E(I)u\|^2 = \int_I |i - \lambda|^{2\sigma} \, d\|E(\lambda)u\|^2 \le C_I \|u\|^2,$$

where $C_I = \sup_{\lambda \in I} |i - \lambda|^{2\sigma}$. Thus $BE(I) = BR(i)^\sigma (i - H)^\sigma E(I)$ is bounded. □

Now we are ready for the

PROOF OF THEOREM 8.3.1. First assume that $b = 0$. Since A is closed, we see by (8.1.2) and (8.3.3) that $R_0(z)f \in D(A)$ for each z and

$$AR_0(z)f = -i\int_0^\infty e^{izt} A e^{-itH_0} f \, dt. \tag{8.3.10}$$

We apply Theorem 8.2.1. To show that $J_I^+(f, g)$ exists, it suffices to show that

$$a\int_I ([R(z) - R_0(z)]f, R(z)g) \, ds \tag{8.3.11}$$

converges as $a \to 0$ for each bounded interval I. Let g be an arbitrary element of $\mathcal{H}$ and let K be a bounded interval containing $\bar{I}$ in its interior. If we replace g by $E(CK)g$, (8.3.11) is bounded by

$$C_I(\|R(i)f\| + \|R_0(i)f\|)\left(a\int_I \|R(z)E(CK)g\|^2 \, ds\right)^{1/2} \tag{8.3.12}$$

(Lemma 8.3.2). By Lemma 8.2.1, the last factor in (8.3.12) converges to $\pi^{1/2}\|\tilde{E}(I)E(CK)g\| = 0$ as $a \to 0$. Thus to show that (8.3.11) converges, it suffices to replace g by $E(K)g$. By Lemma 8.3.1 the integrand equals

$$(AR_0(z)f, BR(\bar{z})R(z)E(K)g)$$
$$= -i\int_0^\infty e^{izt}(Ae^{-itH_0}f, BE(K)R(\bar{z})R(z)g) \, dt.$$

By Lemma 8.3.3, $BE(K)$ is a bounded operator. Hence by Lemma 8.2.1 and its proof,

$$a\int_I e^{izt}(R(z)[BE(K)]^* Ae^{-itH_0}f, R(z)g) \, ds$$

$$\to \pi(Ae^{-itH_0}f, B\tilde{E}(I)e^{-itH}g) \tag{8.3.13}$$

as $a \to 0$. Moreover, the left-hand side of (8.3.13) is bounded by a constant times $\|Ae^{-itH_0}f\|$ (Lemma 8.3.2), which is integrable from 0 to ∞ by hypothesis (8.3.3). Thus (8.3.11) converges to

$$-\pi i\int_0^\infty (Ae^{-itH_0}f, B\tilde{E}(I)e^{-itH}g)\,dt. \tag{8.3.14}$$

This shows that $J_I^+(f,g)$ exists for all $g \in \mathcal{H}$. Next we verify (8.2.4). By Lemma 8.3.1,

$$\begin{aligned}\pi j(z,f,g) &= a\big(R_0(z)f, \big[BR(\bar{z})\big]^* AR_0(z)g\big)\\ &= ai\int_0^\infty e^{-i\bar{z}\sigma}\big(R_0(z)f, \big[BR(\bar{z})\big]^* Ae^{-i\sigma H_0}g\big)\,ds,\end{aligned}$$

provided $R_0(z)g \in D(A)$. Thus by Lemma 8.3.2,

$$\begin{aligned}\int_I |j(z,f,g)|\,ds &\le C_I\|R_0(i)f\|\int_0^\infty \|\big[BR(i)\big]^* Ae^{-i\sigma H_0}g\|\,d\sigma\\ &\le C_I'\|f\|\int_0^\infty \|Ae^{-i\sigma H_0}g\|\,d\sigma.\end{aligned}$$

Set $f_t = e^{-itH_0}f$. Then

$$\begin{aligned}|J_I^+(f_t,f_t)| &\le C_I'\|f\|\int_0^\infty \|Ae^{-i(\sigma+t)H_0}f\|\,d\sigma\\ &= C_I'\|f\|\int_t^\infty \|Ae^{-i\tau H_0}f\|\,d\tau \to 0\end{aligned}$$

by (8.3.3). Thus $f \in M_+$ by Theorem 8.2.1. If $b \neq 0$, the proof just given shows that $e^{ibH_0}f \in M_+$. Hence $W(t)e^{ibH_0}f \to \hat{f}$ as $t \to \infty$. This is the same as $e^{ibH}W(t-b)f \to \hat{f}$ or $W(t-b)f \to e^{-ibH}\hat{f}$, which shows that $f \in M_+$ as well. □

8.4. The Application

In this section we shall show how the theory of Section 8.3 gives an improved version of the results of Section 6.4. In carrying out the argument we shall need a few simple lemmas from spectral theory. By Theorem 1.10.1 we can define any function of a self-adjoint operator H. In particular, we can define $|H|^{1/2}$ by

$$|H|^{1/2}u = \int_{-\infty}^\infty |\lambda|^{1/2}\,dE(\lambda)u \tag{8.4.1}$$

with $D(|H|^{1/2})$ consisting of those u such that

$$\int_{-\infty}^\infty |\lambda|\,d\|E(\lambda)u\|^2 < \infty. \tag{8.4.2}$$

We shall need

Theorem 8.4.1. *Let $h(u)$ be a closed bilinear form bounded from below with dense domain $D(h)$. Assume that h is Hermitian, and let H be the operator* *associated with h. Then H is self-adjoint, $D(|H|^{1/2}) = D(h)$, and*

$$h(u, v) = \int_{-\infty}^{\infty} \lambda \, d(E(\lambda)u, v), \qquad u, v \in D(h). \tag{8.4.3}$$

PROOF. That H is self-adjoint follows from Theorem 5.6.2. We may assume that $h \geq 1$. For we may always add a constant N to h and have the associated operator $H + N$. Then $D(h) = D(h + N)$. Moreover, $D(|H|^{1/2}) = D(|H + N|^{1/2})$ as is easily seen from the inequalities

$$\int |\lambda| \, d\|E(\lambda)u\|^2 \leq \int |\lambda + N| \, d\|E(\lambda)u\|^2 + N\|u\|^2$$
$$\leq \int |\lambda| \, d\|E(\lambda)u\|^2 + 2N\|u\|^2.$$

Now if $u \in D(H)$, we have

$$(Hu, v) = \int_1^{\infty} \lambda \, d(E(\lambda)u, v) \tag{8.4.4}$$

and hence

$$h(u) = \int_1^{\infty} \lambda \, d\|E(\lambda)u\|^2 = \|H^{1/2}u\|^2. \tag{8.4.5}$$

If $u \in D(h)$, there is a sequence $\{u_n\}$ of elements in $D(H)$ such that $h(u_n - u) \to 0$ (see the proof of Theorem 5.6.2). By (8.4.5), $H^{1/2}u_n$ converges to some element f. Also, $u_n \to u$ in $\mathcal{H}$. Since $H^{1/2}$ is self-adjoint, it is closed. Consequently $u \in D(H^{1/2})$ and $H^{1/2}u = f$. Hence $D(h) \subset D(H^{1/2})$ and (8.4.5) holds. Moreover, (8.4.3) holds as well. This can be seen by replacing u by u_n in (8.4.4) and taking the limit. Finally, suppose $u \in D(H^{1/2})$. Then for each n, $E(0, n)u$ is in $D(H)$ as is easily seen from the inequality

$$\int_1^{\infty} \lambda^2 \, d\|E(\lambda)E(0, n)u\|^2 = \int_1^{n} \lambda^2 \, d\|E(\lambda)u\|^2 \leq n^2\|u\|^2.$$

Since

$$\|H^{1/2}E(n, \infty)u\|^2 = \int_n^{\infty} \lambda \, d\|E(\lambda)u\|^2 \to 0 \qquad \text{as} \quad n \to \infty,$$

we have $H^{1/2}E(0, n)u \to f$ in $\mathcal{H}$. By (8.4.5), $h(E(0, n)u - E(0, m)u) \to 0$, showing that $u \in D(h)$. Thus $D(h) = D(H^{1/2})$. □

Theorem 8.4.2. *Let H_0 be a self-adjoint operator on a Hilbert space $\mathcal{H}$ such that*

$$(H_0u, u) \geq 0, \qquad u \in D(H_0), \tag{8.4.6}$$

and let h_0 be the associated bilinear form. Suppose A, B are operators from $\mathcal{H}$ to a Hilbert space $\mathcal{K}$ such that $D(h_0) \subset D(A) \cap D(B)$ with

$$\|Au\|^2 + \|Bu\|^2 \leq \theta h_0(u) + K\|u\|^2, \qquad u \in D(h_0), \tag{8.4.7}$$

where $\theta < 2$. *Then the bilinear form*

$$h(u, v) = h_0(u, v) + (Au, Bv), \; D(h) = D(h_0) \tag{8.4.8}$$

is closed and bounded from below. If it is Hermitian, then the associated operator H *is self-adjoint with* $D(|H|^{1/2}) = D(H_0^{1/2})$.

PROOF. By (8.4.8) and (8.4.7) we have

$$\begin{aligned} h_0(u) = h(u) - (Au, Bu) &\le h(u) + \tfrac{1}{2}\|Au\|^2 + \tfrac{1}{2}\|Bu\|^2 \\ &\le h(u) + \tfrac{1}{2}\theta h_0(u) + \tfrac{1}{2}K\|u\|^2. \end{aligned}$$

Thus

$$\left(1 - \tfrac{1}{2}\theta\right)h_0(u) \le h(u) + \tfrac{1}{2}K\|u\|^2. \tag{8.4.9}$$

Also,

$$\begin{aligned} |h(u)| \le h_0(u) + |(Au, Bu)| &\le h_0(u) + \tfrac{1}{2}\theta h_0(u) + \tfrac{1}{2}K\|u\|^2 \\ &\le \left(1 + \tfrac{1}{2}\theta\right)h_0(u) + \tfrac{1}{2}K\|u\|^2. \end{aligned} \tag{8.4.10}$$

These inequalities imply that h is closed. For if $u_k \in D(h)$, $u_k \to u$, $h(u_j - u_k) \to 0$, then inequality (8.4.9) implies that $h_0(u_j - u_k) \to 0$. Consequently, $H_0^{1/2}u_k$ converges to some element $f \in \mathcal{H}$. Since $H_0^{1/2}$ is a closed operator, we have $u \in D(h_0)$ and $H_0^{1/2}u = f$. Hence $h_0(u_j - u) \to 0$, and consequently, $h(u_j - u) \to 0$. The rest of the theorem follows from Theorem 8.4.1. □

Now we are ready for

Theorem 8.4.3. *Assume that*

$$\int_x^{x+1} |V(y)|\, dy \le C_0 \tag{8.4.11}$$

and that for each real s *there is a* $b(s) > 0$ *such that*

$$V(t, s) \equiv \int_{-\infty}^{\infty} |V(x)|(x - s)^2 \exp\left\{ -\frac{(x - s)^2}{2(1 + t^2)} \right\} dx < \infty \tag{8.4.12}$$

when $t > b(s)$. *Assume further that*

$$\int_{b(s)}^{\infty} t^{-3/2} V(t, s)^{1/2}\, dt < \infty \tag{8.4.13}$$

for each real s. *Let* H *be the operator associated with the bilinear form*

$$h(u, v) = \frac{1}{2m}(u', v') + (Vu, v). \tag{8.4.14}$$

Then $M_+ = L^2$.

PROOF. We apply Theorem 8.3.1. Set $A = |V(x)|^{1/2}$ when $V(x) \ne 0$, $A(x) = e^{-x^2}$ when $V(x) = 0$ and $B(x) = V(x)/A(x)$. Then

$$\|Au\|^2 + \|Bu\|^2 \le h_0(u) + K\|u\|^2, \qquad u \in D(h_0), \tag{8.4.15}$$

by inequality (2.7.7), where h_0 is the bilinear form associated with H_0. H is self-adjoint by Theorem 8.4.2 and (8.3.1) holds because

$$h_0(u) \le 2h(u) + K\|u\|^2 \le \|Hu\|^2 + (K+1)\|u\|^2$$

by (8.4.9) and the fact that $h(u) = (Hu, u)$ for $u \in D(H)$. Clearly, (8.3.2) holds with $\mathcal{K} = L^2$. Let S denote the set of function described in Lemma 6.4.1. By that lemma, S is dense in L^2. Moreover, if $u = \check{\psi}_s$, where $\psi_s(k) = k \exp\{-k^2 - iks\}$, then

$$\|Ae^{-itH_0}u\|^2 = (4 + 4t^2)^{-3/2} V(t, s) \tag{8.4.16}$$

by (6.4.15). Thus (8.3.3) holds for each $f \in S$. Consequently, $S \subset M_+$. Since M_+ is closed and S is dense in L^2, the conclusion follows. □

We conclude this section by showing how the proof of Theorem 8.3.1 can be simplified if we add the assumptions

$$D(|H|^{1/2}) = D(|H_0|^{1/2}), \tag{8.4.17}$$

$$\|Bu\| \le C(\| \, |H|^{1/2}u\| + \|u\|), \qquad u \in D(|H_0|^{1/2}), \tag{8.4.18}$$

and

$$f \in D(|H_0|^{1/2}) \tag{8.4.19}$$

(note that these assumptions make some of the hypotheses of Theorem 8.3.1 redundant). To prove that $f \in M_+$ we set $W(t) = e^{itH}e^{-itH_0}$ and note that

$$\begin{aligned}(W'(t)f, g) &= d(e^{-itH_0}f, e^{-itH}g)/dt \\ &= i(e^{-itH_0}f, He^{-itH}g) - i(H_0e^{-itH_0}f, e^{-itH}g) \\ &= i(Ae^{-itH_0}f, Be^{-itH}g)\end{aligned}$$

by (8.3.2). Thus

$$\begin{aligned}\|[W(t) - W(s)]f\|^2 &= 2\|f\|^2 - 2\,\mathrm{Re}(W(t)f, W(s)f) \\ &= 2\,\mathrm{Re}(W(t)f, [W(t) - W(s)]f) \\ &= 2\,\mathrm{Re}\int_s^t (W(t)f, W'(\sigma)f)\, d\sigma \\ &= 2\,\mathrm{Im}\int_s^t (Be^{-i\sigma H}W(t)f, Ae^{-i\sigma H_0}f)\, d\sigma.\end{aligned}$$

But

$$\begin{aligned}\|Be^{-i\tau H}g\| &\le C(\| \, |H|^{1/2}e^{-i\tau H}g\| + \|g\|) \\ &\le C(\| \, |H|^{1/2}g\| + \|g\|) \le C(\| \, |H_0|^{1/2}g\| + \|g\|).\end{aligned}$$

Thus

$$\|Be^{-i\sigma H}W(t)f\| \le C(\| \, |H_0|^{1/2}f\| + \|f\|),$$

and consequently,

$$\|[W(t)-W(s)]f\|^2 \le C(\|\,|H_0|^{1/2}f\| + \|f\|)\int_s^t \|Ae^{-i\sigma H_0}f\|\, d\sigma,$$

which converges to 0 as $s, t \to \infty$.

Exercises

1. Prove (8.1.1) and (8.1.2).

2. Prove (8.1.9).

3. Explain why the limits (8.1.10) can exist without f being in M_+.

5. Show that the left-hand side of (8.2.4) equals the left-hand side of (8.2.6).

5. Prove (8.2.9) and (8.2.10).

6. Show directly that $\mathcal{H}_c(H)$ is a closed subspace.

7. Show that $\mathcal{H}_c(H)$ consists of those elements which are orthogonal to the eigenvectors of H.

8. Show that $\|BR(z)E(J)\| \le C_I$ when $s \in I$ and I is a positive distance from J.

9. Show that $f_a(\lambda)$ given by (8.2.9) equals 1 when I is the whole real line.

10. Prove (8.3.8).

11. Supply the details in the proof of Lemma 8.3.3.

12. Prove (8.3.10).

13. Prove (8.3.13).

9
Completeness

9.1. Definition

Let H_0, H be self-adjoint operators on a Hilbert space $\mathcal{H}$. A state

$$\psi(t) = e^{-itH}\psi(0) \tag{9.1.1}$$

is called a scattering state if there exist states (called incoming and outgoing asymptotic states)

$$\psi_\pm(t) = e^{-itH_0}\psi_\pm(0) \tag{9.1.2}$$

such that

$$\|\psi(t) - \psi_\pm(t)\| \to 0 \qquad \text{as} \quad t \to \pm\infty \tag{9.1.3}$$

(see Section 6.2). If we set

$$W(t) = e^{itH}e^{-itH_0} \tag{9.1.4}$$

we have

$$W(t)\psi_\pm(0) \to \psi(0) \qquad \text{as} \quad t \to \pm\infty. \tag{9.1.5}$$

In other words,

$$W_\pm\psi_\pm(0) = \psi(0). \tag{9.1.6}$$

Conversely, suppose $\psi_\pm(0) \in M_\pm$ satisfy (9.1.6). Then (9.1.5) holds, and if we define $\psi(t)$, $\psi_\pm(t)$ by (9.1.1), (9.1.2), respectively, then we obtain (9.1.3). Hence $\psi(t)$ is a scattering state and $\psi_\pm(t)$ are its asymptotic states. This shows us that ψ is the value at $t = 0$ of an incoming asymptotic state for a scattering state iff $\psi \in M_-$ and $W_-\psi \in R_+$. Similarly, it is the value at $t = 0$ of an outgoing asymptotic state for a scattering state iff $\psi \in M_+$ and

$W_+\psi \in R_-$. From the point of view of experimental physics it is important that there be as many scattering states as possible. On the other hand, we saw in Section 6.2 that it is unrealistic to expect scattering for states which are eigenelements. These considerations suggest the following definitions which describe the situations when one has the maximum amount of scattering. The wave operators are said to be (*weakly asymptotically*) *complete* if every $\psi \in \mathcal{H}_c(H_0)$ is the value at $t = 0$ of an incoming asymptotic state for a scattering state and the value at $t = 0$ of an outgoing asymptotic state for a scattering state. This is equivalent to saying

$$\mathcal{H}_c(H_0) \subset M_\pm \quad \text{and} \quad R_+ \cap \mathcal{H}_c(H) = R_- \cap \mathcal{H}_c(H). \tag{9.1.7}$$

To see this we note

Lemma 9.1.1. *If $\psi \in M_\pm$, then*

$$W_\pm\psi \in \mathcal{H}_c(H) \quad \text{iff} \quad \psi \in \mathcal{H}_c(H_0). \tag{9.1.8}$$

The proof of the lemma will be given in Section 9.13. The definition of completeness given above guarantees that all elements in $\mathcal{H}_c(H_0)$ form incoming and outgoing asymptotic states for scattering states. But it does not concern itself with the scattering states as such. In particular, it does not say anything about those elements in $\mathcal{H}$ which give rise to scattering states. Thus one can formulate a stronger definition of completeness. We shall say that the wave operators $W_\pm$ are *strongly* (*asymptotically*) *complete* if they are complete and every $\psi \in \mathcal{H}_c(H)$ is the value at $t = 0$ of a scattering state. This is equivalent to

$$\mathcal{H}_c(H_0) \subset M_\pm, \qquad \mathcal{H}_c(H) \subset R_\pm. \tag{9.1.9}$$

We can look at strong completeness from a slightly different angle. To emphasize the roles of the operators let us denote the wave operators as we have defined them by

$$W_\pm(H, H_0)f = \lim_{t\to\pm\infty} W(t)f, \tag{9.1.10}$$

where $W(t)$ is given by (9.1.4). The subspaces $M_\pm$ and $R_\pm$ will be denoted by $M_\pm(H, H_0)$ and $R_\pm(H, H_0)$, respectively. This notation has the convenience of allowing one to consider wave operators for different pairs of self-adjoining operators. In particular, we can look at the operators $W_\pm(H_0, H)$. The following relationships are convenient.

Lemma 9.1.2. *We have*

$$R_\pm(H, H_0) = M_\pm(H_0, H) \tag{9.1.11}$$

and

$$W_\pm(H_0, H) = W_\pm(H, H_0)^{-1}. \tag{9.1.12}$$

PROOF. We note that

$$\|W(t)^*g - f\| = \|g - W(t)f\|. \tag{9.1.13}$$

Thus, $f \in M_\pm(H, H_0)$, $g \in R_\pm(H, H_0)$ and $W_\pm(H, H_0)f = g$ iff $g \in M_\pm(H_0, H)$, $f \in R_\pm(H_0, H)$ and $W_\pm(H_0, H)g = f$. □

Using Lemma 9.1.2, we see that strong completeness is equivalent to

$$\mathcal{H}_c(H_0) \subset M_\pm(H, H_0), \qquad \mathcal{H}_c(H) \subset M_\pm(H_0, H) \tag{9.1.14}$$

Thus the operators $W_\pm(H_0, H)$ are strongly complete iff the same is true for the operators $W_\pm(H, H_0)$.

In this chapter we shall be concerned with finding sufficient conditions for completeness both in abstract setting and for the Hamiltonian that we are studying. When there is no chance of confusion we shall revert to our old notation of $W_\pm$, $M_\pm$, $R_\pm$.

Before we continue we must issue the following warning. Until this point we have tried to avoid the use of measure theory and Lebesgue integration. However, in the present chapter it is forced upon us in Section 9.10. Moreover, there we shall need to use several theorems for analytic functions of a complex variable. Although most of the results of this chapter can be understood without Section 9.10, our main theorems, Theorems 9.11.2 and 9.14.1, require it.

9.2. The Abstract Theory

In order to find conditions which imply completeness we return momentarily to the considerations of Section 8.2. First we note the following corollary of Theorem 8.2.1. For sets L, M we shall write $L \subset\subset M$ to mean that the closure of L is bounded and contained in M. If M is open, this implies that the boundaries of L and M have nothing in common.

Theorem 9.2.1. *Suppose $f \in \mathcal{H}_c(H_0)$ and there is an open interval Λ such that for each interval $I \subset\subset \Lambda$.* (a) *There is a dense subset S_I of $\mathcal{H}$ such that $J_I^+(f, g)$ exists for each $g \in S_I$ and* (b)

$$J_I^+\left(e^{-itH_0}E_0(I)f, e^{-itH_0}E_0(I)f\right) \to 0 \qquad \text{as} \quad t \to \infty. \tag{9.2.1}$$

Then $E_0(\Lambda)f \in M_+$.

PROOF. Take $I \subset\subset \Lambda$. From (8.2.15) and (8.2.20) we see that

$$J_I^+(E_0(I)f, g) = J_I^+(f, g) \tag{9.2.2}$$

exists for all $g \in \mathcal{H}$. On the other hand, if the interval K does not intersect I, we have

$$\pi \int_K |j(z, E_0(I)f, g)| \, ds$$

$$\leq \left(a \int_K \|R_0(z)E_0(I)f\|^2 \, ds \right)^{1/2} \left(a \int_K \|[R(z) - R_0(z)]g\|^2 \, ds \right)^{1/2}$$

and this tends to 0 by Lemmas 8.2.1 and 8.2.2. Thus

$$J_K^+(E_0(I)f, g) = 0 \tag{9.2.3}$$

for all $g \in \mathcal{H}$. Consequently, $J_L^+(E_0(I)f, g)$ exists for any interval L and any element $g \in \mathcal{H}$. In particular, if K does not intersect I, we have by (9.2.3)

$$J_K^+(E_0(I)f_t, E_0(I)f_t) = 0, \tag{9.2.4}$$

where $f_t = e^{-itH_0}f$. Thus by (9.2.1) and (9.2.4),

$$J_L^+(E_0(I)f_t, E_0(I)f_t) \to 0 \quad \text{as} \quad t \to \infty$$

for any interval L. Thus all of the hypotheses of Theorem 8.2.1 are satisfied with f replaced by $E_0(I)f$. Hence $E_0(I)f \in M_+$ for each $I \subset\subset \Lambda$. Suppose $\Lambda = (a, b)$. Then $E_0(a + \varepsilon, b - \varepsilon)f \in M_+$ for each $\varepsilon > 0$. Since $f \in \mathcal{H}_c(H_0)$, we have $E_0(a + \varepsilon, b - \varepsilon)f \to E_0(\Lambda)f$ as $\varepsilon \to 0$. Hence $E_0(\Lambda)f \in M_+$. □

It is easy to show that every open set on the real line is the denumerable union of disjoint open intervals. Using this fact, we have

Corollary 9.2.1. *If Λ is an open set and hypotheses* (a) *and* (b) *of Theorem* 9.2.1 *hold, then $E_0(\Lambda)f \in M_+$.*

Before we continue we shall need several lemmas which we prove in the next section.

9.3. Some Identities

In this section we let H be a self-adjoint operator on a Hilbert space $\mathcal{H}$ with resolvent $R(z)$ and spectral family $\{E(\lambda)\}$. We let A be a closed operator from $\mathcal{H}$ to a Hilbert space $\mathcal{K}$. If $z = s + ia$, $a > 0$, we have by (8.1.6) and (8.1.12)

$$AR(z)u = -i\int_0^\infty e^{izt}Ae^{-itH}u \, dt$$

$$= -i(2\pi)^{1/2}\bar{F}\{\chi_{(0,\infty)}e^{-at}Ae^{-itH}u\} \tag{9.3.1}$$

and

$$AR(\bar{z})u = i\int_{-\infty}^{0} e^{i\bar{z}t}Ae^{-itH}u\,dt$$
$$= i(2\pi)^{1/2}\bar{F}\{\chi_{(-\infty,0)}e^{at}Ae^{-itH}u\}. \tag{9.3.2}$$

If we apply Parseval's identity, we have

$$\int_{-\infty}^{\infty} \|AR(z)u\|^2\,ds = 2\pi\int_0^{\infty} e^{-2at}\|Ae^{-itH}u\|^2\,dt, \tag{9.3.3}$$

$$\int_{-\infty}^{\infty} \|AR(\bar{z})u\|^2\,ds = 2\pi\int_{-\infty}^{0} e^{2at}\|Ae^{-itH}u\|^2\,dt \tag{9.3.4}$$

and

$$\int_{-\infty}^{\infty} (AR(z)u, AR(\bar{z})v)\,ds = 0. \tag{9.3.5}$$

Thus,

$$\int_{-\infty}^{\infty} \|A[R(z) - R(\bar{z})]u\|^2\,ds = 2\pi\int_{-\infty}^{\infty} e^{-2a|t|}\|Ae^{-itH}u\|^2\,dt. \tag{9.3.6}$$

Moreover, if we set

$$u_+(s) = -i\int_0^{\infty} e^{ist}Ae^{-itH}u\,dt, \tag{9.3.7}$$

$$u_-(s) = i\int_{-\infty}^{0} e^{ist}Ae^{-itH}u\,dt, \tag{9.3.8}$$

then

$$\int_{-\infty}^{\infty} \|AR(z)u - u_+(s)\|^2\,ds = 2\pi\int_0^{\infty} (e^{-at} - 1)^2\|Ae^{-itH}u\|^2\,dt \tag{9.3.9}$$

and

$$\int_{-\infty}^{\infty} \|AR(\bar{z})u - u_-(s)\|^2\,ds = 2\pi\int_{-\infty}^{0} (e^{at} - 1)^2\|Ae^{-itH}u\|^2\,dt. \tag{9.3.10}$$

Thus we have

Lemma 9.3.1. *Suppose $e^{-itH}u \in D(A)$ for $t > 0$ and*

$$C_+ \equiv \int_0^{\infty} \|Ae^{-itH}u\|^2\,dt < \infty. \tag{9.3.11}$$

Then $R(z)u \in D(A)$ for $a = \operatorname{Im} z > 0$ and

$$\int_{-\infty}^{\infty} \|AR(z)u\|^2\,ds \le C_+. \tag{9.3.12}$$

Moreover,

$$\int_{-\infty}^{\infty} \|AR(z)u - u_+(s)\|^2\, ds \to 0 \qquad as \quad a \to 0. \tag{9.3.13}$$

A similar statement holds if the interval $(0, \infty)$ *is replaced by* $(-\infty, 0)$.

We also have

Lemma 9.3.2. *Suppose*

$$a\|AR(s + ia)\|^2 \le C_0, \qquad a > 0, \quad -\infty < s < \infty. \tag{9.3.14}$$

Then

$$\int_{-\infty}^{\infty} \|Ae^{-itH}u\|^2\, dt \le 2C_0\|u\|^2. \tag{9.3.15}$$

The same conclusion holds if (9.3.14) *is replaced by*

$$a\|AR(s - ia)\|^2 \le C_0, \qquad a > 0, \quad -\infty < s < \infty. \tag{9.3.16}$$

PROOF. By 9.3.6,

$$2\pi\int_{-\infty}^{\infty} e^{-2a|t|}\|Ae^{-itH}u\|^2\, dt = 4a^2\int_{-\infty}^{\infty} \|AR(z)R(\bar{z})u\|^2\, ds. \tag{9.3.17}$$

If (9.3.14) holds, this is

$$\le 4aC_0\int_{-\infty}^{\infty} \|R(\bar{z})u\|^2\, ds = 4\pi C_0\|u\|^2$$

by Lemma 8.2.2. If (9.3.16) holds, (9.3.17) is

$$\le 4aC_0\int_{-\infty}^{\infty} \|R(z)u\|^2\, ds = 4\pi C_0\|u\|^2.$$

Letting $a \to 0$, we obtain (9.3.15). □

Lemma 9.3.3. *Suppose*

$$a\|AR(z)\|^2 \le C_1, \qquad s \in I, \quad a > 0, \tag{9.3.18}$$

for some set I. *Then*

$$a\|AR(z)E(I)\|^2 \le 4C_1, \qquad -\infty < s < \infty. \tag{9.3.19}$$

PROOF. Clearly, (9.3.18) holds for $s \in \bar{I}$. If $s \in \bar{I}$, we have

$$a\|AR(z)E(I)u\|^2 \le C_1\|E(I)u\|^2 \le C_1\|u\|^2.$$

Hence, (9.3.19) holds. If $s \notin \bar{I}$, let σ be the closest point of $\bar{I}$ to s. Put $\zeta = \sigma + ia$. By the first resolvent equation [see (8.3.8)],

$$\|AR(z)E(I)\| = \|A[R(\zeta) + (\zeta - z)R(\zeta)R(z)]E(I)\| \le \|AR(\zeta)E(I)\|$$
$$+ |s - \sigma|\, \|AR(\zeta)E(I)R(z)E(I)\|.$$

Now,

$$\|R(z)E(I)u\|^2 = \int_I |z - \lambda|^{-2}\, d\|E(\lambda)u\|^2$$
$$\leq |s - \sigma|^{-2}\|E(I)u\|^2.$$

Hence

$$\|AR(z)E(I)\| \leq 2\|AR(\zeta)E(I)\|.$$

This implies (9.3.19). □

The proof of Lemma 9.3.3 also gives

Corollary 9.3.1. *If*

$$a\|AR(\bar{z})\|^2 \leq C_1, \qquad s \in I, \quad a > 0, \tag{9.3.20}$$

then (9.3.19) *holds with z replaced by* $\bar{z}$.

9.4. Another Form

In this section we specialize to the situation when there are operators A, B from $\mathcal{H}$ to a Hilbert space $\mathcal{K}$ such that $D(H_0) \subset D(A)$, $D(H) \subset D(B)$, and

$$(u, Hv) = (H_0u, v) + (Au, Bv)_{\mathcal{K}}, \qquad u \in D(H_0), \qquad v \in D(H). \tag{9.4.1}$$

We have

Theorem 9.4.1. *Assume that there are an open set Λ and an $f \in \mathcal{H}_c(H_0)$ such that for each interval $I \subset\subset \Lambda$,*

$$\int_0^\infty \|Ae^{-itH_0}E_0(I)f\|^2\, dt < \infty \tag{9.4.2}$$

and

$$a\|BR(s - ia)\|^2 \leq C_I, \qquad s \in I, \quad a > 0. \tag{9.4.3}$$

Then $E_0(\Lambda)f \in M_+$.

PROOF. By Corollary 9.3.1,

$$a\|BR(s - ia)E(\bar{I})\|^2 \leq 4C_I. \tag{9.4.4}$$

In view of Lemma 9.3.2, this implies

$$\int_{-\infty}^\infty \|Be^{-itH}E(\bar{I})g\|^2\, dt \leq 8C_I\|g\|^2. \tag{9.4.5}$$

Using (9.4.2) and (9.4.5), we see by Lemma 9.3.1 that there are functions

$f_+(s)$, $g_\pm(s)$ such that

$$\int_{-\infty}^{\infty} \|AR_0(z)E_0(I)f - f_+(s)\|^2\, ds \to 0 \quad \text{as} \quad a \to 0, \quad (9.4.6)$$

$$\int_{-\infty}^{\infty} \|BR(s \pm ia)E(\bar{I})g - g_\pm(s)\|^2\, ds \to 0 \quad \text{as} \quad a \to 0. \quad (9.4.7)$$

Moreover, by Lemma 8.3.1,

$$\begin{aligned} a\int_I &\big([R(z) - R_0(z)]E_0(I)f, R(z)E(\bar{I})g\big)\, ds \\ &= a\int_I \big(AR_0(z)E_0(I)f, BR(z)R(\bar{z})E(\bar{I})g\big)\, ds \\ &= -\frac{1}{2}i\int_I \big(AR_0(z)E_0(I)f, B[R(z) - R(\bar{z})]E(\bar{I})g\big)\, ds \\ &\to -\frac{1}{2}i\int_I (f_+(s), [g_+(s) - g_-(s)])\, ds \end{aligned}$$

as $a \to 0$. On the other hand,

$$\begin{aligned} a\int_I &|([R(z) - R_0(z)]E_0(I)f, R(z)E(C\bar{I})g)|\, ds \\ &\le \left(a\int_I \|[R(z) - R_0(z)]E_0(I)f\|^2\, ds\right)^{1/2}\left(a\int_I \|R(z)E(C\bar{I})g\|^2\, ds\right)^{1/2}, \end{aligned}$$

which converges to 0 as $a \to 0$ by Lemma 8.2.1 and 8.2.2. Thus $J_I^+(E_0(I)f, g)$ exists for each $g \in \mathcal{H}$. Next set $f_t = e^{-itH_0}f$. Then

$$\begin{aligned} a\int_I &|(R_0(z)E_0(I)f_t, [R(z) - R_0(z)]E_0(I)f_t)|\, ds \\ &= a\int_I |(BR(\bar{z})R_0(z)E_0(I)f_t, AR_0(z)E_0(I)f_t)|\, ds \\ &\le \left(a^2\int_I \|BR(\bar{z})R_0(z)E_0(I)f_t\|^2\, ds\right)^{1/2}\left(\int_I \|AR_0(z)E_0(I)f_t\|^2\, ds\right)^{1/2} \\ &\le C_I^{1/2}\left(a\int_I \|R_0(z)E_0(I)f_t\|^2\, ds\right)^{1/2}\left(\int_0^{\infty} \|Ae^{-i\sigma H_0}E_0(I)f_t\|^2\, d\sigma\right)^{1/2} \\ &\le \pi^{1/2}C_I^{1/2}\|E_0(I)f\|\left(\int_t^{\infty} \|Ae^{-i\tau H_0}E_0(I)f\|^2\, d\tau\right)^{1/2} \end{aligned}$$

by (9.4.3). This shows that (9.2.1) holds. Let Λ_0 be any component of Λ. Then all of the hypotheses of Theorem 2.1 are satisfied for Λ_0. Hence $E_0(\Lambda_0)f \in M_+$. Since this is true for each component Λ_0 of Λ, we see that $E_0(\Lambda)f \in M_+$. □

Now we come to our first completeness result.

Theorem 9.4.2. *Suppose* (9.4.1) *holds and there is an open set* Λ *such that* $E_0(\Lambda)$ *is the projection onto* $\mathcal{H}_c(H_0)$ *and*

$$a\|AR_0(s \pm ia)\|^2 + a\|BR(s \pm ia)\|^2 \le C_I, \qquad s \in I, \quad a > 0 \quad (9.4.8)$$

for each $I \subset\subset \Lambda$. *Then*

$$E(\Lambda)\mathcal{H}_c(H) = R_\pm \cap \mathcal{H}_c(H). \qquad (9.4.9)$$

The wave operators are complete.

In proving Theorem 9.4.2 we shall make use of the following lemma, which is proved in Section 9.13.

Lemma 9.4.1. *For each interval* I *and each* $f \in M_\pm$ *we have*

$$E(I)W_\pm f = W_\pm E_0(I)f. \qquad (9.4.10)$$

Now we give the

PROOF OF THEOREM 9.4.2. Suppose $f \in \mathcal{H}_c(H_0)$. By Corollary 9.3.1, (9.4.8) implies

$$a\|AR_0(z)E_0(I)\|^2 + a\|BR(z)E(I)\|^2 \le 8C_I, \qquad -\infty < s < \infty.$$

Lemma 9.3.2 then implies

$$\int_{-\infty}^{\infty} \|Ae^{-itH_0}E_0(I)f\|^2\,dt < \infty.$$

All of the hypotheses of Theorem 9.4.1 are satisfied. Thus $E_0(\Lambda)f \in M_+$. Since $f \in \mathcal{H}_c(H_0)$, $E_0(\Lambda)f = f$ by hypothesis. If we consider the pair $-H_0$, $-H$ in place of H_0, H we see that $f \in M_-$ as well. Next we note that hypothesis (9.4.8) is symmetric in H_0 and H. Thus we can interchange A and B, H_0 and H in applying Theorem 9.4.1. This gives $E(\Lambda)g \in M_\pm(H_0, H) = R_\pm$ for each $g \in \mathcal{H}_c(H)$. Now suppose $h \in \mathcal{H}_c(H)$ is orthogonal to $E(\Lambda)\mathcal{H}_c(H)$. Then $E(\Lambda)h = 0$. If $h = W_\pm f, f \in M_\pm$, then by Lemma 9.4.1

$$E(\Lambda)h = E(\Lambda)W_\pm f = W_\pm E_0(\Lambda)f,$$

which implies $E_0(\Lambda)f = 0$. Thus $f \perp \mathcal{H}_c(H_0)$. But $f \in \mathcal{H}_c(H_0)$ by Lemma 9.1.1. Thus f and h vanish. This shows that (9.4.9) holds. The wave operators are complete in view of (9.1.7). □

9.5. The Unperturbed Resolvent Operator

We want to discuss the completeness of our Hamiltonian operator. To obtain optimal results we shall write $V(x)$ as the product of two functions $A(x)$ and $B(x)$ so that (9.4.1) will hold. Then we want to apply Theorem

9.4.2. First we concentrate on verifying

$$a\|AR_0(z)\|^2 \le C_I, \qquad z = s \pm ia, \quad s \in I \tag{9.5.1}$$

[see (9.4.9)]. We shall prove

Theorem 9.5.1. *Let* Λ *be the real axis with the origin removed. If* $A \in L^2$, *then* (9.5.1) *holds for each* $I \subset\subset \Lambda$.

For convenience we shall assume $2m = \hbar = 1$ from this point on. In proving Theorem 9.5.1 we shall use

Theorem 9.5.2. *If* $\eta = \operatorname{Im} \kappa > 0$ *and* $\kappa^2 = z$, *then*

$$R_0(z)f(x) = \frac{1}{2\kappa i}\int_{-\infty}^{\infty} e^{i\kappa|x-y|}f(y)\,dy. \tag{9.5.2}$$

PROOF. We could tell you to merely differentiate (9.5.2) twice to verify that it is indeed a solution of

$$\kappa^2 u + u'' = f, \tag{9.5.3}$$

but this would not tell you how it was obtained. Let us give a derivation. Solving (9.5.3) for u is equivalent to solving

$$(\kappa + D)u = v, \qquad (\kappa - D)v = f \tag{9.5.4}$$

for u, v, where $D = d/i\,dx$. Now,

$$D(e^{-i\kappa x}v) = e^{-i\kappa x}(D - \kappa)v = -e^{-i\kappa x}f. \tag{9.5.5}$$

Hence

$$e^{-i\kappa x}v(x) = v(0) - i\int_0^x e^{-i\kappa y}f(y)\,dy \tag{9.5.6}$$

Taking absolute values we have

$$|v(x)| = e^{-\eta x}|v(0) - i\int_0^x e^{-i\kappa y}f(y)\,dy| \tag{9.5.7}$$

Since $f \in L^2$, the integral

$$\int_{-\infty}^{0} e^{-i\kappa y}f(y)\,dy \tag{9.5.8}$$

exists. Thus the only way the right-hand side of (9.5.7) can be in L^2 is if the expression in the absolute value signs tends to 0 as $x \to -\infty$. Thus,

$$v(0) = -i\int_{-\infty}^{0} e^{-i\kappa y}f(y)\,dy. \tag{9.5.9}$$

This gives

$$v(x) = -i\int_{-\infty}^{x} e^{i\kappa(x-y)}f(y)\,dy. \tag{9.5.10}$$

Note that $v(x)$ given by (9.5.10) is in L^2. For by Schwarz's inequality,

$$|v(x)|^2 \leq \int_{-\infty}^{x} e^{\eta(y-x)}\,dy \int_{-\infty}^{x} e^{\eta(y-x)}|f(y)|^2\,dy$$

and consequently

$$\int_{-\infty}^{\infty} |v(x)|^2\,dx \leq \frac{1}{\eta}\int_{-\infty}^{\infty}\int_{-\infty}^{x} e^{\eta(y-x)}|f(y)|^2\,dy\,dx$$

$$= \frac{1}{\eta}\int_{-\infty}^{\infty}\left[\int_{y}^{\infty} e^{\eta(y-x)}\,dx\right]|f(y)|^2\,dy = \eta^{-2}\|f\|^2. \tag{9.5.11}$$

The same reasoning gives

$$D(e^{i\kappa x}u) = e^{i\kappa x}(D+\kappa)u = e^{i\kappa x}v \tag{9.5.12}$$

and consequently

$$e^{i\kappa x}u(x) = u(0) + i\int_0^x e^{i\kappa y}v(y)\,dy. \tag{9.5.13}$$

Since $v \in L^2$ and

$$|u(x)| = e^{\eta x}\left|u(0) + i\int_0^x e^{i\kappa y}v(y)\,dy\right|,$$

the only way $u(x)$ can be in L^2 is if

$$u(0) = -i\int_0^{\infty} e^{i\kappa y}v(y)\,dy. \tag{9.5.14}$$

Thus we must have

$$u(x) = -i\int_x^{\infty} e^{i\kappa(y-x)}v(y)\,dy. \tag{9.5.15}$$

As before, it is simple to verify that $v \in L^2$ implies $u \in L^2$. The next step is to find u in terms of f. For this purpose we substitute (9.5.10) into (9.5.15). This gives

$$u(x) = -\int_x^{\infty} e^{i\kappa(y-x)}\int_{-\infty}^{y} e^{i\kappa(y-t)}f(t)\,dt\,dy$$

$$= -\int_x^{\infty} e^{2i\kappa y}\,dy\int_{-\infty}^{x} e^{-i\kappa(x+t)}f(t)\,dt$$

$$-\int_x^{\infty}\left[\int_t^{\infty} e^{2i\kappa y}\,dy\right]e^{-i\kappa(x+t)}f(t)\,dt$$

$$= (2i\kappa)^{-1}\left(\int_{-\infty}^{x} e^{i\kappa(x-t)}f(t)\,dt + \int_x^{\infty} e^{i\kappa(t-x)}f(t)\,dt\right). \tag{9.5.16}$$

This gives (9.5.2). □

Equipped with Theorem 9.5.2 we give the

PROOF OF THEOREM 9.5.1. By (9.5.2),

$$|R_0(z)f(x)|^2 \leq \frac{1}{4|\kappa|^2}\int_{-\infty}^{\infty} e^{-2\eta|x-y|}\,dy\int_{-\infty}^{\infty}|f(y)|^2\,dy$$
$$= \|f\|^2/4|\kappa|^2\eta. \tag{9.5.17}$$

Thus

$$\|AR_0(t)f\|^2 \leq \|A\|^2\|f\|^2/4|\kappa|^2\eta. \tag{9.5.18}$$

Now if $z = s \pm ia$ and $\kappa = \sigma + i\eta$, then $\pm a = 2\sigma\eta$ and consequently $a \leq 2|\sigma|\eta$. Thus we have

$$a\|AR_0(z)f\|^2 \leq \|A\|^2\|f\|^2/2|\kappa|. \tag{9.5.19}$$

This proves the theorem. Note that we get an explicit bound for C_I. □

For future reference we shall need some additional formulas and estimates.

Lemma 9.5.1. *If $A \in L^2$, then*

$$[AR_0(z)]^*f(y) = \frac{i}{2\bar{\kappa}}\int_{-\infty}^{\infty} e^{-i\bar{\kappa}|x-y|}A(x)^*f(x)\,dx. \tag{9.5.20}$$

PROOF. For $v \in L^2$ we have

$$(f, AR_0(z)v) = \frac{i}{2\bar{\kappa}}\iint f(x)A(x)^*e^{-i\bar{\kappa}|x-y|}v(y)^*\,dy\,dx.$$

This gives (9.5.20). □

Lemma 9.5.2.

$$2iDR_0(z)f(x) = \int_{-\infty}^{x} e^{i\kappa(x-y)}f(y)\,dy - \int_{x}^{\infty} e^{i\kappa(y-x)}f(y)\,dy \tag{9.5.21}$$

and

$$|DR_0(z)f(x)|^2 \leq \|f\|^2/2\eta. \tag{9.5.22}$$

PROOF. First assume that f is continuous as well as in L^2. Then by (9.5.2),

$$-2\kappa DR_0(z)f(x) = f(x) + i\kappa\int_{-\infty}^{x} e^{i\kappa(x-y)}f(y)\,dy$$
$$-f(x) - i\kappa\int_{x}^{\infty} e^{i\kappa(y-x)}f(y)\,dy.$$

This gives (9.5.21) for such f. Schwarz's inequality then gives (9.5.22). Now if f is any function in L^2, we can approximate it by continuous functions f_n in L^2 (Theorem 7.8.2). We apply (9.5.21) to the approximating functions f_n and then we let them converge to f. The right-hand side of (9.5.21) converges to the corresponding expression for f in L^2. Moreover, $R_0(z)f_n$ converges to $R_0(z)f$ in L^2. Since D is a closed operator, we obtain (9.5.21). □

Lemma 9.5.3. *If $A \in L^2$, then*

$$\|ADR_0(z)f\|^2 \leq \|A\|^2\|f\|^2/2\eta \tag{9.5.23}$$

and

$$2[ADR_0(z)]^*f(y) = i\int_y^{\infty} e^{i\kappa(y-x)}A(x)^*f(x)\,dx - i\int_{-\infty}^{y} e^{i\bar{\kappa}(x-y)}A(x)\,dx. \tag{9.5.24}$$

PROOF. The inequality (9.5.23) follows from (9.5.22). If $v \in L^2$, we have by (9.5.21)

$$2(f, ADR_0(z)v) = i\int_{-\infty}^{\infty}\int_{-\infty}^{x} f(x)A(x)^*e^{-i\bar{\kappa}(x-y)}v(y)^*\,dy\,dx$$
$$-i\int_{-\infty}^{\infty}\int_{x}^{\infty} f(x)A(x)^*e^{-i\bar{\kappa}(y-x)}v(y)^*\,dy\,dx.$$

This gives (9.5.24). □

9.6. The Perturbed Operator

Our next job is to examine the inequality

$$a\|BR(z)\|^2 \leq C_I, \qquad s \in I. \tag{9.6.1}$$

Here our task becomes more complicated. For in our case as well as in most applications, H is a perturbation of H_0 in some sense. Usually H_0 and V are given and H is defined by means of some abstract method. While one is often able to construct $R_0(z)$ (as we did in Section 9.5), it is usually a very formidable task to find $R(z)$. Thus it becomes a serious problem to verify (9.6.1) when one cannot find an explicit expression for $R(z)$. We are going to present a method which allows one to prove (9.6.1) using only assumptions on H_0, A, and B. It is a rather involved procedure, but well worth the effort. We shall need some elementary facts concerning linear algebra and analytic functions of a complex variable. They will be stated as needed.

The basic idea is as follows. Assume that (9.4.1) holds. By Lemma 8.3.1,

$$BR(z) = BR_0(z) + B[AR_0(\bar{z})]^*BR(z)$$

and consequently

$$(1 - B[AR_0(\bar{z})]^*)BR(z) = BR_0(z). \tag{9.6.2}$$

(In this outline we are not being careful about domains. We shall take care of these details later.) We see from this that (9.6.1) will hold if we can prove

1. $a\|BR_0(z)\|^2 \le C_I$, $s \in I$, and
2. $G_0(z) = 1 - B[AR_0(\bar{z})]^*$ has an inverse for each z near I which is uniformly bounded for $s \in I$.

Notice that both these conditions depend only on H_0, A, and B. Now condition (1) is similar to (9.5.1) and can be verified in the same way. However, condition (2) does not look easy at all. Suppose we can show by some means that $G_0(z)$ indeed has an inverse for each nonreal z. What we need is that this inverse should be uniformly bounded for z near I. A moment's reflection will convince us that this is not an easy thing to prove. It is the heart of the problem.

We proceed as follows. Suppose $G_0(s + ia)$ converges in norm as $a \to 0$ to a bounded operator $G_{0+}(s)$ uniformly on the interval I. Suppose further that $G_{0+}(s)$ has a bounded inverse $G_+(s)$ for each $s \in I$. If we define

$$G_{0+}(z) = G_0(z), \qquad a > 0, \quad s \in I, \tag{9.6.3}$$

$$G_+(z) = G_0(z)^{-1}, \qquad a > 0, \quad s \in I, \tag{9.6.4}$$

then $G_{0+}(z)$ is uniformly continuous in the closure of

$$\omega_I = \{z = s + ia | s \in I, 0 < a < b\} \tag{9.6.5}$$

and its inverse equals $G_+(z)$. It is then a simple matter to show that the inverse of an operator which is uniformly continuous in a parameter is also uniformly continuous in the parameter. In particular, $G_+(z)$ is uniformly bounded in ω_I, and our goal will be achieved.

This is an outline of the program. We shall carry it out in the next few sections. We conclude this section by proving

Lemma 9.6.1. *Let $F(z)$ be a bounded operator on a Hilbert space $\mathcal{H}$ for z in a closed bounded set Σ in the complex plane. Assume that $F(z)$ is continuous in z for $z \in \Sigma$ and it has a bounded inverse on $\mathcal{H}$ for each z in Σ. Then its inverse $G(z)$ is continuous in z and uniformly bounded for $z \in \Sigma$.*

PROOF. First we show that there is a constant C such that

$$\|G(z)\| \le C, \qquad z \in \Sigma. \tag{9.6.6}$$

Suppose (9.6.6) did not hold. Then there would be a sequence z_k of points in Σ such that

$$\|G(z_k)\| \to \infty. \tag{9.6.7}$$

Since Σ is bounded and closed, there is a subsequence which converges. Thus we may assume $z_k \to z_0 \in \Sigma$. Set $L_k = G(z_k)/\|G(z_k)\|$. Then $\|L_k\| = 1$ and $\|F(z_k)L_k\| = 1/\|G(z_k)\| \to 0$ as $k \to \infty$ by (9.6.6). Thus,

$$\|F(z_0)L_k\| \le \|F(z_k)L_k\| + \|[F(z_0) - F(z_k)]L_k\| \to 0 \quad \text{as} \quad k \to \infty.$$

This implies

$$1 = \|L_k\| = \|G(z_0)F(z_0)L_k\| \le \|G(z_0)\| \, \|F(z_0)L_k\| \to 0 \qquad \text{as} \quad k \to \infty,$$

providing a contradiction. Thus (9.6.6) holds. Once we know this we have

$$G(z) - G(z') = G(z)[F(z') - F(z)]G(z')$$

and consequently

$$\|G(z) - G(z')\| \le C^2 \|F(z') - F(z)\|. \qquad \square$$

9.7. Compact Operators

We shall need some information concerning compact operators on a Hilbert space $\mathcal{H}$. In particular we shall need

Theorem 9.7.1. *Let K be a compact operator on $\mathcal{H}$. Then $\sigma(K)$ consists of at most a denumerable set of points in the complex plane with no nonzero accumulation point. Every nonzero point $\lambda \in \sigma(K)$ is an eigenvalue with dim $N(\lambda - K) < \infty$.*

In proving Theorem 9.7.1 we shall use the following lemmas.

Lemma 9.7.1. *If $T = 1 - K$, then* dim $N(T) < \infty$.

PROOF. Clearly $N(T)$ is a subspace of $\mathcal{H}$. If it were infinite dimensional, it would have an orthonormal sequence $\{\varphi_n\}$ (Lemma 3.5.2). Since K is compact, $\{K\varphi_n\}$ has a convergent subsequence. By discarding the rest we may assume that $K\varphi_n \to \psi$. Thus $\varphi_n = T\varphi_n + K\varphi_n = K\varphi_n \to \psi$, contradicting the fact that $\{\varphi_n\}$ is an orthonormal sequence [cf. (3.6.3)]. $\square$

Lemma 9.7.2. *There is a constant C such that*

$$\|u\| \le C\|Tu\|, \qquad u \perp N(T). \tag{9.7.1}$$

PROOF. If (9.7.1) did not hold, there would be a sequence $\{u_n\} \subset N(T)^\perp$ such that

$$\|u_n\| = 1, \qquad Tu_n \to 0. \tag{9.7.2}$$

Thus $\{Ku_n\}$ has a convergent subsequence. Again we may assume that $Ku_n \to v$. Hence $u_n = Tu_n + Ku_n \to v$, showing that $Tv = 0$. Since $u_n \perp N(T)$, the same is true of v. Thus we must have $v = 0$. But $\|v\| = \lim\|u_n\| = 1$. This contradiction proves the lemma. $\square$

Lemma 9.7.3. *$R(T)$ is closed.*

PROOF. Suppose $Tx_n \to y$. We can write $x_n = x_n' + x_n''$, where $x_n' \perp N(T)$ while $x_n'' \in N(T)$ (Corollary 3.5.1 and Theorem 3.5.1). Thus $Tx_n' \to y$. By Lemma 9.7.2, $\{x_n'\}$ is a Cauchy sequence in $\mathcal{H}$ and hence it converges to an element x. Since T is bounded, $Tx'_n \to Tx$, and consequently $Tx = y$. Hence $y \in R(T)$. □

Lemma 9.7.4. *If $N(T) = \{0\}$, then $R(T) = \mathcal{H}$.*

PROOF. Note that $R(T^n)$ is a closed subspace for each $n > 0$, and

$$R(T^{n+1}) \subset R(T^n). \tag{9.7.3}$$

Suppose there is an $n > 0$ such that

$$R(T^{n+1}) = R(T^n). \tag{9.7.4}$$

Let x be any element of $\mathcal{H}$. Then $T^n x \in R(T^n)$. Thus there is a $y \in \mathcal{H}$ such that $T^n x = T^{n+1} y$. Since T is one-to-one, we have $x = Ty$. Consequently, $\mathcal{H} = R(T)$. Thus in order for the lemma to be false, we must have

$$R(T^{n+1}) \neq R(T^n) \tag{9.7.5}$$

for each n. By Corollary 3.5.3, there is a sequence $\{y_n\}$ of elements such that

$$y_n \in R(T^n), \qquad y_n \perp R(T^{n+1}), \qquad \|y_n\| = 1. \tag{9.7.6}$$

If $n > m$, then

$$Ky_m - Ky_n = y_m - (y_n + Ty_m - Ty_n) = y_m - y,$$

where $y \in R(T^{m+1})$. Hence $y_m \perp y$ and consequently

$$\|Ky_m - Ky_n\|^2 = \|y_m\|^2 + \|y\|^2 \geq 1.$$

This shows that $\{Ky_n\}$ cannot have a convergent subsequence, contradicting the fact that K is compact. Hence there is an n such that (9.7.4) holds. This proves the lemma. □

Now we can give the

PROOF OF THEOREM 9.7.1. First we note that if $\lambda \neq 0$ is not an eigenvalue of K, then it must be in $\rho(K)$. For if $\dim N(\lambda - K) = 0$, then $R(1 - \lambda^{-1}K) = \mathcal{H}$ (Lemma 9.7.4). Thus $(1 - \lambda^{-1}K)$ has an inverse defined everywhere on $\mathcal{H}$, and it is bounded by Lemma 9.7.2. Thus $\lambda \in \rho(K)$. To complete the proof we show that for each $\varepsilon > 0$ the set $|z| \geq \varepsilon$ in the complex plane contains at most a finite number of eigenvalues of K such that $|\lambda_m| \geq \varepsilon$. Then for each n there is an $x_n \neq 0$ such that

$$(\lambda_n - K)x_n = 0. \tag{9.7.7}$$

I claim that for any n, the elements $x_1, \ldots, x_n$ are linearly independent. Otherwise, these would be an m such that

$$x_m = \sum_{j=1}^{m-1} \alpha_j x_j, \tag{9.7.8}$$

while the $x_1, \ldots, x_{m-1}$ are independent. By (9.7.8),

$$Kx_m = \sum_{j=1}^{m-1} \alpha_j \lambda_j x_j.$$

But

$$Kx_m = \lambda_m x_m = \sum_{j=1}^{m-1} \alpha_j \lambda_m x_j$$

and consequently

$$\sum_{j=1}^{m-1} \alpha_j (\lambda_j - \lambda_m) x_j = 0.$$

Since $\lambda_j \neq \lambda_m$ for $j \neq m$ and the $x_1, \ldots, x_{m-1}$ are linearly independent, we see that all the α_j must vanish. But this would make x_m vanish by (9.7.8), contradicting the way it was chosen.

Next let M_m be the subspace spanned by $x_1, \ldots, x_m$. Then $M_n \subset M_{n+1}$ but $M_n \neq M_{n+1}$ for each n. Again by Corollary 3.5.3 there is a sequence $\{y_n\}$ such that

$$y_n \in M_n, \qquad y_n \perp M_{n-1}, \qquad \|y_n\| = 1.$$

Moreover, we note that K maps M_n into itself. For if

$$x = \sum_1^n \alpha_j x_j, \tag{9.7.9}$$

then

$$Kx = \sum_1^n \alpha_j \lambda_j x_j \in M_n.$$

This gives

$$(K - \lambda_n)x = \sum_1^{n-1} \alpha_j (\lambda_j - \lambda_n) x_j \in M_{n-1},$$

showing that $K - \lambda_n$ maps M_n into M_{n-1}. Now if $n > m$,

$$Ky_n - Ky_m = \lambda_n \left[y_n - \lambda_n^{-1}(Ky_m + (\lambda_n - K)y_n) \right] = \lambda_n (y_n - y),$$

where $y \in M_{n-1}$. Thus

$$\|Ky_n - Ky_m\|^2 = |\lambda_n|^2 (\|y_n\|^2 + \|y\|^2) \geq \varepsilon^2.$$

This shows that $\{Ky_n\}$ cannot have a convergent subsequence, contradicting the fact that K is compact. Thus K can have only a finite number of eigenvalues in $|z| \geq \varepsilon$, and the theorem is proved. □

We shall also use

Theorem 9.7.2. *K is compact iff it has the property that $Ku_n \to 0$ whenever $\{u_n\}$ is a bounded sequence converging weakly to* 0.

PROOF. Suppose K is compact and $\{u_n\}$ is such a sequence. If Ku_n did not converge to 0 there would be a subsequence (also denoted by $\{u_n\}$) and a $\delta > 0$ such that

$$\|Ku_n\| \geq \delta. \tag{9.7.10}$$

Since K is compact, there is a subsequence (again denoted by $\{u_n\}$) such that $Ku_n \to w$. Since $u_n \to 0$ weakly, we have for any $v \in \mathcal{H}$, $(w, v) = \lim(Ku_n, v) = \lim(u_n, K^*v) = 0$. Hence $w = 0$. On the other hand, (9.7.10) shows that $\|w\| \geq \delta$. This contradiction shows that $Ku_n \to 0$. Conversely, suppose K has the property described and $\{u_n\}$ is a bounded sequence. By Theorem 3.6.3, it has a subsequence (also denoted by $\{u_n\}$) converging weakly to some element $u \in \mathcal{H}$. Thus $u_n - u \to 0$ weakly, and thus by hypothesis $K(u_n - u) \to 0$ strongly. This shows that K is compact. □

Theorem 9.7.3. *If $\{K_n\}$ is a sequence of compact operators and $\|K_n - K\| \to 0$ as $n \to \infty$, then K is compact.*

PROOF. Suppose $\{u_k\}$ satisfies $\|u_k\| \leq C$ and converges weakly to 0. Let $\varepsilon > 0$ be given, and take n so large that $\|K_n - K\| < \varepsilon/2C$. Then

$$\|Ku_k\| \leq \|(K - K_n)u_k\| + \|K_n u_k\| \leq \tfrac{1}{2}\varepsilon + \|K_n u_k\|.$$

Then take N so large that

$$\|K_n u_k\| < \tfrac{1}{2}\varepsilon, \qquad k > N.$$

Thus

$$\|Ku_k\| < \varepsilon, \qquad k > N.$$

This means that $Ku_k \to 0$. Hence K is compact by Theorem 9.7.2. □

It can be shown that every weakly convergent sequence is bounded, but we shall not need this fact here.

9.8. Analytic Dependence

We shall need to consider operators which depend on a parameter in various ways. We have been doing this all along without paying any attention to it. For instance, the resolvent operator $R(z)$ depends on a complex parameter z.

Let Ω be an open set in the complex plane, and let $T(z)$ be a bounded operator on a Hilbert space $\mathcal{H}$ for each $z \in \overline{\Omega}$. We shall say that $T(z)$

depends continuously on z in $\bar{\Omega}$ if

$$\|T(z) - T(z_0)\| \to 0 \quad \text{as} \quad z \to z_0 \tag{9.8.1}$$

for each $z_0 \in \bar{\Omega}$. We shall say that it depends analytically on z in Ω if the difference quotient

$$[T(z) - T(z_0)]/(z - z_0) \tag{9.8.2}$$

converges in norm as $z \to z_0$ to a bounded operator $T'(z_0)$ for each $z_0 \in \Omega$. It is not difficult to check that all of the basic theorems of complex function theory carry over to this case as well. In particular, the Cauchy integral theorem and formula are valid. Moreover, for each $z_0 \in \Omega$, $T(z)$ can be expanded in a Taylor series of the form

$$T(z) = \sum_0^\infty (z - z_0)^k T_k, \tag{9.8.3}$$

where the T_k are bounded operators and

$$\sum_0^\infty |z - z_0|^k \|T_k\| < \infty \tag{9.8.4}$$

for z in a neighborhood of z_0. An important example is given by

Lemma 9.8.1. *Let A be a closed operator on $\mathcal{H}$ with nonempty resolvent set $\rho(A)$. Then the resolvent operator $R(z) = (z - A)^{-1}$ depends analytically on z in $\rho(A)$.*

PROOF. Let z_0 be any point in $\rho(A)$. Then I claim that

$$R(z) = \sum_1^\infty (z_0 - z)^{k-1} R(z_0)^k. \tag{9.8.5}$$

This series converges for $|z - z_0| \, \|R(z_0)\| < 1$. To show that it equals $R(z)$, note that

$$(z - A)\left[R(z) - \sum_1^n (z_0 - z)^{k-1} R(z_0)^k\right] = (z_0 - z)^n R(z_0)^n, \tag{9.8.6}$$

which converges to 0 in norm for z close to z_0. Apply $R(z)$ to both sides of (9.8.6) and let $n \to \infty$. Once we have (9.8.5) it is not difficult to show, as in the scalar case, that $R(z)$ has a derivative at z_0. □

Our next aim is to prove

Theorem 9.8.1. *Let K be a compact operator on $\mathcal{H}$ and let D be a disk with boundary γ in the complex plane such that $\bar{D}$ does not contain the origin and γ does not contain any points of $\sigma(K)$. Then the operator*

$$P = \frac{1}{2\pi i} \int_\gamma (z - K)^{-1} \, dz \tag{9.8.7}$$

is a bounded projection onto a finite-dimensional subspace and

$$PK = KP. \tag{9.8.8}$$

If z_0 is the center of D and $z_0 - K$ is one-to-one on $R(P)$, then $z_0 \in \rho(K)$.

In proving the theorem, we shall make use of the simple

Lemma 9.8.2. *If a projection is compact, then its range is finite dimensional.*

PROOF. If P is a projection and $R(P)$ is infinite dimensional, then it has an orthonormal sequence $\{\varphi_n\}$ (Lemma 3.5.2). Since $P\varphi_n = \varphi_n$, we see that $\{P\varphi_n\}$ has no convergent subsequence. Hence P cannot be compact. ☐

Now we give the

PROOF OF THEOREM 9.8.1. The integral over γ exists because $R(z) = (z - k)^{-1}$ is analytic on $\rho(K)$. To show that P is a projection, let γ_1 be a circle with center z_0 and radius slightly smaller than that of γ and such that there are no points of $\sigma(K)$ on γ_1 or between γ and γ_1. Then by the Cauchy integral theorem

$$P = \frac{1}{2\pi i} \int_{\gamma_1} R(\zeta)\, d\zeta.$$

Hence

$$P^2 = (2\pi i)^{-2} \int_{\gamma} \int_{\gamma_1} R(z)R(\zeta)\, dz\, d\zeta$$

$$= (2\pi i)^{-2} \int_{\gamma} \int_{\gamma_1} [R(z) - R(\zeta)](\zeta - z)^{-1}\, dz\, d\zeta$$

$$= (2\pi i)^{-2} \int_{\gamma} R(z) \int_{\gamma_1} (\zeta - z)^{-1}\, d\zeta\, dz$$

$$- (2\pi i)^{-2} \int_{\gamma_1} R(\zeta) \int_{\gamma} (\zeta - z)^{-1}\, dz\, d\zeta = (2\pi i)^{-1} \int_{\gamma_1} R(\zeta)\, d\zeta = P$$

where we have used the first resolvent equation [cf. (8.3.8)]. Note that the points ζ are inside γ while the points z are outsides γ_1. To show that $R(P)$ is finite dimensional we note that

$$R(z) - z^{-1} = Kz^{-1}R(z). \tag{9.8.9}$$

Hence

$$P = (2\pi i)^{-1} \int_{\gamma} [z^{-1} + Kz^{-1}R(z)]\, dz$$

$$= (2\pi i)^{-1} K \int_{\gamma} z^{-1}R(z)\, dz \tag{9.8.10}$$

since the origin is not in D. This shows that P is a compact operator. Hence its range must be finite dimensional by Lemma 9.8.2. Note that (9.8.8) follows immediately from the definition (9.8.7).

Next set

$$S = (2\pi i)^{-1} \int_{\gamma} (z - z_0)^{-1} R(z)\, dz. \tag{9.8.11}$$

Then

$$\begin{aligned} S(z_0 - K) &= (z_0 - K)S \\ &= (2\pi i)^{-1} \int_{\gamma} \left[(z - z_0)^{-1} - R(z) \right] dz = 1 - P. \end{aligned} \tag{9.8.12}$$

Moreover, if $z_0 - K$ is one-to-one on the finite-dimensional subspace $R(P)$, then there is a bounded operator L on $R(P)$ such that

$$(z_0 - K)LP = L(z_0 - K)P = P \tag{9.8.13}$$

(here we are using the fact that linear operators on a finite-dimensional space which are one-to-one are also onto and the fact that all linear operators on a finite-dimensional space are bounded). Set $T = S + LP$. Then by (9.8.12) and (9.8.13), $(z_0 - K)T = T(z_0 - K) = 1$. Since T is a bounded operator, we see that $z \in \rho(K)$. □

9.9. Projections

We shall also need a few results concerning projections. The first is

Theorem 9.9.1. *Let P, Q be bounded projections on $\mathcal{H}$ such that $\|P - Q\| < 1$. Then there is a bounded invertible operator S on H such that $SP = QS$. In particular,* $\dim R(P) = \dim R(Q)$.

PROOF. Set

$$R = (P - Q)^2 = P + Q - PQ - QP.$$

Then

$$PR = P - PQP = RP, \qquad QR = Q - QPQ = RQ.$$

Since $\|R\| \le \|P - Q\|^2 < 1$, the operator $1 - R$ is invertible. Set

$$S = (1 - R)^{-1}\left[QP + (1 - Q)(1 - P) \right],$$

$$T = (1 - R)^{-1}\left[PQ + (1 - P)(1 - Q) \right].$$

Then

$$ST = TS = (1 - R)^{-1}. \tag{9.9.1}$$

This shows that S is invertible. Moreover, $SP = (1 - R)^{-1}QP = QS$. To prove the last statement, let $x_1, \ldots, x_n$ be elements of $\mathcal{H}$. They are linearly independent iff $Sx_1, \ldots, Sx_n$ are. Now

$$Px_k = x_k \qquad \text{iff} \quad QSx_k = Sx_k.$$

Thus the x_k are in $R(P)$ iff the Sx_k are in $R(Q)$. Thus dim $R(P) = n$ iff dim $R(Q) = n$. □

Next we have

Theorem 9.9.2. *Let Ω be a bounded open set in the plane and $P(z)$ a projection on $\mathcal{H}$ for $z \in \overline{\Omega}$ depending analytically on z for $z \in \Omega$ and continuously on z for $z \in \overline{\Omega}$. Let z_0 be any point of $\overline{\Omega}$. Then there are a neighborhood N of z_0 and an invertible operator $X(z)$ depending analytically on z in $N \cap \Omega$ and continuously on z in $N \cap \overline{\Omega}$ such that*

$$P(z)X(z) = X(z)P(z_0). \tag{9.9.2}$$

PROOF. Since $P^2 = P$, we have

$$P'P + PP' = P' \tag{9.9.3}$$

and consequently

$$PP'P = 0. \tag{9.9.4}$$

Set

$$Q = P'P - PP'. \tag{9.9.5}$$

Then

$$PQ = -PP', \qquad QP = P'P, \qquad QP - PQ = P'. \tag{9.9.6}$$

Suppose we can solve

$$X' = QX, \qquad X(z_0) = 1, \qquad Y' = -YQ, \qquad Y(z_0) = 1 \tag{9.9.7}$$

in $N \cap \overline{\Omega}$, where N is some neighborhood of z_0, with $X(z)$ analytic in $N \cap \Omega$, continuous in $N \cap \overline{\Omega}$, and invertible. Then

$$(YX)' = Y'X + YX' = -YQX + YQX = 0.$$

Since $Y(z_0)X(z_0) = 1$, we see that $Y(z)X(z) = 1$ in $N \cap \Omega$. Since X is invertible, this shows that $Y = X^{-1}$. Moreover,

$$(YPX)' = Y'PX + YP'X + YPX' = 0$$

by (9.9.3), (9.9.6), and (9.9.7). Thus

$$Y(z)P(z)X(z) = P(z_0). \tag{9.9.8}$$

This gives the desired result. Therefore it remains only to solve (9.9.7) with X having the desired properties. Set $X_0(z) = 1$ and define $X_n(z)$ inductively by

$$X_n(z) = 1 + \int_{z_0}^{z} Q(\zeta)X_{n-1}(\zeta)\, d\zeta. \tag{9.9.9}$$

By induction, each $X_n(z)$ is analytic in Ω and the integral is independent of the path. If the $X_n(z)$ converge uniformly in some neighborhood of z_0, then the limit $X(z)$ satisfies

$$X(z) = 1 + \int_{z_0}^{z} Q(\zeta)X(\zeta)\,d\zeta. \tag{9.9.10}$$

Clearly this is a solution of (9.9.7) with the desired properties. It will be invertible if we take the neighborhood so small that

$$\left\| \int_{z_0}^{z} Q(\zeta)X(\zeta)\,d\zeta \right\| < 1. \tag{9.9.11}$$

Thus it suffices to show that the $X_n(z)$ converge uniformly in some neighborhood of z_0. Let N_δ be the intersection of $\bar{\Omega}$ with the disk $|z - z_0| < \delta$. Then by (9.9.9),

$$\|X_{n+1}(z) - X_n(z)\| \le \delta C_0 \sup_{N_\delta} \|X_n(\zeta) - X_{n-1}(\zeta)\|$$

for $z \in N_\delta$, where

$$C_0 = \sup_{\Omega} \|Q(z)\|.$$

Repeating n times and setting

$$M = \sup_{N_\delta} \|X_1(z) - X_0(z)\|,$$

we obtain

$$\|X_{n+1}(z) - X_n(z)\| \le \delta^n C_0^n M.$$

Hence if $n > m$,

$$\|X_n(z) - X_m(z)\| \le M \sum_{k=m}^{n-1} \delta^k C_0^k \le M\delta^m C_0^m / (1 - \delta C_0).$$

Thus we see that if $\delta C_0 < 1$, then the $X_n(z)$ converge uniformly in N_δ. Thus we have solved (9.9.7) for X. A similar argument will obtain a solution Y as well. □

9.10. An Analytic Function Theorem

The following theorem is easily proved for scalar functions of a complex variable. For any interval I let the rectangle ω_I be defined by (9.6.5), where b is a given positive number.

Theorem 9.10.1. *Suppose $f(z)$ is an analytic function in ω_I and does not vanish there. If $f(z)$ is continuous in $\bar{\omega}_I$, then it can vanish on I only on a set of measure* 0.

We shall prove this theorem at the end of the section. However, for our purposes we shall need an extension of this theorem to operators. Such a theorem would be easy to prove if we were interested in the subset of I where the operator vanished. This is not what we have in mind. All we want to find is the subset of I where the operator is not invertible. This is a more complicated situation and involves strong restrictions. We state it as

Theorem 9.10.2. *Let $T(z)$ be a bounded operator on $\mathcal{H}$ for each $z \in \bar{\omega}_I$ which depends analytically on z in ω_I and continuously on z in $\bar{\omega}_I$. Suppose that $K(z) = 1 - T(z)$ is compact for $z \in \omega_I$ and that $T(z)$ has a bounded inverse for each $z \in \omega_I$. Then the set of those $s \in I$ for which $T(s)$ has no bounded inverse has measure* 0.

PROOF. First we note that it suffices to show that for any $s_0 \in \bar{I}$ there is a neighborhood N of s_0 such that the subset of $N \cap I$ where $T(s)$ does not have a bounded inverse has measure 0.

Let s_0 be any point of $\bar{I}$. Since $K(z)$ is continuous in $\bar{\omega}_I$ and compact in ω_I, it is compact in the whole of $\bar{\omega}_I$ (Theorem 9.7.3). Thus $K(s_0)$ is a compact operator. Since the points of $\sigma(K(s_0))$ are isolated (Theorem 9.7.1), there is a circle γ with center 1 and radius < 1 which does not intersect $\sigma(K(s_0))$. By Lemma 9.6.1, there is a constant C_0 such that

$$\|(\zeta - K(s_0))^{-1}\| \leq C_0, \qquad \zeta \in \gamma. \tag{9.10.1}$$

Now,

$$\zeta - K(z) = [\zeta - K(s_0)]\big(1 - [\zeta - K(s_0)]^{-1}[K(z) - K(s_0)]\big). \tag{9.10.2}$$

Since $K(z)$ is continuous in z, there is a $\delta > 0$ such that

$$\|K(z) - K(s_0)\| < 1/C_0 \tag{9.10.3}$$

for $|z - s_0| < \delta$. This will make the last operator in (9.10.2) have a bounded inverse. Hence $\zeta \in \rho(K(z))$ when $\zeta \in \gamma$ and

$$|z - s_0| < \delta. \tag{9.10.4}$$

For such z set

$$P(z) = (2\pi i)^{-1}\int_\gamma [\zeta - K(z)]^{-1}\, d\zeta. \tag{9.10.5}$$

Clearly, $P(z)$ depends continuously on z. Moreover, one easily checks that

$$P'(z) = (2\pi i)^{-1}K'(z)\int_\gamma [\zeta - K(z)]^{-2}\, d\zeta. \tag{9.10.6}$$

Hence $P(z)$ depends analytically on z. By Theorem 9.8.1, $P(z)$ is a bounded projection onto a finite-dimensional subspace. The dimension of this subspace does not change with z (Theorem 9.9.1). We apply Theorem 9.9.2 to Ω equal to the intersection of ω_I and the disk $|z - s_0| < \frac{1}{2}\delta$. Thus there is a neighborhood N of s_0 and an invertible operator $X(z)$ depending analytically on z in $N \cap \Omega$ and continuously on z in $N \cap \bar{\Omega}$ such that

$$P(z)X(z) = X(z)P(s_0). \tag{9.10.7}$$

Consider the operator

$$S(z) = P(s_0)X(z)^{-1}T(z)X(z)P(s_0). \tag{9.10.8}$$

It is a mapping of $R(P(s_0))$ into itself which depends analytically on z in $N \cap \Omega$ and continuously on z in $N \cap \bar{\Omega}$. If $x_1, \ldots, x_m$ is a basis for $R(P(s_0))$, we can write $S(z)$ in the form

$$S(z)\sum \alpha_j x_j = \sum \alpha_j a_{jk}(z)x_k, \tag{9.10.9}$$

where the $a_{jk}(z)$ are scalar functions which are analytic in $N \cap \Omega$ and continuous in $N \cap \bar{\Omega}$. Thus $S(z)$ is invertible in $R(P(s_0))$ iff

$$f(z) = \det(a_{jk}(z)) \neq 0. \tag{9.10.10}$$

Now suppose there were a $z \in N \cap \Omega$ such that $f(z) = 0$. This would mean that $S(z)$ is not invertible. Hence there would be a v such that $P(s_0)v \neq 0$ and $S(z)v = 0$. By (9.10.7), this means that

$$X(z)^{-1}P(z)T(z)P(z)X(z)v = 0.$$

This is equivalent to

$$T(z)P(z)X(z)v = 0.$$

But

$$P(z)X(z)v = X(z)P(s_0)v \neq 0.$$

Consequently, $T(z)$ would not be one-to-one and could not have an inverse, contrary to hypothesis. Thus $f(z)$ cannot vanish in $N \cap \Omega$. We can now apply Theorem 9.10.1 to conclude that $f(s)$ can vanish on $N \cap \bar{\Omega} \cap I$ only on a set of measure 0. This implies that $S(s)$ can fail to have an inverse only on such a set.

Finally, we note that $T(s)$ has a bounded inverse whenever $S(s)$ does. To see this suppose u is such that $P(s)u \neq 0$. Then

$$\begin{aligned} T(s)P(s)u &= P(s)T(s)P^2(s)u \\ &= X(s)P(s_0)X(s)^{-1}T(s)X(s)P(s_0)X(s)^{-1}P(s)u \\ &= X(s)S(s)X(s)^{-1}P(s)u. \end{aligned}$$

If $S(s)$ is invertible, this implies that $T(s)P(s)u \neq 0$. Thus $T(s)$ is one-to-one on $R(P(s))$. This in turn implies that $T(s)$ has a bounded inverse (Theorem 9.8.1).

To summarize, we have shown that $f(s)$ can vanish on $N \cap \bar{\Omega} \cap I$ only on a set of measure 0 and that $T(s)$ has a bounded inverse whenever $f(s) \neq 0$. Hence $T(s)$ can fail to have a bounded inverse only on a set of measure 0. □

It remains to prove Theorem 9.10.1. We base the proof on

Theorem 9.10.3. *Let $F(z)$ be an analytic function in $|z| < 1$ which does not vanish there. If $F(z)$ is continuous in $|z| \le 1$, then*

$$\int_0^{2\pi} \left|\log|F(e^{i\theta})|\,\right| d\theta < \infty. \tag{9.10.11}$$

PROOF. For $r \le 1$ set

$$u_r(\theta) = K - \log|F(re^{i\theta})|,$$

where

$$K = \max_{|z| \le 1} \log|F(z)|.$$

The constant K is finite because $F(z)$ is continuous in $|z| \le 1$. For each $r < 1$ the functions $u_r(\theta)$ are nonnegative and integrable on $(0, 2\pi)$, and they converge pointwise to $u_1(\theta)$ as $r \to 1$. Moreover, $\log|F(z)| = \text{Re} \log F(z)$ is harmonic in $|z| < 1$ since $F(z)$ does not vanish there. By the mean value theorem for harmonic functions,

$$2\pi \log|F(0)| = \int_0^{2\pi} \log|F(re^{i\theta})|\, d\theta.$$

Hence

$$\int_0^{2\pi} u_r(\theta)\, d\theta = 2\pi(K - \log|F(0)|).$$

Now we can apply Fatou's lemma to conclude that $u_1(\theta)$ is integrable from 0 to 2π. This gives (9.10.11). □

Note that (9.10.11) implies that $F(e^{i\theta}) \neq 0$ almost everywhere. Using this fact we give the

PROOF OF THEOREM 9.10.1. Map ω_I conformally onto the unit disk $|w| < 1$ by a function $w = g(z)$. This mapping can be extended to be continuous on the boundary of ω_I. Note that $\log g(z)$ has a single-valued branch in ω_I and that $|g(z)| \to 1$ as $\text{Im}\, z \to 0$. Thus $\text{Re} \log g(z) = \log|g(z)| \to 0$ as $\text{Im}\, z \to 0$. It follows by the reflection principle that $\log g(z)$ and consequently $g(z)$ is analytic on I. Moreover, by the Cauchy–Riemann equations,

$$\partial \arg g(s + it)/\partial s = -\partial \log|g(s + it)|/\partial t.$$

Thus

$$\partial \arg g(s)/\partial s \ge 0, \qquad s \in I. \tag{9.10.12}$$

If $g'(s)$ vanished at any point $s_0 \in I$, then the two segments of I to the right and left of s_0 would be mapped onto curves making an angle $< \pi$ at $g(s_0)$. This cannot happen in view of (9.10.12). Thus $g'(s) \neq 0$ for $s \in I$.

Next set $F(w) = f(g^{-1}(w))$. Then $F(w)$ satisfies all of the hypotheses of Theorem 9.10.3. Consequently, the set of points where it vanishes must have measure 0. Since $g'(s) \neq 0$ for $s \in I$, the set of points of I mapped into this set must also have measure 0. □

9.11. The Combined Results

Now we are ready to formulate conditions which will imply condition (9.6.1). We are willing to make assumptions on $R_0(z)$ and on the perturbation, but we do not want to make any assumptions which require information about $R(z)$. In some expressions we shall denote the closure of an operator L by $[L]$. The use of this symbol will mean that the operator L is known to be closable. First we have

Theorem 9.11.1. *Suppose* (9.4.1) *holds with* $D(H_0) \subset D(A) \cap D(B)$, $D(H) \subset D(B)$, *and that* $Q_0(z) = [B(AR_0(\bar{z}))^*]$ *is bounded and everywhere defined on* $\mathcal{H}$ *for* $\operatorname{Im} z \neq 0$. *Assume that* $G_0(z) = 1 - Q_0(z)$ *has a bounded inverse for* $\operatorname{Im} z \neq 0$ *and that there is an open set* Λ *such that* $Q_0(z)$ *can be extended to be continuous in norm in* $\bar{\omega}_I$ *for each* $I \subset\subset \Lambda$. *Assume further that there is a* $z_1 \in \rho(H_0)$ *such that* $[BR_0(z)(AR_0(\bar{z}))^*]$ *is compact for* $\operatorname{Im} z \neq 0$ *and that*

$$|\operatorname{Im} z| \, \|BR_0(z)\|^2 \leq C_I, \qquad z \in \omega_I, \quad I \subset\subset \Lambda. \tag{9.11.1}$$

Then there is an open subset Γ *of* Λ *such that* $\Lambda - \Gamma$ *has measure* 0 *and*

$$|\operatorname{Im} z| \, \|BR(z)\|^2 \leq C_I', \qquad z \in \omega_I, \quad I \subset\subset \Gamma. \tag{9.11.2}$$

PROOF. Set $G(z) = G_0(z)^{-1}$, and let Γ be the set of those $s \in \Lambda$ such that $G_0(s)$ has a bounded inverse. Assume that $\operatorname{Im} z_1 \neq 0$. Then

$$G(z_1)G_0(z) = 1 + K(z), \tag{9.11.3}$$

where

$$K(z) = (z - z_1)G(z_1)\left[BR_0(z)(AR_0(\bar{z}_1))^*\right]. \tag{9.11.4}$$

To see this, note that the domain of $(AR_0(\bar{z}_1))^*$ is equal to that of $(z - z_1)R_0(z)(AR_0(\bar{z}_1))^*$, which is contained in that of

$$(z_1 - z)(AR_0(\bar{z}_1)R_0(\bar{z}))^* = \left(A\left[R_0(\bar{z}) - R_0(\bar{z}_1)\right]\right)^*.$$

Hence the domains of $(AR_0(\bar{z}_1))^*$ and $(AR_0(\bar{z}))^*$ coincide. Once this is known, (9.11.4) follows by an easy computation. By hypothesis, $K(z)$ is compact for $\operatorname{Im} z \neq 0$. Since $G_0(z)$ is continuous in $\bar{\omega}_I$, the same is true of $K(z)$ by (9.11.3). Moreover, (9.11.3) also shows that $1 + K(z)$ has a bounded inverse for $\operatorname{Im} z \neq 0$. Hence we can apply Theorem 9.10.2 to conclude that the subset of I where $1 + K(s)$ fails to have a bounded inverse has measure 0. By (9.11.3), the same must be true of $G_0(s)$. Thus $\Lambda - \Gamma$ has measure 0. If $I \subset\subset \Gamma$, then $G_0(z)$ has a bounded inverse for each $z \in \bar{\omega}_I$. Thus $G(z)$ is bounded in $\bar{\omega}_I$ by Lemma 9.6.1. Inequality (9.11.2) follows from (9.11.1) and (9.6.2).

In case z_1 happens to be real, we cannot define $K(z)$ by (9.11.4) because we do not know that $G(z_1)$ exists. We circumvent this in the following way. Let z_0 be a nonreal point outside $\bar{\omega}_I$. Then for $z \neq z_0$ we have

$$\begin{aligned} BR_0(z)(AR_0(\bar{z}_0))^* = {} & BR_0(z)(AR_0(\bar{z}_1))^* \\ & + (z_1 - z_0)(z_0 - z)^{-1} \\ & \times B[R_0(z) - R_0(z_0)](AR_0(\bar{z}_1))^*. \quad (9.11.5) \end{aligned}$$

This shows $[BR_0(z)(AR_0(\bar{z}_0))^*]$ is compact for nonreal $z \neq z_0$. In particular, this holds for $z \in \omega_I$. Hence we may substitute z_0 for z_1 in (9.11.3) and (9.11.4) and proceed as above. □

If we combine Theorems 9.4.2 and 9.11.1, we obtain

Theorem 9.11.2. *Suppose* (9.4.1) *holds with* $D(H_0) \subset D(A) \cap D(B)$, $D(H) \subset D(B)$, *and that* $Q_0(z) = [B(AR_0(\bar{z}))^*]$, $G_0(z) = 1 - Q_0(z)$, *and* $G(z) = G_0(z)^{-1}$ *are bounded and everywhere defined for* Im $z \neq 0$. *Assume that there is an open set* Λ *such that* $E_0(\Lambda)$ *is the (orthogonal) projection onto* $\mathcal{H}_c(H_0)$, *and that* $Q_0(z)$, $Q_0(\bar{z})$ *are uniformly continuous in* ω_I *for each* $I \subset\subset \Lambda$. *Assume also that there is a* z_1 *such that* $[BR_0(z)(AR_0(z_1))^*]$ *is compact when* Im $z \neq 0$ *and that*

$$|\text{Im } z|(\|AR_0(z)\|^2 + \|BR_0(z)\|^2) \leq C_I, \qquad z, \bar{z} \in \omega_I, \quad I \subset\subset \Lambda. \tag{9.11.6}$$

Assume further that $E_0(\Gamma) = E_0(\Lambda)$ *for each open subset* $\Gamma \subset \Lambda$ *such that* $\Lambda - \Gamma$ *has measure* 0. *Then the wave operators are complete.*

PROOF. All of the hypotheses of Theorem 9.4.2 are assumed except

$$|\text{Im } z| \; \|BR(z)\|^2 \leq C_I, \qquad z, \bar{z} \in \omega_I, \quad I \subset\subset \Lambda. \tag{9.11.7}$$

Since $Q_0(z)$ is uniformly continuous in ω_I, it can be extended to be continuous in the whole of $\bar{\omega}_I$. Then we can apply Theorem 9.11.1 to conclude that there is an open subset $\Gamma \subset \Lambda$ such that $\Lambda - \Gamma$ has measure 0 and (9.11.7) holds with Λ replaced by Γ. However, $E_0(\Gamma) = E_0(\Lambda)$ by hypothesis, so we can apply Theorem 9.4.2 using Γ in place of Λ. □

An important hypothesis in Theorems 9.11.1 and 9.11.2 is the existence of a bounded inverse of $G_0(z)$ for Im $z \neq 0$. Now we give a criterion for the existence of such an inverse.

Theorem 9.11.3. *Suppose* $D(H_0) \cup D(H) \subset D(A) \cap D(B)$ *and*

$$\begin{aligned} (u, Hv) - (H_0 u, v) = (Au, Bv)_{\mathcal{K}} = (Bu, Av)_{\mathcal{K}}, & \\ u \in D(H_0), \quad v \in D(H), & \qquad (9.11.8) \end{aligned}$$

and that $D(A^*)$ is dense. Assume also that $Q_0(z) = [B(AR_0(\bar{z}))^*]$, $Q(z) = [B(AR(\bar{z}))^*]$ are bounded and everywhere defined on $\mathcal{K}$. Then $G_0(z) = 1 - Q_0(z)$ has a bounded inverse $G(z)$ on $\mathcal{K}$ given by $G(z) = 1 + Q(z)$.

In proving Theorem 9.11.3 we shall make use of

Lemma 9.11.1. *Let T be a bounded operator everywhere defined and*

$$R(T) \subset D(A), \qquad R(T^*) \subset D(B). \tag{9.11.9}$$

If $D(A^*)$ is dense and $B(AT)^*$ is bounded, then

$$B(AT)^*u = BT^*A^*u, \qquad u \in D(A^*). \tag{9.11.10}$$

Moreover,

$$[B(AT)^*] = [BT^*A^*] \tag{9.11.11}$$

and they are everywhere defined.

PROOF. If $u \in D(A^*)$, then

$$(u, ATv) = (A^*u, Tv) = (T^*A^*u, v)$$

This shows that $u \in D[(AT)^*]$ and

$$(AT)^*u = T^*A^*u. \tag{9.11.12}$$

Since $R(T^*) \subset D(B)$, this shows that $(AT)^*u \in D(B)$ and (9.11.10) holds. Now suppose u is any element. Then there is a sequence $\{u_n\}$ of elements in $D(A^*)$ converging to u. Thus

$$BT^*A^*u_n = B(AT)^*u_n \to [B(AT)^*]u.$$

This gives (9.11.11). □

PROOF OF THEOREM 9.11.3. By (9.11.8),

$$R(z) - R_0(z) = [AR_0(\bar{z})]^*BR(z) = [AR(\bar{z})]^*BR_0(z). \tag{9.11.13}$$

Suppose $u \in D(A^*)$. Then by Lemma 9.11.1,

$$Q_0(z)u = BR_0(z)A^*u, \qquad Q(z) = BR(z)A^*u.$$

If we define $G(z)$ to be $1 + Q(z)$, we have

$$\begin{aligned} G_0(z)G(z)u - u &= Q(z)u - Q_0(z)u - Q_0(z)Q(z)u \\ &= B[R(z) - R_0(z)]A^*u - Q_0(z)BR(z)A^*u \\ &= \{B[AR_0(\bar{z})]^* - Q_0(z)\}BR(z)A^*u. \end{aligned}$$

This vanishes because the range of $R(z) - R_0(z)$ is contained in $D(B)$ and consequently $Q_0(z)$ equals $B[AR_0(\bar{z})]^*$ on the range of $BR(z)$. Similar reasoning gives

$$G(z)G_0(z)u = u.$$

Since both $G_0(z)$ and $G(z)$ are bounded and everywhere defined and $D(A^*)$ is dense, it follows that

$$G(z)G_0(z) = G_0(z)G(z) = 1. \qquad \square$$

9.12. Absolute Continuity

In order to apply Theorem 9.11.2 we must find an open set Λ such that

a. $E_0(\Lambda)$ is the projection onto $\mathcal{H}_c(H_0)$ and
b. $E_0(\Lambda) = E_0(\Gamma)$ for each open subset Γ of Λ such that $\Lambda - \Gamma$ has measure 0.

In our application, H_0 is the free Hamiltonian as defined in Section 2.5. In this section we shall verify that (a) and (b) do indeed hold for the free Hamiltonian. In order to do this we shall need a representation for the spectral family $\{E_0(\lambda)\}$ of H_0. Let f be any function in L^2. Then it is easily checked that

$$[R_0(z)f]\hat{\ } = (z - k^2)^{-1}\hat{f}(k). \tag{9.12.1}$$

Thus if $z = s + ia$ and I is an open interval,

$$a\int_I \|R_0(z)f\|^2\, ds = \int_{-\infty}^{\infty}\left[\int_I \frac{a\, ds}{(s - k^2)^2 + a^2}\right]|\hat{f}(k)|^2\, dk. \tag{9.12.2}$$

The inner integral converges to π if $k^2 \in I$, $\frac{1}{2}\pi$ if k^2 is an endpoint of I and 0 if $k^2 \notin \bar{I}$. Thus by Lemma 8.2.1,

$$(\tilde{E}_0(I)f, f) = \int_{k^2 \in I} |\hat{f}(k)|^2\, dk. \tag{9.12.3}$$

Note that it does not matter whether the integral in (9.12.3) is taken over the set $k^2 \in I$, $k^2 \in \bar{I}$, or a set in between. Suppose $I = (b, c)$. Set

$$E_0(\lambda -)u = \lim_{0<\delta\to 0} E(\lambda - \delta)u. \tag{9.12.4}$$

Then

$$E_0(\bar{I}) = E_0(c) - E_0(b -), \qquad E_0(I) = E_0(c -) - E_0(b), \tag{9.12.5}$$

and consequently

$$2\tilde{E}_0(I) = E_0(c) + E_0(c -) - E_0(b) - E_0(b -). \tag{9.12.6}$$

Now if $b' < b$ and $b \to b'$, then

$$E_0(b)f \to E_0(b')f, \qquad E_0(b -)f \to E_0(b')f, \tag{9.12.7}$$

with a similar statement for c. Thus if $b' < b < c' < c$ and $b \to b'$, $c \to c'$, then we have

$$([E_0(c') - E_0(b')]f, f) = \int_{k^2 \in I'} |\hat{f}(k)|^2 \, dk,$$

where $I' = (b', c')$. Since b and c were arbitrary, this gives

$$(E_0(I)f, f) = \int_{k^2 \in I} |\hat{f}(k)|^2 \, dk. \tag{9.12.8}$$

Since the right-hand side of (9.12.8) converges to 0 with the length of I, we see that $(E_0(\lambda)f, f)$ is continuous for each $f \in L^2$. Hence we have

Theorem 9.12.1. *For the free Hamiltonian H_0, $\mathcal{H}_c(H_0)$ is the whole of L^2.*

Next we note that any open set Λ of the real line R is a denumerable union of nonoverlapping open intervals. Thus

$$(E_0(\Lambda)f, f) = \int_{k^2 \in \Lambda} |\hat{f}(k)|^2 \, dk. \tag{9.12.9}$$

From this we see that if the set $(0, \infty) - \Lambda$ has measure 0, then

$$(E_0(\Lambda)f, f) = \int_{-\infty}^{\infty} |\hat{f}(k)|^2 \, dk = \|f\|^2, \tag{9.12.10}$$

from which we see that

$$E_0(\Lambda) = 1. \tag{9.12.11}$$

Thus we have

Theorem 9.12.2. *For the free Hamiltonian if Λ is any open set such that $(0, \infty) - \Lambda$ has measure 0, then $E_0(\Lambda) = 1$.*

Further consequences of (9.12.8) are

$$E_0(\lambda)f = 0, \qquad \lambda < 0, \tag{9.12.12}$$

$$(E_0(\lambda)f, f) = \int_{-\lambda^{1/2}}^{\lambda^{1/2}} |\hat{f}(k)|^2 \, dk, \qquad \lambda \geq 0. \tag{9.12.13}$$

Another useful property can be observed as follows. Suppose

$$\Lambda = \bigcup_n I_n, \tag{9.12.14}$$

where the intervals I_n do not overlap (the union can be finite or denumerable). If we set

$$|\Lambda| = \sum_n |I_n|,$$

where $|I_n|$ denotes the length of I_n, it follows from (9.12.9) that

$$E_0(\Lambda)f \to 0 \qquad \text{as} \quad |\Lambda| \to 0. \tag{9.12.15}$$

Whenever (9.12.15) holds we say that f is *absolutely continuous* with respect to H_0. In our case we have shown that all functions in L^2 are absolutely continuous with respect to the free Hamiltonian H_0.

It is not difficult to show that for any self-adjoint operator A the set of elements which are absolutely continuous with respect to A form a closed subspace. It is called the *subspace of absolute continuity* of A and is denoted by $\mathcal{H}_{ac}(A)$. Its orthogonal complement is called the *subspace of singularity* and denoted by $\mathcal{H}_s(A)$. It contains the eigenvectors of A. Thus we have

Theorem 9.12.3. *For the free Hamiltonian H_0 we have*

$$\mathcal{H}_c(H_0) = \mathcal{H}_{ac}(H_0) = L^2, \qquad \mathcal{H}_s(H_0) = \varnothing.$$

9.13. The Intertwining Relations

In this section we shall prove Lemmas 9.1.1 and 9.4.1. First we note

Lemma 9.13.1. *If* $\operatorname{Im} z \neq 0$, *then*

$$W_\pm R_0(z) = R(z) W_\pm. \tag{9.13.1}$$

PROOF. By Theorem 6.3.3 and (8.1.2), we have for $\operatorname{Im} z > 0$

$$R(z) W_\pm f = -i \int_0^\infty e^{it(z-H)} W_\pm f \, dt = -i \int_0^\infty W_\pm e^{it(z-H_0)} f \, dt \tag{9.13.2}$$

when $f \in M_\pm$. This shows that $R_0(z)f \in M_\pm$ and

$$R(z) W_\pm f = W_\pm R_0(z) f. \tag{9.13.3}$$

A similar argument works in the case $\operatorname{Im} z < 0$. Conversely, if $R_0(z)f \in M_\pm$, then $(z - H_0)R_0(z)f \in M_\pm$ (Theorem 6.3.8). Thus (9.13.3) holds in this case as well. □

Now we can give the

PROOF OF LEMMA 9.4.1. We have by Lemma 9.13.1

$$(R(z) W_\pm f, R(z)g) = (R_0(z)f, R_0(z) W_\pm^* g)$$

for $f \in M_\pm$ and $g \in R_\pm$. Thus by Lemma 8.2.1,

$$(\tilde{E}(I) W_\pm f, g) = (\tilde{E}_0(I)f, W_\pm^* g)$$

for any interval I. Hence

$$\tilde{E}(I) W_\pm f = W_\pm \tilde{E}_0(I) f, \qquad f \in M_\pm. \tag{9.13.4}$$

Using (9.12.4)–(9.12.7) and reasoning as in the proof of (9.12.8), we obtain (9.4.10). □

The formulas (9.4.10) are called the *intertwining relations* because they intertwine $E_0(I)$, $E(I)$, and the $W_\pm$. Another consequence of (9.13.4) is

$$E(\bar{I})W_\pm f = W_\pm E_0(\bar{I})f, \qquad f \in M_\pm. \tag{9.13.5}$$

This can be used to give the

PROOF OF LEMMA 9.1.1. If $\psi \in \mathcal{H}_c(H_0)$, then $E_0(\bar{I})\psi = E_0(I)\psi$ for each interval I. Hence by (9.4.10) and (9.13.5),

$$\left[E(\bar{I}) - E(I)\right]W_\pm\psi = W_\pm\left[E_0(\bar{I}) - E_0(I)\right]\psi = 0.$$

This means that $W_\pm\psi \in \mathcal{H}_c(H)$. The converse is obtained by reasoning in the opposite direction. □

A consequence of Lemma 9.4.1 is

Lemma 9.13.2. *If $M_\pm \subset \mathcal{H}_{ac}(H_0)$, then $R_\pm \subset \mathcal{H}_{ac}(H)$.*

PROOF. Let Λ be given by (9.12.14). Then by (9.4.10) and Theorem 9.12.3,

$$E(\Lambda)W_\pm f = W_\pm E_0(\Lambda)f \to 0 \qquad \text{as} \quad |\Lambda| \to 0.$$ □

9.14. The Application

Now we return to our Hamiltonian H and find a sufficient condition on the potential that will insure that the wave operators are complete. We shall assume

$$\int_{-\infty}^{\infty} |V(x)|\,dx < \infty \tag{9.14.1}$$

and define H by means of bilinear forms. Specifically, we shall take $A(x) = |V(x)|^{1/2}$ when $V(x) \neq 0$, $A(x) = e^{-x^2}$ when $V(x) = 0$ and $B(x) = V(x)/A(x)$. Then both A and B are in L^2, and

$$\|Au\|^2 + \|Bu\|^2 \leq h_0(u) + K\|u\|^2 \tag{9.14.2}$$

holds by (2.7.7). Once this is known we can apply Theorem 8.4.2 to show that the operator H associated with

$$h(u, v) = (u', v') + (Vu, v) \tag{9.14.3}$$

is self-adjoint and satisfies $D(|H|^{1/2}) = D(H_0^{1/2})$. The aim of this section (and chapter) is to prove

Theorem 9.14.1. *If* (9.14.1) *holds, then the wave operators $W_\pm$ are complete.*

The proof of this theorem will be accomplished by a series of lemmas.

Lemma 9.14.1. *If A and B are in L^2, then*

$$\|[B[AR_0(z)]^*]\| \le \tfrac{1}{2}|z|^{-1/2}\|B\|\,\|A\|. \tag{9.14.4}$$

PROOF. By Lemma 9.5.1, we have for $u, v \in L^2$

$$|(B[AR_0(z)]^*u, v)| \le \tfrac{1}{2}|z|^{-1/2}\int\int|B(x)A(y)u(y)v(x)|\,dx\,dy.$$

The double integral is bounded by

$$\left(\int\int|B(x)u(y)|^2\,dx\,dy\right)^{1/2}\left(\int\int|A(y)v(x)|^2\,dx\,dy\right)^{1/2} = \|B\|\,\|u\|\,\|A\|v.$$

This gives (9.14.4). □

Lemma 9.14.2. *If*

$$\int_{-\infty}^{\infty}(1+|x|)^2(|A(x)|^2+|B(x)|^2)\,dx < \infty, \tag{9.14.5}$$

then $Q_0(z) = [B(AR_0(z))^*]$ is uniformly continuous in ω_I for any bounded interval I which is bounded away from the origin.

PROOF. Let z, z' be points of ω_I and let κ, κ' be related to them as in Theorem 9.5.2. Set

$$g(\kappa) = \kappa^{-1}e^{i\kappa|x|}.$$

Then

$$g'(\kappa) = (i\kappa|x|-1)e^{i\kappa|x|}/\kappa^2.$$

Suppose $I = (c, d)$ with $c > 0$. If $\kappa = \sigma + i\eta$ and $z = s + ia$, then $s = \sigma^2 - \eta^2$ and $a = 2\sigma\eta$. Thus $\sigma > c^{1/2}$ and consequently

$$|g'(\kappa)| \le (1+|\kappa x|)/c.$$

If we set $\kappa(\theta) = (1-\theta)\kappa' + \theta\kappa$, $0 \le \theta \le 1$, then $\operatorname{Re}\kappa(\theta) \ge c^{1/2}$ in ω_I and

$$\begin{aligned} g(\kappa) - g(\kappa') &= g(\kappa(1)) - g(\kappa(0)) \\ &= \int_0^1 g'(\kappa(\theta))\kappa'(\theta)\,d\theta = (\kappa-\kappa')\int_0^1 g'(\kappa(\theta))\,d\theta. \end{aligned}$$

Hence

$$|g(\kappa) - g(\kappa')| \le |\kappa-\kappa'|(1+K|x|)/c,$$

where K is an upper bound for $|z|^{1/2}$ in ω_I. Thus by (9.5.2),

$$|R_0(z)f - R_0(z')f| \le \tfrac{1}{2}c^{-1}|\kappa-\kappa'|\int_{-\infty}^{\infty}(1+K|x-y|)|f(y)|\,dy.$$

Consequently,

$$\begin{aligned} &|([Q_0(z) - Q_0(z')]u, v)| \\ &\qquad \le \tfrac{1}{2}c^{-1}|\kappa-\kappa'|\int\int(1+K|x-y|)|B(x)A(y)u(y)v(x)|\,dx\,dy. \end{aligned}$$

The integral is bounded by

$$\left(\int\int (1 + K|x|)^2 |B(x)u(y)|^2 \, dx \, dy\right)^{1/2}$$

$$\times \left(\int\int (1 + K|y|)^2 |A(y)v(x)|^2 \, dx \, dy\right)^{1/2}$$

$$\leq \left(\int (1 + K|x|)^2 |B(x)|^2 \, dx \int (1 + K|y|)^2 |A(y)|^2 \, dy\right)^{1/2} \|u\| \, \|v\|.$$

Hence (9.12.5) implies

$$\|Q_0(z) - Q_0(z')\| \leq C|\kappa - \kappa'|. \tag{9.14.6}$$

This gives uniform continuity of $Q_0(z)$ in ω_I. □

Lemma 9.14.3. *If $A \in L^2$ and $\operatorname{Im} z \neq 0$, then $AR_0(z)$ is a compact operator on L^2.*

PROOF. By Theorem 3.7.5, A is H_0-compact. Let $\{f_n\}$ be a bounded sequence in L^2. Then $\{R_0(z)f_n\}$ satisfies

$$\|R_0(z)f_n\| + \|H_0 R_0(z)f_n\| \leq C.$$

Thus $\{AR_0(z)f_n\}$ has a convergent subsequence. This shows that $AR_0(z)$ is compact. □

Lemma 9.14.4. *If A and B are in L^2, then $Q_0(z)$ is uniformly continuous in any ω_I for I bounded and away from the origin.*

PROOF. For each n set

$$A_n(x) = A(x), \qquad B_n(x) = B(x), \qquad |x| \leq n,$$
$$A_n(x) = B_n(x) = 0, \qquad |x| > n,$$

and $Q_n(z) = [B_n(A_n R_0(\bar{z}))^*]$. Then by Lemma 9.14.2,

$$\|Q_n(z) - Q_0(z)\| \leq C(\|A\| \, \|B_n - B\| + \|A_n - A\| \, \|B\|).$$

Thus $Q_n(z)$ converges to $Q_0(z)$ uniformly in ω_I. On the other hand, A_n and B_n satisfy (9.14.5) for each n. Hence $Q_n(z)$ is uniformly continuous in ω_I for each n (Lemma 9.14.2). Combining these two facts we obtain the desired result. □

Lemma 9.14.5. *If $A(x)$ is real valued and locally in L^2, then the operator of multiplication by A is self-adjoint.*

PROOF. Since $Au \in L^2$ for $u \in C_0^\infty$, we see that $D(A)$ is dense. Now suppose u, f are in L^2 and

$$(u, Av) = (f, v), \qquad v \in D(A). \tag{9.14.7}$$

Then

$$\int [u(x)A(x) - f(x)]v(x)^* \, dx = 0, \qquad v \in C_0^\infty. \qquad (9.14.8)$$

Note that Au is locally integrable, being the product of a function in L^2 and one locally in L^2. Let φ be a function in C_0^∞ and let $j_n(x)$ be defined by (7.8.3). Then (9.14.8) gives

$$\int [u(x)A(x) - f(x)]\varphi(x)j_n(x-y) \, dx = 0$$

for each real y. This means

$$J_n[(uA - f)\varphi] \equiv 0$$

for each n. But $(uA - f)\varphi$ is in L^1, and consequently $J_n[(uA - f)\varphi]$ converges to $(uA - f)\varphi$ in L^1 by (7.8.16). Thus $(uA - f)\varphi = 0$ for each $\varphi \in C_0^\infty$. But for each bounded interval I we can find a $\varphi \in C_0^\infty$ which equals 1 on I (Theorem 7.8.3). This shows that $Au = f \in L^2$. Hence $u \in D(A)$. Thus $D(A^*) \subset D(A)$. Since A is Hermitian, the result follows. □

Now we are ready for the

PROOF OF THEOREM 9.14.1. We apply Theorem 9.11.2. Clearly, (9.11.8) holds. Since

$$h_0(u) = (H_0u, u) \le \tfrac{1}{2}(\|H_0u\|^2 + \|u\|^2), \qquad u \in D(H_0)$$

and

$$|h(u)| = |(Hu, u)| \le \tfrac{1}{2}(\|Hu\|^2 + \|u\|^2), \qquad u \in D(H),$$

inequality (9.14.2) and the fact that $D(|H|^{1/2}) = D(|H_0|^{1/2})$ imply that

$$\|Au\|^2 + \|Bu\|^2 \le \tfrac{1}{2}\|H_0u\|^2 + K'\|u\|^2, \qquad u \in D(H_0), \quad (9.14.9)$$

$$\|Au\|^2 + \|Bu\|^2 \le \tfrac{1}{2}\|Hu\|^2 + K'\|u\|^2, \qquad u \in D(H). \quad (9.14.10)$$

Thus $D(H_0) \cup D(H) \subset D(A) \cap D(B)$. By Lemma 9.14.5, A and B are self-adjoint operators. Moreover, (9.14.9) and (9.14.10) imply that $BR_0(z)$ and $BR(z)$ are bounded and everywhere defined. Hence $BR_0(z)A$ and $BR(z)A$ are densely defined. Lemmas 9.11.1 and 9.14.1 show that $Q_0(z)$ is bounded and everywhere defined for $\operatorname{Im} z \neq 0$. Let $\Lambda = (0, \infty)$. Then $E_0(\Lambda) = 1$ and $E_0(\Gamma) = 1$ for each open set Γ such that $\Lambda - \Gamma$ has measure 0 (Theorem 9.12.2). Since $BR_0(z)$ is a compact operator for $\operatorname{Im} z \neq 0$ (Lemma 9.14.3), the same is true of $[BR_0(z)(AR_0(\bar{z}_1))^*]$ for any nonreal z_1. Inequality (9.11.6) follows from Theorem 9.5.1. Moreover, $Q_0(z)$, $Q_0(\bar{z})$ are uniformly continuous in ω_I for each $I \subset\subset \Lambda$ by Lemma 9.14.4. Next we note that for $\operatorname{Im} z \neq 0$, the operator $|i - H|^{1/2}R(z)$ maps L^2 into

$D(|H|^{1/2})$. In fact, we have

$$(|i - H|^{1/2}R(z)u, |i - H|^{1/2}v) = \int |i - \lambda|(z - \lambda)^{-1}\, d(E(\lambda)u, v) \tag{9.14.11}$$

and this is bounded by a constant times $\|u\|\,\|v\|$. Thus there is an $f \in L^2$ such that the right-hand side of (9.14.11) equals (f, v) for all $u, v \in L^2$. Since

$$B(AR(\bar{z}))^* = B|R(i)|^{1/2}(A|i - H|^{1/2}R(\bar{z}))^*,$$

we see that $Q(z)$ is bounded and everywhere defined. The same is true of $G_0(z)^{-1}$ by Theorem 9.11.3. Thus all of the hypotheses of Theorem 9.11.2 are verified, and the conclusion follows. □

We can even give a stronger version of Theorem 9.14.1.

Theorem 9.14.2. *If* (9.14.1) *holds, then* $R_\pm = \mathcal{H}_{ac}(H)$.

PROOF. In the proof above we showed that there is an open set Λ such that $C\Lambda$ has measure 0 and (9.4.8) holds for $I \subset\subset \Lambda$. Thus (9.4.9) holds by Theorem 9.4.2. By Theorem 9.12.3 and Lemma 9.13.2 we know that $R_\pm \subset \mathcal{H}_{ac}(H)$. Hence $R_\pm = E(\Lambda)\mathcal{H}_c(H)$. Thus if g is any function in $\mathcal{H}_{ac}(H)$, then $E(\Lambda)g \in R_\pm$. This implies that $g \in R_\pm$. To see this, cover $C\Lambda$ with an open set Γ. Since $g \in \mathcal{H}_{ac}(H)$, we have $E(\Gamma)g \to 0$ as $|\Gamma| \to 0$. Hence $E(C\Lambda)g = 0$. Consequently, $g = E(\Lambda)g \in R_\pm$. □

Exercises

1. Using Lemma 9.1.1, show that (9.1.7) is equivalent to the definition of completeness.

2. Show that (9.1.9) is equivalent to strong completeness.

3. Prove that every open set on the real line is a denumerable union of disjoint open intervals.

4. Prove Corollary 9.2.1.

5. Prove Lemma 9.3.1.

6. Prove Corollary 9.3.1.

7. Show that (9.5.2) gives the unique solution of (9.5.3).

8. Prove (9.5.11).

9. Show that $u(x)$ given by (9.5.15) is in L^2.

10. Prove (9.5.16).

11. Prove (9.6.4).

12. If K is compact and $T = 1 - K$, show that $R(T^n)$ is a closed subspace for each $n \geq 0$.

13. Prove (9.8.6).

14. Show that (9.8.6) implies that $R'(z_0)$ exists.

15. Prove that every linear operator on a finite-dimensional space which is one-to-one is also onto.

16. Show that all linear operators on a finite-dimensional space are bounded.

17. Prove (9.9.1).

18. Prove Theorem 9.10.1.

19. Prove (9.10.6).

20. Show that the functions $a_{jk}(z)$ in (9.10.9) are analytic in $N \cap \Omega$.

21. Let Ω be an open set and $T(z)$ a bounded operator which depends uniformly continuously in z for $z \in \Omega$. Show that $T(z)$ can be defined for $z \in \bar{\Omega}$ to depend uniformly continuously on z.

22. Using (9.11.9), show that (9.11.8) implies that $Q(z)$ is bounded.

23. Prove Theorem 9.14.3 for I an interval of negative numbers.

24. Assume that $T_n(z)$ is uniformly continuous in Ω for each n and $T_n(z) \to T(z)$ in norm as $n \to \infty$ uniformly in Ω. Show that $T(z)$ is uniformly continuous in Ω.

25. Show how (9.13.2) implies (9.13.3).

26. Prove (9.13.8).

27. Show that (9.1.7) implies $R_+ = R_-$.

28. Prove the last statement made in the proof of Theorem 9.4.1.

29. If L and M are two dense open subsets of R, show that $L \cap M$ is also dense in R.

30. Prove (9.12.1).

31. Show that (9.1.7) holds iff $\mathcal{H}_c(H_0) \subset M_{\pm}$ and $R_+ = R_-$.

32. Prove (9.11.13).

33. Prove the last statement made in the proof of Theorem 9.11.3.

34. Prove that the limit (9.12.4) exists.

35. Prove (9.12.7).

36. Prove (9.12.9).

37. Fill in the details in the proof of Theorem 9.10.1.

38. Show that (9.12.10) implies (9.12.11).

39. Prove (9.13.5).

10
Strong Completeness

10.1. The More Difficult Problem

In Chapter 9 we found a method of proving completeness of the wave operators and applied it to our Hamiltonian. When the method applies, it shows that $\mathcal{H}_c(H_0) \subset M_\pm$ and that $R_+ = R_-$. However, it does not show that $\mathcal{H}_c(H) \subset R_\pm$ (see the first section of that chapter). From the point of view of physics it is desired that every $\psi \in \mathcal{H}_c(H)$ be the value at $t = 0$ of a scattering state, that is, that strong completeness hold.

The difficulty in proving strong completeness can be seen from Theorem 9.4.2. There we assumed that $E_0(\Lambda)$ is the projection onto $\mathcal{H}_c(H_0)$ and that hypothesis (9.4.8) holds. The proof of that theorem shows that $E_0(\Lambda)\mathcal{H}_c(H) \subset R_\pm$. However, we do not know that $E(\Lambda)$ is the projection onto $\mathcal{H}_c(H)$ and cannot conclude that $\mathcal{H}_c(H) \subset R_\pm$. Now the assumption that $E_0(\Lambda)$ is the projection onto $\mathcal{H}_c(H_0)$ is reasonable for the unperturbed operator and was readily verified for our free Hamiltonian in Section 9.12, provided $(0, \infty) - \Lambda$ has measure 0. However, an assumption that $E(\Lambda)$ is the projection onto $\mathcal{H}_c(H)$ concerns the perturbed Hamiltonian and is very difficult to verify in practice. The knowledge that the complement of Λ has measure 0 does not guarantee that $E(\Lambda)$ is the projection onto $\mathcal{H}_c(H)$. Thus the theorems of Section 9.2 do not help in this respect.

In this section we present a theory which gives strong completeness. Then we shall apply the theory to our Hamiltonian to find sufficient conditions on V for strong completeness.

10.2. The Abstract Theory

As we mentioned in Section 10.1, the fact that the complement $C\Lambda = R - \Lambda$ of Λ has measure 0 does not necessarily imply that

$$E(\Lambda)f = f, \qquad f \in \mathcal{H}_c(H). \tag{10.2.1}$$

This leads to the question whether there is a property of $C\Lambda$ which will guarantee (10.2.1). An answer to this is given by

Lemma 10.2.1. *If* $C\Lambda$ *is a denumerable set, then* (10.2.1) *holds.*

PROOF. Let $\{\lambda\}$ denote the set consisting of the point λ only. Then

$$\begin{aligned} E(C\{\lambda\}) &= E(-\infty, \lambda) + E(\lambda, \infty) \\ &= E(\lambda -) + 1 - E(\lambda). \end{aligned}$$

Consequently,

$$P(\lambda) = E(\{\lambda\}) = E(\lambda) - E(\lambda -). \tag{10.2.2}$$

Suppose

$$C\Lambda = \bigcup_{k=1}^{\infty} \{\lambda_k\}.$$

Then

$$E(C\Lambda)f = \sum_{k=1}^{\infty} [E(\lambda_k) - E(\lambda_k -)]f = 0$$

since $f \in \mathcal{H}_c(H)$. Thus (10.2.1) holds. □

Using Lemma 10.2.1, we have the following counterpart of Theorem 9.4.2.

Theorem 10.2.1. *Assume that there are operators A, B from $\mathcal{H}$ to $\mathcal{K}$ such that*

$$D(H_0) \subset D(A), \qquad D(H) \subset D(B) \tag{10.2.3}$$

and

$$(u, Hv) = (H_0 u, v) + (Au, Bv)_{\mathcal{K}}, \qquad u \in D(H_0), \quad v \in D(H). \tag{10.2.4}$$

Suppose Λ is an open subset of the real line such that $C\Lambda$ is denumerable and

$$a\|AR_0(s \pm ia)\|^2 + a\|BR(s \pm ia)\|^2 \le C_I, \qquad s \in I, \quad a > 0 \tag{10.2.5}$$

holds for all $I \subset\subset \Lambda$. Then the wave operators are strongly complete.

PROOF. The hypotheses of this theorem are symmetric in H_0 and H. By Theorem 9.4.1, $E_0(\Lambda)f \in M_+$ for each $f \in \mathcal{H}_c(H_0)$, and $E(\Lambda)g \in R_+$ for each $g \in \mathcal{H}_c(H_0)$. Now we apply Lemma 10.2.1. □

As before, we shall look for hypotheses on H_0, A, B which will imply (10.2.5). Our main result in this direction is

Theorem 10.2.2. *Suppose that A, B are closed and there is a number θ such that $0 \le \theta \le 1$,*

$$D(|H_0|^\theta) = D(|H|^\theta) = D_\theta \subset D(A), \tag{10.2.6}$$

$$D(|H_0|^{1-\theta}) = D(|H|^{1-\theta}) = D_{1-\theta} \subset D(A), \tag{10.2.7}$$

and (10.2.4) holds. Set $Q_0(z) = A[BR_0(\bar{z})]^*$, $G_0(z) = 1 - Q_0(z)$. (They are bounded operators on $\mathcal{H}$ for $\operatorname{Im} z \neq 0$.) Assume that there is a $z_1 \in \rho(H_0)$ such that $AR_0(z)[BR_0(z_1)]^*$ is a compact operator on $\mathcal{H}$ for all nonreal z. Assume further that there is an open set Γ such that $C\Gamma$ is denumerable and such that $G_0(s \pm ia) \to G_{0\pm}(s)$ in norm for $s \in \Gamma$, where the $G_{0\pm}(s)$ are continuous in s, and

$$a\|AR_0(s \pm ia)\|^2 + a\|BR_0(s \pm ia)\|^2 \le C_I, \qquad s \in I, \quad a > 0, \tag{10.2.8}$$

for each $I \subset\subset \Gamma$. Assume also that $D(B^*)$ is dense, and for each $g \in N[G_{0\pm}(s)]$ there is a function $\sigma(\delta) \to 0$ as $\delta \to 0$ such that

$$\|[BE_0(I)R_0(s+it)]^*g\| \le \sigma(|I|) \tag{10.2.9}$$

when s is the midpoint of $I \subset\subset \Gamma$. Finally, we assume that there is a locally bounded function $C(s)$ in Γ and functions $\tau_j(\delta) \to 0$ as $\delta \to 0$ such that

$$\begin{aligned}&\|[BE_0(I)R_0(s+it)]^*Au\|\\ &\quad \le C(s)[\tau_1(|I|) + \tau_2(|s-\lambda|/|I|)\|u\|, u \in N(H-s)]\end{aligned} \tag{10.2.10}$$

where λ is the center of the interval $I \subset\subset \Gamma$. Then the wave operators are strongly complete.

One should compare this theorem with Theorem 9.11.2. The basic change is the addition of hypotheses (10.2.9) and (10.2.10). The proof of Theorem 10.2.2 is not difficult and will be given in the next sections. Then we shall apply it to our Hamiltonian.

10.3. The Technique

In proving Theorem 10.2.2 we shall make use of the following lemmas.

Lemma 10.3.1. *If $g \in N[G_{0\pm}(s)]$, then there is a $u \in D(H)$ such that $g = Au$ and $(H - s)u = 0$.*

PROOF. Let $\varepsilon > 0$ be given, and let $I \subset\subset \Gamma$ be an interval with center s such that $\sigma(|I|) < \varepsilon$. Then by (10.2.9),

$$\|\{B[R_0(s + it) - R_0(s + ir)]\}^*g\| \le 2\varepsilon$$
$$+\|\{BE_0(CI)[R_0(s + it) - R_0(s + ir)]\}^*g\|.$$

Thus $[BR_0(s + it)]^*g$ converges to some element $u \in \mathcal{H}$ as $t \to 0$. Moreover,

$$A[BR_0(s \pm ia)]^*g \to g - G_{0\pm}(s)g = g \tag{10.3.1}$$

as $a \to 0$. Since A is closed, we see that $u \in D(A)$ and $Au = g$. Now,

$$([s \pm ia - H_0]w, [BR_0(s \pm ia)]^*g) = (Bw, Au)$$

for all $w \in D(H_0)$. Letting $a \to 0$, we get

$$([s - H_0]w, u) = (Bw, Au), \qquad w \in D(H_0). \tag{10.3.2}$$

This implies

$$|([i - H_0]w, u)| \le C\| \, |i - H_0|^{1-\theta}w\|, \qquad w \in D(H_0) \tag{10.3.3}$$

or

$$|(|i - H_0|^\theta h, u)| \le C\|h\|, \qquad h \in D_\theta. \tag{10.3.4}$$

This shows that $u \in D_\theta$. I claim that (10.3.2) holds not only for all $w \in D(H_0)$ but for all $w \in D_{1-\theta}$. To see this, let w be any element in $D_{1-\theta}$, and set $w_k = E_0(-k, k)w$. Then

$$w_k \in D(H_0) \qquad \text{and} \qquad |i - H_0|^{1-\theta}(w_k - w) \to 0, \qquad k \to \infty. \tag{10.3.5}$$

Apply (10.3.2) with w replaced by w_k and then let $k \to \infty$. The same reasoning shows that (10.2.4) holds for $u \in D_\theta$, $v \in D_{1-\theta}$. Thus we have

$$([s - H]w, u) = 0, \qquad w \in D_{1-\theta}.$$

Since $D(H) \subset D_{1-\theta}$, we see that $u \in D(H)$ and $[s - H]u = 0$. □

Lemma 10.3.2. *The set of points $s \in \Gamma$ for which $N[G_{0\pm}(s)] \neq \{0\}$ has no limit point in Γ.*

PROOF. Suppose $g_k \in N[G_{0\pm}(\lambda_k)]$, $g_k \neq 0$, with the λ_k different and $\lambda_k \to \lambda \in \Gamma$. Then by Lemma 10.3.1, the λ_k are eigenvalues of H with eigenvectors u_k satisfying $Au_k = g_k$ obtained as in the proof of Lemma 10.3.1. We may assume that $\|u_k\| = 1$ for each k. Since

$$g_k = (i + |\lambda_k|^\theta)A(i + |H|^\theta)^{-1}u_k, \tag{10.3.6}$$

we see that the norms of the g_k are uniformly bounded. Hence there is a subsequence (also denoted by $\{g_k\}$) which converges weakly (Theorem 3.6.3).

Next we note that

$$(v, Hu) = (H_0 v, u) + (Bv, Au)_{\mathcal{K}}, \qquad v \in D_{1-\theta}, \quad u \in D_\theta. \tag{10.3.7}$$

[This is just the conjugate of (10.2.4).] Thus (9.11.8) holds. Moreover, since $A|R(z)|^\theta$ and $B|R(z)|^{1-\theta}$ are bounded operators everywhere defined for $\operatorname{Im} z \neq 0$, we see that the same is true of $Q(z)$. Hence we may apply Theorem 9.11.3 to conclude that $G(z) = G_0(z)^{-1}$ is bounded and everywhere defined for $\operatorname{Im} z \neq 0$. [Note that $D(B^*)$ is dense.] Set

$$K(z) = (z - z_1)G_0(z_1)^{-1} A R_0(z)[BR_0(\bar{z}_1)]^*, \tag{10.3.8}$$

where z_1 is the point mentioned in the hypothesis of Theorem 10.2.2. If z_1 is real, we use the trick employed in the proof of Theorem 9.11.1. By hypothesis, $K(z)$ is compact for $\operatorname{Im} z \neq 0$. Moreover,

$$G_0(z) = G_0(z_1)[1 + K(z)]. \tag{10.3.9}$$

Thus, $K(s \pm ia) \to K_\pm(s)$ in norm as $a \to 0$ uniformly on intervals $I \subset\subset \Gamma$. Thus, $K_\pm(s)$ is compact for each $s \in \Gamma$ (Theorem 9.7.3) and depends continuously on s (the uniform limit of continuous functions). By (10.3.9),

$$G_{0\pm}(s) = G_0(z_1)[1 + K_\pm(s)]. \tag{10.3.10}$$

Hence

$$g_k = [K_\pm(\lambda) - K_\pm(\lambda_k)]g_k - K_\pm(\lambda)g_k.$$

Since the g_k converge weakly and the $K_\pm(s)$ are compact and depend continuously on s, we see that the g_k converge strongly (Theorem 9.7.2). Let $\varepsilon > 0$ be given and let I be an interval with center λ containing the λ_k in its interior and such that $C(\lambda_k)\tau_i(|I|) < \varepsilon$ for all k [this is possible because $C(s)$ is locally bounded]. Then take the λ_k so close to λ that $C(\lambda_k)\tau_2(|\lambda - \lambda_k|/|I|) < \varepsilon$ for all k. Then

$$\begin{aligned}\|u_j - u_k\| \le\ & \|u_j - [BR_0(\lambda_j \pm ia)]^* g_j\| \\ & + \|[BR_0(\lambda_j \pm ia)]^* g_j - [BR_0(\lambda_k \pm ia)]^* g_k\| \\ & + \|[BR_0(\lambda_k \pm ia)]^* g_k - u_k\|.\end{aligned}$$

The middle term on the right is bounded by

$$\begin{aligned}&\|[BE_0(I)R_0(\lambda_j \pm ia)]^* g_j\| + \|[BE_0(I)R_0(\lambda_k \pm ia)]^* g_k\| \\ &+ \|[BE_0(CI)R_0(\lambda_j \pm ia)]^* g_j - [BE_0(CI)R_0(\lambda_k \pm ia)]^* g_k\|\end{aligned}$$

Letting $a \to 0$ we find that

$$\|u_j - u_k\| \le 4\varepsilon + \|[BE_0(CI)R_0(\lambda_j)]^* g_j - [BE_0(CI)R_0(\lambda_k)]^* g_k\|.$$

This shows that the u_k form a Cauchy sequence. But they are orthonormal, being the eigenelements of a self-adjoint operator corresponding to different eigenvectors. In fact we have

$$(\lambda_j - \lambda_k)(u_j, u_k) = (Hu_j, u_k) - (u_j, Hu_k) = 0$$

This contradiction gives the lemma. □

Now we can give the

PROOF OF THEOREM 10.3.2. Let $e_\pm$ be the set of those $s \in \Gamma$ such that $N[G_{0\pm}(s)] \neq \{0\}$, and set $e = e_+ \cup e_-$. By Lemma 10.3.2, $e_\pm$ has no limit point in Γ. Thus the same is true of e. If we set $\Lambda = \Gamma - e$, then $C\Lambda = C\Gamma \cup e$ is denumerable. Moreover, (10.3.10) and the fact that the $K_\pm(s)$ are compact show us that $G_{0\pm}(s)$ has a bounded inverse for $s \in \Lambda$ (Lemma 9.7.4). Hence the same is true of $G_0(z)$ for $z, \bar{z} \in \omega_I$, $I \subset\subset \Lambda$ (Lemma 9.6.1). Now it is easily checked that

$$Q_0(z) = [BR_0(\bar{z})A^*]^* = (B[AR_0(z)]^*)^* \tag{10.3.11}$$

(see Lemma 9.11.1). Thus

$$G_0(\bar{z})^* BR(z) = BR_0(z) \tag{10.3.12}$$

Since $G_0(z)$ has a bounded inverse for $z, \bar{z} \in \omega_I$ when $I \subset\subset \Lambda$, the same is true of $G_0(z)^*$. Thus (10.2.5) follows from (10.2.8) and we can apply Theorem 10.2.1. □

One should note that A and B are interchanged in the definition of $Q_0(z)$ as compared with Chapter 9.

10.4. Verification for the Hamiltonian

Now we return to our Hamiltonian H. We shall assume that

$$\int_{-\infty}^{\infty} |V(x)|\, dx < \infty \tag{10.4.1}$$

as we did in Chapter 9. Thus all of the theorems of that chapter will hold. In particular, the wave operators are complete. We define the Hamiltonian and we have

$$D(|H|^{1/2}) = D(|H_0|^{1/2}). \tag{10.4.2}$$

This suggests that we take $\theta = \frac{1}{2}$ in Theorem 10.2.2. If we do this and take $A(x) = |V(x)|^{1/2}$, $B(x) = V(x)/A(x)$ when $V(x) \neq 0$, $A(x) = e^{-x^2}$, $B(x) = 0$ when $V(x) = 0$, and $\Gamma = R - \{0\}$ (the real numbers minus the origin), we find that all of the hypotheses except (10.2.9) and (10.2.10) are verified (see the lemmas of Section 9.14). The $G_0(z)$ are uniformly continuous in each ω_I with $I \subset\subset \Gamma$ and consequently can be extended to be uniformly continuous in $\bar{\omega}_I$. Thus it remains to find assumptions on V so that (10.2.9) and (10.2.10) will hold.

For any $f \in L^2$ the Fourier transform of $R_0(z)f$ is $(z - k^2)^{-1}\hat{f}(k)$. Thus by (9.12.8),

$$([BE_0(I)R_0(z)]^*g, f) = (E_0(I)Bg, R_0(z)f)$$
$$= \int_{k^2 \in I} (\bar{z} - k^2)^{-1}\widehat{Bg}(k)\hat{f}(k)^* \, dk. \quad (10.4.3)$$

Hence

$$\|[BE_0(I)R_0(z)]^*g\|^2 = \int_{k^2 \in I} |z - k^2|^{-2}|\widehat{Bg}(k)|^2 \, dk. \quad (10.4.4)$$

If $z = s + it$, we want this to be bounded independently of t and tend to 0 as $|I| \to 0$ when $g \in N[G_{0\pm}(s)]$. If $s < 0$, then I must be contained in the negative real axis and the right-hand side of (10.4.4) vanishes. Thus we only need consider the case $s > 0$. In that case the right-hand side of (10.4.4) cannot be expected to be bounded as $t \to 0$ unless $\widehat{Bg}(k)$ vanishes in some sense when $k = \pm s^{1/2}$. That this is indeed true follows from

Lemma 10.4.1. *Suppose $s > 0$ and $h(k)$ is a bounded function continuous at $k = s^{1/2}$ and at $k = -s^{1/2}$. Then*

$$a\int_{-\infty}^{\infty} \frac{h(k)\, dk}{(k^2 - s)^2 + a^2} \to \tfrac{1}{2}\pi s^{-1/2}[h(s^{1/2}) + h(-s^{1/2})] \qquad \text{as} \quad a \to 0.$$
$$(10.4.5)$$

PROOF. Split the integral up into the sum of the integrals over $(-\infty, 0)$ and $(0, \infty)$. In each of these introduce the variable $t = k^2$. Then the left-hand side of (10.4.5) becomes

$$\tfrac{1}{2}a\int_0^{\infty} [h(t^{1/2}) + h(-t^{1/2})][(t - s)^2 + a^2]^{-1}t^{-1/2}\, dt.$$

Let $b > 0$ be such that $b < s$. The integrals over the intervals $(0, s - b)$ and $(s + b, \infty)$ converge to 0 because $t - s$ is bounded away from zero in these intervals. In the interval $(s - b, s + b)$ introduce the variable $y = (t - s)/a$. The integral becomes

$$\tfrac{1}{2}\int_{-b/a}^{b/a} [h(\sqrt{ay + s}\,) - h(-\sqrt{ay + s}\,)](y^2 + 1)^{-1}(ay + s)^{-1/2}\, dy.$$

Since $ay + s \geq s - b > 0$, the integrand is bounded by a constant divided by $y^2 + 1$. Moreover, it converges to

$$\tfrac{1}{2}[h(s^{1/2}) + h(-s^{1/2})]s^{-1/2}(y^2 + 1)^{-1}$$

for each y. The Lebesgue dominated convergence theorem now gives the result. □

From this we get

Lemma 10.4.2. *Suppose* $G_{0+}(s)g = 0$, $s > 0$, *and* $\hat{Bg}(k)$ *is bounded and continuous at* $k = s^{1/2}$ *and* $k = -s^{1/2}$. *Then* $\hat{Bg}(\pm s^{1/2}) = 0$. *A similar statement holds for* $G_{0-}(s)$.

PROOF. If $G_{0+}(s)g = 0$, then

$$AR_0(s + ia)Bg \to g \qquad \text{in } L^2 \quad \text{as } a \to 0. \tag{10.4.6}$$

Set $C(x) = \operatorname{sgn} V(x) = V(x)/|V(x)|$ when $V(x) \neq 0$ and vanishing when $V(x) = 0$. Then $B = CA$ and (10.4.6) implies

$$BR_0(s \pm ia)Bg \to Cg.$$

Consequently,

$$\int_{-\infty}^{\infty} (s + ia - k^2)^{-1} |\hat{Bg}(k)|^2 \, dk = (BR_0(s + ia)Bg, g) \to (Cg, g).$$

Taking imaginary parts, we have

$$a \int_{-\infty}^{\infty} \left[(s - k^2)^2 + a^2 \right]^{-1} |\hat{Bg}(k)|^2 \, dk \to 0.$$

since C is real valued. In view of Lemma 10.4.1 this gives

$$|\hat{Bg}(s^{1/2})|^2 + |\hat{Bg}(-s^{1/2})|^2 = 0,$$

the desired result. □

The vanishing of $\hat{Bg}$ at the two points is necessary for (10.4.4) to be bounded as $t = \operatorname{Im} z \to 0$, but it is not sufficient. We must therefore look for a sufficient condition. One such condition can be expressed in terms of Hölder continuity.

A function $u(x)$ defined on an open set Ω is said to be (locally) *Hölder continuous* in Ω if, for each $x_0 \in \Omega$, it satisfies an inequality of the form

$$|u(x) - u(x_0)| \leq C|x - x_0|^\alpha$$

for x near x_0, where $0 < \alpha \leq 1$. The constants C and α may depend on x_0. If $\alpha = 1$, $u(x)$ is said to be *Lipschitz continuous*. The constant α is called the Hölder exponent at x_0. Clearly a function which is Hölder continuous is continuous. We shall use

Lemma 10.4.3. *If* $\hat{Bg}(k)$ *is bounded and Hölder continuous in* Γ *with exponent* $> \frac{1}{2}$, *then* (10.2.9) *holds for each* $g \in N[G_{0\pm}(s)]$.

PROOF. Let s be any positive number, and let $I = (a, b)$ be any interval with $a > 0$. Set $h = \hat{Bg}$. Then $h(s^{1/2}) = h(-s^{1/2}) = 0$ by Lemma 10.4.2 and there is an $\alpha > \frac{1}{2}$ such that

$$|h(k)| = |h(k) - h(\pm s^{1/2})| \leq C|k \mp s^{1/2}|^\alpha.$$

The right-hand side of (10.4.4) consists of two integrals, one over $(a^{1/2}, b^{1/2})$ and the other over $(-b^{1/2}, -a^{1/2})$. The former is bounded by

$$C\int_{a^{1/2}}^{b^{1/2}} |s^{1/2}+k|^{-2}|s^{1/2}-k|^{2\alpha-2}\,dk, \tag{10.4.7}$$

which exists since $\alpha > \frac{1}{2}$. Moreover, it tends to 0 as $|I| \to 0$. A similar estimate holds for the other integral. Thus (10.2.9) holds. □

In view of Lemma 10.4.3, our next step should be to find conditions on $V(x)$ which will imply that $\hat{B}g$ is bounded and Hölder continuous. For this purpose we shall use

Lemma 10.4.4. *If $W(x)$, $w(x)$ are in L^2, then $h = \widehat{Ww}$ is bounded and satisfies*

$$|h(k)| \le (2\pi)^{-1/2}\|W\|\,\|w\|. \tag{10.4.8}$$

If $W\rho^\beta$ and w are in L^2 for some $\beta > 0$, then h is Hölder continuous with exponent β and satisfies

$$|h(k) - h(k')| \le C|k-k'|^\beta\|W\rho^\beta\|\,\|w\| \tag{10.4.9}$$

PROOF. Inequality (10.4.8) follows from

$$|h(k)| \le (2\pi)^{-1/2}\int |W(x)w(x)|\,dx.$$

We also have

$$|h(k)-h(t)| \le (2\pi)^{-1/2}\int |e^{-ikx}-e^{-itx}|\,|W(x)w(x)|\,dx$$

$$\le C|k-t|^\beta \int |x|^\beta |W(x)w(x)|\,dx,$$

which implies (10.4.9). □

Combining Lemmas 10.4.3 and 10.4.4 we have

Lemma 10.4.5. *If*

$$C_\alpha = \int |V(x)|\rho(x)^\alpha\,dx < \infty \tag{10.4.10}$$

holds for some $\alpha > 1$, then (10.2.9) holds.

PROOF. If we take $\beta = \frac{1}{2}\alpha$, then $B\rho^\beta$ is in L^2. Consequently, $\hat{B}g$ is bounded and Hölder continuous with exponent β (Lemma 10.4.4). Since $\beta > \frac{1}{2}$, we can apply Lemma 10.4.3 to obtain the desired conclusion. □

Now we turn our attention to (10.2.10). For this purpose we shall need

Lemma 10.4.6. *If $(H-s)v = 0$, then $G_{0\pm}(s)\,Av = 0$.*

PROOF. By (10.2.4), we have

$$([s - H_0]u, v) = (Bu, Av), \qquad u \in D(H_0). \tag{10.4.11}$$

Thus, if $z = s \pm ia$,

$$([s - H_0]R_0(\bar{z})Aw, v) = (BR_0(\bar{z})Aw, Av), \qquad w \in D(A).$$

Consequently,

$$([1 \pm iaR_0(\bar{z})]Aw, v) = (Q_0(z)^* w, Av)$$

and

$$(Aw, [1 \mp iaR_0(z)]v) = (w, Q_0(z)Av).$$

Now, by (10.2.8),

$$a^2\|AR_0(z)\|^2 \le aC_I \to 0 \qquad \text{as} \quad a \to 0,$$

where I is any interval containing s in its interior. Hence

$$(Aw, v) = (w, Q_{0\pm}(s)Av), \qquad w \in D(A).$$

This gives the lemma. □

Next suppose $V(x)$ satisfies (10.4.10) for some $\alpha > 1$, $s \in I \subset\subset \Gamma$, and $(H - s)u = 0$. Set $g = Au$. Then $G_{0\pm}(s)g = 0$ by Lemma 10.4.6. Now,

$$\|[BE_0(I)R_0(s + it)]^* g\|^2 = \int_{k^2 \in I} |s + it - k^2|^{-2} |\hat{B}g(k)|^2 \, dk. \tag{10.4.12}$$

By (10.4.9),

$$|\hat{B}g(k)| \le C|k \pm s^{1/2}|^{\alpha/2}\|g\|, \tag{10.4.13}$$

where C depends only on C_α. If $I = (a, b)$ with $a > 0$, then the right-hand side of (10.4.12) is the sum of

$$\int_{a^{1/2}}^{b^{1/2}} |s + it - k^2|^{-2} |\hat{B}g(k)|^2 \, dk \tag{10.4.14}$$

and a similar integral over $(-b^{1/2}, -a^{1/2})$. By (10.4.13) this integral is bounded by

$$C^2\|g\|^2 \int_{a^{1/2}}^{b^{1/2}} |s^{1/2} + k|^{-2} |s^{1/2} - k|^{\alpha - 2} \, dk$$

$$\le C^2\|g\|^2 s^{-1/2}(\alpha - 1)^{-1}\left[(b^{1/2} - s^{1/2})^{\alpha - 1} + (s^{1/2} - a^{1/2})^{\alpha - 1}\right]$$

$$\le C' s^{-1/2}(\alpha - 1)^{-1}(b - a)^{\alpha - 1}\|g\|^2. \tag{10.4.15}$$

The same estimate holds for the other integral. Since

$$g = (i + s^{1/2})A(i + |H|^{1/2})^{-1}u, \tag{10.4.16}$$

there is a constant independent of s, u such that

$$\|g\| \le C(1+s)^{1/2}\|u\|. \tag{10.4.17}$$

Thus (10.2.10) holds with

$$\tau_1(\delta) = \delta^{\alpha-1}, \qquad \tau_2(\delta) = 0,$$

and

$$C(s) = C''(1+s)^{1/2}s^{-1/2}.$$

Hence we have proved

Lemma 10.4.7. *If V satisfies* (10.4.10) *for some $\alpha > 1$, then* (10.2.10) *holds.*

Combining all of the results so far we have

Theorem 10.4.1. *If* (10.4.10) *holds for some $\alpha > 1$, then the wave operators are strongly complete.*

10.5. An Extension

In this section we shall show how to weaken the hypothesis of Theorem 10.4.1. We shall prove

Theorem 10.5.1. *Assume that*

$$\int |V(x)|^p \rho(x)^\alpha \, dx < \infty \tag{10.5.1}$$

for some α, p such that

$$1 \le p \le 2, \qquad \alpha > 1. \tag{10.5.2}$$

The the wave operators are strongly complete.

In proving this theorem we shall use the L^p norms given by

$$\|f\|_{p,I} = \left(\int_I |f(x)|^p \, dx\right)^{1/p}, \qquad 1 \le p < \infty, \tag{10.5.3}$$

$$\|f\|_{\infty,I} = \sup_I |f(x)|, \tag{10.5.4}$$

where I is any subset of the real line. The set of those f for which $\|f\|_{p,I}$ is finite is denoted by $L^p(I)$. For $1 \le p \le \infty$ we set $p' = p/(p-1)$, $\infty' = 1$, and $\|f\|_p = \|f\|_{p,R}$.

An important property of the Fourier transform is given by the following theorem proved in Appendix A.

Theorem 10.5.2. *If* $1 \le p \le 2$, *then*

$$\|\hat{f}\|_{p'} \le (2\pi)^{(1/2)-1/p}\|f\|_p. \tag{10.5.5}$$

We shall be using the following lemmas.

Lemma 10.5.1. *If* $A\rho^\beta \in Lr$, $q \le r$, *and* $\gamma + q^{-1} < \beta + r^{-1}$, *then* $A\rho^\gamma$ *is in* L^q.

PROOF. If $r < \infty$, we have

$$\int |A(x)\rho(x)^\gamma|^q\, dx \le \left(\int |A(x)|^{tq}\rho(x)^{\beta r}\, dx\right)^{1/t} \times \left(\int \rho(x)^{(t\gamma q - \beta r)t'/t}\, dx\right)^{1/t'} \tag{10.5.6}$$

by Hölder's inequality [see (7.3.27)]. We take $t = r/q$ and verify that the last integral is finite because the power of ρ is < -1. The case $r = \infty$ is left as an exercise. □

Lemma 10.5.2. *If*

$$A\rho^\beta \in L^r \tag{10.5.7}$$

for some β, r *satisfying*

$$r \le 2, \qquad \beta + r^{-1} > \tfrac{1}{2}, \tag{10.5.8}$$

and $w \in L^2$, *then* $(A^q w)\hat{}$ *is Hölder continuous with any exponent* $< q(\beta + r^{-1}) - \frac{1}{2}$ *provided* $1 \le q \le \frac{1}{2}r$.

PROOF. Set $h(k) = (A^q w)\hat{}$. Then by 10.4.9

$$|h(k) - h(k')| \le C|k - k'|^\tau \|A^q\rho^\tau\|\, \|w\| \tag{10.5.9}$$

for any $\tau \ge 0$. Now Lemma 10.5.1 tells us that $A\rho^{\tau/q} \in L^{2q}$ if $(\tau/q) + (q/2) < \beta + (1/r)$. This is the same as saying that $A^q\rho^\tau$ is in L^2. Hence $h(k)$ is Hölder continuous with exponent τ under these conditions. □

Lemma 10.5.3. *If* L *is a one-to-one self-adjoint operator, then* L^{-1} *with* $D(L^{-1}) = R(L)$ *is self-adjoint.*

PROOF. Set $M = L^{-1}$ with $D(M) = R(L)$. Suppose u, f are such that

$$(u, Mv) = (f, v), \qquad v \in D(M).$$

Then

$$(u, w) = (f, Lw), \qquad w \in D(L).$$

Since L is self-adjoint, this implies that $f \in D(L)$ and $Lf = u$. Consequently, $u \in D(M)$ and $Mu = f$. □

For any number θ we define A^θ to be the operator of multiplication by the function $A(x)^\theta$. If $0 \le \theta \le 1$, then $A(x)^\theta$ is locally in L^2 [since $A(x)$ is], and consequently it is self-adjoint (Lemma 9.14.5). In view of Lemma 10.5.3, $A^{-\theta}$ is self-adjoint as well. We define an operator T by

$$(T\varphi, w) = \int_I \varphi(k)\widehat{Aw}(k)^*\,dk \tag{10.5.10}$$

and prove

Lemma 10.5.4. *If $A \in L^r$ and*

$$1 \le p \le 2, \qquad \theta = 1 - r\left(p^{-1} - \tfrac{1}{2}\right), \tag{10.5.11}$$

then T maps $L^p(I)$ into $R(A^\theta)$.

PROOF. By Theorem 10.5.2 we have

$$\begin{aligned}|(T\varphi, A^{-\theta}w)| &\le \|\varphi\|_{p,I}\|(A^{1-\theta}w)^\wedge\|_{p'}\\ &\le C\|\varphi\|_{p,I}\|A^{1-\theta}w\|_p.\end{aligned}$$

Moreover, by Hölder's inequality

$$\|A^{1-\theta}w\|_p \le \|A^{1-\theta}\|_t\|w\|, \qquad t^{-1} + \tfrac{1}{2} = p^{-1}. \tag{10.5.12}$$

If we take $t = r/(1-\theta)$, we have

$$|(T\varphi, A^{-\theta}w)| \le C\|\varphi\|_{p,I}\|w\|, \qquad w \in L^2. \tag{10.5.13}$$

By the Riesz representation theorem (Theorem 1.9.1) for each $\varphi \in L^p(I)$ there is an $f \in L^2$ such that

$$(T\varphi, A^{-\theta}w) = (f, w), \qquad w \in L^2. \tag{10.5.14}$$

In particular, (10.5.14) holds for all w in $D(A^{-\theta})$. Since $A^{-\theta}$ is self-adjoint, it follows that $T\varphi \in D(A^{-\theta}) = R(A^\theta)$. □

Lemma 10.5.5. *Suppose* (10.5.7) *holds for some β, r satisfying* (10.5.8) *and $\beta r > 1$. Then for each $s \ne 0$ we have $N(G_{0\pm}(s)) \subset R(A)$.*

PROOF. If $s < 0$, then $R_0(s)$ exists since $s \in \rho(\hat{H}_0)$. Consequently, $Q_0(s)$ maps L^2 into $R(A)$. Thus the conclusion certainly holds in this case. Now suppose $s > 0$ and let I be an interval such that $I \subset\subset (0, \infty)$ with s in its interior. Then

$$\begin{aligned}BE_0(CI)R_0(z) = BR_0(i) - BR_0(i)E_0(I) + (i-z)BR_0(i)E_0(CI)R_0(i)\\ + (i-z)^2BR_0(i)E_0(CI)R_0(z)R_0(i),\end{aligned} \tag{10.5.15}$$

where $z = s \pm ia$. This shows that the operator $[BE_0(CI)R_0(s)]^*$ maps L^2 into $D(A)$ (note that this operator exists because s is an interior point of I).

Set

$$Q_{0\pm I}(s)v = Q_{0\pm}(s)v - A[BE_0(CI)R_0(s)]^*v$$
$$= \lim_{0<a\to 0} A[BE_0(I)R_0(s \pm ia)]^*v. \qquad (10.5.16)$$

Let g be any function in $N[G_{0\pm}(s)]$, and set

$$\varphi(k) = \hat{B}g(k)/(s - k^2). \qquad (10.5.17)$$

Then by (10.4.3)

$$(Q_{0\pm I}(s)g, w) = (T\varphi, w), \qquad w \in L^2, \qquad (10.5.18)$$

where T is the operator defined by (10.5.10). Now suppose $g \in R(A^{\nu})$ for some ν satisfying $0 \leq \nu \leq \frac{1}{2}r - 1$. Then $g = A^{\nu}h$, where $h \in L^2$. Thus $Bg = BA^{\nu}h$ is Hölder continuous with any exponent $< (1 + \nu)(\beta + r^{-1}) - \frac{1}{2}$ by Lemma 10.5.2. This implies that $\varphi(k)$ given by (10.5.17) is in $L^p(I)$ for any $p \leq 2$ satisfying

$$1 - p^{-1} < (1 + \nu)(\beta + r^{-1}) - \tfrac{1}{2}. \qquad (10.5.19)$$

Now (10.5.7) implies that $A \in L^{\sigma}$ for any σ satisfying

$$\sigma^{-1} < \beta + r^{-1} \qquad (10.5.20)$$

(Lemma 10.5.1). This in turn implies that $T\varphi \in R(A^{\theta})$ for

$$\theta = 1 - \sigma\left(p^{-1} - \tfrac{1}{2}\right) \qquad (10.5.21)$$

(Lemma 10.5.4). If the right-hand side of (10.5.19) is $> \frac{1}{2}$, then we can take $p = 1$ in (10.5.21) giving $\theta = 1$. Thus $T\varphi \in R(A)$. By (10.5.16) and (10.5.18), we see that $g = Q_{0\pm}(s)g \in R(A)$, the desired result. On the other hand, if the right-hand side of (10.5.19) is $\leq \frac{1}{2}$, we can still take θ to be any number satisfying

$$\theta < 2 + \nu - (\beta + r^{-1})^{-1} \qquad (10.5.22)$$

by (10.5.19)–(10.5.21). By (10.5.8), we see that $\theta - \nu$ can be taken greater than a positive number depending only on β and r. By Lemma 10.5.4, $T\varphi \in R(A^{\theta})$. If $\theta < \frac{1}{2}r - 1$, we can make θ our new ν and reapply the same argument to obtain a larger θ. We can repeat this process until $\nu \geq \frac{1}{2}r - 1$. Applying the argument one final time we get $T\varphi \in R(A^{\theta})$ for any θ satisfying

$$\theta < 1 + \tfrac{1}{2}r - (\beta + r^{-1})^{-1}. \qquad (10.5.23)$$

Now we use the assumption $\beta r > 1$ to allow us to take $\theta = 1$ in (10.5.23). Hence $T\varphi \in R(A)$ and consequently $g \in R(A)$ as before. □

Lemma 10.5.6. *If w, $Aw \in L^2$, and $0 < \nu < 1$, then $A^{\nu}w \in L^2$.*

PROOF. Set $q = 1/\nu$. Then by Hölder's inequality,

$$\int |A(x)^\nu u(x)|^2\,dx = \int |A(x)u(x)|^{2\nu}|u(x)|^{2-2\nu}\,dx$$

$$\le \left(\int |A(x)u(x)|^{2\nu q}\,dx\right)^\nu \left(\int |u(x)|^{2q'(1-\nu)}\,dx\right)^{1-2}$$

$$= \|Au\|^{2\nu}\|u\|^{2-2\nu}. \quad \square \tag{10.5.24}$$

Now we can give the

PROOF OF THEOREM 10.5.1. Assumption (10.5.1) is equivalent to $A\rho^\beta \in L^r$ with $2 \le r \le 4$ and $\beta r > 1$. Thus (10.5.8) is satisfied. Let g be any element in $N[G_{0\pm}(s)]$. Then $g = Au$ for some $u \in D(A)$ by Lemma 10.5.5. Set $p = \frac{1}{2}r$ and $w = A^{2-p}u$. Then $w \in L^2$ by Lemma 10.5.6. Moreover, by (10.4.17) and (10.5.24), we have

$$\|w\| \le C(1+s)^{\nu/2}\|u\|. \tag{10.5.25}$$

By Lemma 10.5.2, $(A^p w)\hat{}$ is Hölder continuous with exponent $> \frac{1}{2}$. Hence the same is true of $\hat{B}g$. If we use Lemma 10.4.2 and (10.5.9), we get

$$|\hat{B}g(k)| \le C|k \pm s^{1/2}|^\tau \|w\| \tag{10.5.26}$$

for some $\tau > \frac{1}{2}$. Reasoning as in (10.4.12), (10.4.14), and (10.4.15), we see that (10.2.9) holds. The same reasoning works if we start with $u \in N(H - s)$ and we use Lemma 10.4.6. Thus we see that (10.2.10) holds as well. The conclusion follows from Theorem 10.2.2. $\square$

A useful corollary of Theorem 10.5.1 is

Corollary 10.5.1. *If $V\rho^\alpha \in L^p$ with*

$$2 < p \le \infty, \qquad \alpha > 1 - p^{-1}, \tag{10.5.27}$$

then the wave operators are strongly complete.

PROOF. The hypothesis is equivalent to $A\rho^{\alpha/2} \in L^{2p}$. This implies $A\rho^\beta \in L^4$ for some $\beta > \frac{1}{4}$ (Lemma 10.5.1). Now we can apply Theorem 10.5.1. $\square$

Combining Theorem 10.5.1 and Corollary 10.5.1 we have

Theorem 10.5.3. *If $V\rho^\alpha \in L^p$ for some α, p satisfying*

$$1 \le p \le \infty, \qquad \alpha > \max(p^{-1}, 1 - p^{-1}),$$

then the wave operators are strongly complete.

10.6. The Principle of Limiting Absorption

We have actually proved more than what Theorems 10.2.2 and 10.5.3 state. In fact we have

Theorem 10.6.1. *Under the hypotheses of Theorem* 10.2.2, *let* $s \in \Gamma$ *be a point which is not an eigenvalue of* H. *If* $v \in R(B^*)$, *set* $u(z) = R(z)v$, *where* $z = s \pm ia$. *Then* $u(z) \in D(A)$ *and* $Au(z) \to h_{\pm}(s)$ *in* $\mathcal{H}$ *as* $a \to 0$, *where* $h_{\pm}(s)$ *depends continuously on* s.

PROOF. That $u(z) \in D(A)$ is known from the fact that $D(H_0) \subset D_\theta \subset D(A)$. Now

$$G_0(z)AR(z) = AR_0(z) \tag{10.6.1}$$

[see (9.6.2)]. Thus, if $v = B^*w$, we have

$$Au(z) = G_0(z)^{-1}Q_0(z)w.$$

Since s is not an eigenvalue of H, we have $N[G_{0\pm}(s)] = \{0\}$ (Lemma 10.3.1). Hence $G_0(z)^{-1}$ is bounded and continuous in a neighborhood of s (Lemma 9.6.1). Since the same is true of $Q_0(z)$, the result follows. □

Theorem 10.6.2. *Let* A_1, B_1 *be closed operators from* $\mathcal{H}$ *to* $\mathcal{H}$ *such that* $D_\theta \subset D(A_1)$, $D_{1-\theta} \subset D(B_1)$ *and the operators* $A_1[BR_0(\bar{z})]^*$, $A[B_1R(\bar{z})]^*$, $A_1[B_1R_0(\bar{z})]^*$ *satisfy the hypotheses of* $Q_0(z)$ *in Theorem* 10.2.2. *If* $s \in \Gamma$ *is not an eigenvalue of* H, *then the operators* $A_1[B_1R_0(s \pm ia)]^*$ *converge in norm to limits which depend continuously on* s.

PROOF. By (10.6.1) and Lemma 8.3.1

$$R(z) = R_0(z) + \left[BR_0(\bar{z})\right]^* G_0(z)^{-1}AR_0(z).$$

Apply A_1 on the left and B_1 on the right. Since $G_0(z)$ is invertible in the neighborhood of s, the result follows. □

Theorems 10.6.1 and 10.6.2 are forms of the limiting absorption principle, which states that under certain conditions $A_1[B_1R(z)]^*$ can have boundary values as z approaches the real axis even though $R(z)$ itself may not. In the case of Theorem 10.5.3 we can prove

Theorem 10.6.3. *Under the hypotheses of Theorem* 10.5.3 *we have*

a. $\mathcal{H}_c(H) = \mathcal{H}_{ac}(H) = R_{\pm}$,
b. $\mathcal{H}_s(H) = \mathcal{H}_p(H)$, *the closure of the set of eigenvectors, and*
c. *the eigenvalues of* H *are isolated with* 0, ∞ *as the only possible limit points.*

PROOF. In applying Theorem 10.2.2 we took $\Gamma = R - \{0\}$. By Lemmas 10.4.6 and 10.3.2, the eigenvalues of H have no limit points in Γ. Since H is bounded from below, $-\infty$ cannot be a limit point for the eigenvalues of H. This proves (c). To prove (a) note that

$$G_0(\bar{z})^* BR(z) = BR_0(z). \tag{10.6.2}$$

Let v be any function in $D(B)$ and let I be any bounded interval not containing the origin or any eigenvalue of H. Then

$$\int_I \|R(z)Bv\|^2\,ds = \int_I \left\|\left[BR_0(\bar{z})\right]^* G_0(z)^{-1} v\right\|^2 ds. \tag{10.6.3}$$

Set $h(z) = [BE_0(I)]^* G_0(z)^{-1} v$. Then

$$\begin{aligned} a\int_I \left\|\left[BR_0(\bar{z})E_0(I)\right]^* G_0(z)^{-1} v\right\|^2 ds &= a\int_I \|E_0(I)R_0(z)h(z)\|^2\,ds \\ &= \pi \int_I \int_{k^2\in I} \delta_a(s-k^2)|(h(z))\hat{\ }(k)|^2\,dk\,ds \\ &\to \pi \int_{k^2\in I} |(h(k^2))\hat{\ }(k)|^2\,dk \qquad \text{as} \quad a\to 0 \end{aligned} \tag{10.6.4}$$

by (9.12.8). On the other hand,

$$a\int_I \left\|\left[BR_0(\bar{z})E_0(CI)\right]^* G_0(z)^{-1} v\right\|^2 ds \to 0 \qquad \text{as} \quad a \to 0$$

since the integral is bounded. Thus by Lemma 8.2.1, we have in view of (10.6.3)

$$\|\tilde{E}(I)Bv\|^2 = \int_{k^2\in I} |(h(k^2))\hat{\ }(k)|^2\,dk. \tag{10.6.5}$$

Thus, if I is a union of such intervals which do not intersect, we have

$$\|E(\Lambda)Bv\| \to 0 \qquad \text{as} \quad |\Lambda| \to 0.$$

Since $R(B)$ is dense, we have

$$\|E(\Lambda)f\| \to 0 \qquad \text{as} \quad |\Lambda| \to 0 \tag{10.6.6}$$

for any f. Since the eigenvalues of H are isolated, we see that (10.6.6) holds for any $f \in \mathcal{H}_c(H)$ and any union Λ of nonoverlapping open intervals. Thus we have shown that $\mathcal{H}_c(H) \subset \mathcal{H}_{ac}(H)$. Since the opposite inclusion is always the case, we see that the two subspaces coincide. They are both contained in $R_\pm$ by Theorem 10.5.3. On the other hand, H_0 has no eigenvalues, so that the $R_\pm$ cannot contain any other elements. This proves (a). Finally, we note that (b) follows from (a) since $\mathcal{H}_s(H) = \mathcal{H}_{ac}(H)^\perp$ and $\mathcal{H}_p(H) = \mathcal{H}_c(H)^\perp$ by definition. □

Exercises

1. Prove (10.3.3) and (10.3.4).

2. Prove (10.3.5).

3. Show that the g_k given by (10.3.6) have uniformly bounded norms.

4. Prove (10.3.9).

5. Show that $Q_0(z)$ in Theorem 10.2.2 is bounded for Im $z \neq 0$.

6. Prove (10.3.11) and (10.3.12).

7. Carry out the details of the proof of (10.4.5).

8. Prove (10.4.9).

9. Prove (10.4.13).

10. Prove (10.4.15) for the other interval.

11. Prove (10.4.16) and (10.4.17).

12. Show that the expressions (10.5.3) and (10.5.4) are norms.

13. Verify that the last integral in (10.5.6) is finite.

14. Prove (10.5.12).

15. Prove (10.5.15) and show how it implies
$$R([BE_0(CI)R_0(s)]^*) \subset D(A).$$

16. If $\varphi(k)$ is given by (10.5.17) and $\hat{B}g(k)$ is Hölder continuous with exponent τ in I, show that $\varphi \in L^p(I)$ for any p satisfying $1 - p^{-1} < \tau$.

17. Prove (10.5.24).

18. Show that $\theta = 1$ is a solution of (10.5.25) if $\beta r > 1$.

19. Prove Lemma 10.5.1 when $r = \infty$.

20. Prove (10.6.1) and (10.6.2).

21. Prove (10.6.4).

22. Show that (10.6.6) holds for any $f \in \mathcal{H}_c(H)$ and any union Λ of nonintersecting open intervals.

11

Oscillating Potentials

11.1. A Surprise

So far in our study of scattering theory for various potentials we have been able to show that the wave operators exist everywhere ($M_\pm = L^2$) and are complete if the potential $V(x)$ is in $L^1(-\infty, \infty)$ (Theorem 9.14.1). On the other hand, we showed in Chapter 7 that they do not exist ($M_\pm = \{0\}$) when $V(x) = c/|x|$. Since this potential just misses being in L^1, one might think that we have the complete picture more or less. This is true if one does not take into account any oscillations that the potentials may have. On the other hand, we are going to show that for potentials of the form

$$V(x) = q(x)e^x \sin e^x \tag{11.1.1}$$

the wave operators are complete if $q \in L^1 \cap L^2$ and $q' \in L^1$. Clearly such potentials need not satisfy any of the sufficient conditions we have given for the existence of the wave operators. The problem in dealing with potentials of the form (11.1.1) is how to take the oscillations into account. We shall show that this can be done by writing $V(x)$ in the form

$$V(x) = w'(x) + V_0(x). \tag{11.1.2}$$

For then we have

$$\begin{aligned}(w'u, v) &= ([wu]', v) - (wu', v) \\ &= -(wu, v') - (wu', v).\end{aligned} \tag{11.1.3}$$

This shows us that multiplication by $V(x)$ is equivalent to applying the first-order differential operator

$$M = i\,Dw - iw\,D + V_0, \tag{11.1.4}$$

where $D = d/i\,dx$. In other words, the problem is reduced to the study of scattering theory for the operator

$$H = H_0 + M, \tag{11.1.5}$$

where M is a Hermitian differential operator of order one. The advantage of replacing V by M is that in treating (11.1.5) we need only make hypotheses concerning w and not w'. On the other hand, the presence of the derivatives does not cause any great difficulty. In fact we shall see that most of the theorems already proved will apply to this situation as well.

We present the simple theory in the next section and then we shall examine various types of oscillating potentials. There will be even more surprises in store for us then.

As before we shall simplify the formulas by assuming $2m = \hbar = 1$.

11.2. The Hamiltonian

The first step in treating the problem is to define the Hamiltonian operator H precisely. This is not difficult. In fact we have

Theorem 11.2.1. *If*

$$\int_a^{a+1} (|w(x)|^2 + |V_0(x)|)\, dx \le C_1, \tag{11.2.1}$$

then for every $\varepsilon > 0$ there is a K such that

$$|(Mu, v)| \le \varepsilon \|u'\|^2 + K\|u\|^2. \tag{11.2.2}$$

PROOF. By Theorem 2.7.1,

$$\|wu\| + \|\, |V_0|^{1/2} u\| \le \varepsilon \|u'\| + K\|u\|.$$

Hence

$$\begin{aligned} |(wu, u')| &\le \varepsilon \|u'\|^2 + K\|u\|\, \|u'\| \\ &\le 2\varepsilon \|u'\|^2 + K\varepsilon^{-1}\|u\|^2 \end{aligned}$$

The result now follows from (11.1.3). □

Once we have Theorem 11.2.1 we have no difficulty defining a self-adjoint operator corresponding to (11.1.5). In fact if we set

$$h(u, v) = (u', v') + (Mu, v), \tag{11.2.3}$$

then we see that (4.2.10) holds. By Theorem 4.2.1, we see that the operator H associated with $h(u, v)$ is self-adjoint. It is the forms extension of (11.1.5), and we take it to be our Hamiltonian. The main theorem we are going to prove is

Theorem 11.2.2. *Assume that V is given by* (11.1.2) *with $w \in L^1 \cap L^2$ and $V_0 \in L^1$. Then the operator $H_0 + V$ has a forms extension H. The wave operators are complete.*

In proving Theorem 11.2.2 we shall make use of Theorem 9.11.2. Our first step is to find the operators A, B used there. For this purpose we shall use

Lemma 11.2.1. *A function $w(x)$ can be written in the form $w(x) = w_1(x)w_2(x)$, where $w_1 \in L^2$ and $w_2 \in L^2 \cap L^\infty$ iff $w \in L^1 \cap L^2$.*

PROOF. If w is the product of such functions, then it is in L^1 and in L^2 (being the product of a function in L^2 and a function in $L^2 \cap L^\infty$). Conversely, suppose $w \in L^1 \cap L^2$. Set

$$w_2 = \begin{cases} |w|^{1/2} & \text{when} \quad |w| \le 1 \\ 1 & \text{when} \quad |w| > 1 \end{cases}$$

and set $w_1 = w/w_2$. Clearly $w_2 \in L^\infty$. Since it satisfies $|w_2|^2 \le |w|$, it is also in L^2. On the other hand, $|w_1| \le |w| + |w|^{1/2} \in L^2$ by hypothesis. □

Now by (11.1.3) and Lemma 11.2.1 we have

$$(w'u, v) = -(w_1 u, w_2 v') - (w_2 u', w_1 v). \tag{11.2.4}$$

To define A and B, we let $\mathcal{K}$ be the direct sum of three copies of L^2. By this we mean that elements of $\mathcal{K}$ consist of vectors of the form $\{u_1, u_2, u_3\}$, $u_j \in L^2$, with

$$\|\{u_1, u_2, u_3\}\|^2 = \|u_1\|^2 + \|u_2\|^2 + \|u_3\|^2.$$

With this norm $\mathcal{K}$ becomes a Hilbert space. If we set

$$Au = \{w_1 u, -w_2 u', |V_0|^{1/2} u\}$$

and

$$Bv = \{-w_2 v', w_1 v, V_0 |V_0|^{-1/2} v\},$$

where w_1 and w_2 are determined by Lemma 11.2.1, we have

$$(Vu, v) = (Au, Bv)_{\mathcal{K}} = (Bu, Av)_{\mathcal{K}} \tag{11.2.5}$$

and

$$\|Au\|^2 + \|Bu\|^2 \le C(\|u'\|^2 + \|u\|^2). \tag{11.2.6}$$

This last inequality shows that

$$D(|H_0|^{1/2}) \subset D(A) \cap D(B). \tag{11.2.7}$$

Moreover, since H is a forms extension, we have $D(|H_0|^{1/2}) = D(|H|^{1/2})$. Next we note that

$$[AR_0(z)]^*\{u_1, u_2, u_3\} = [w_1 R_0(z)]^* u_1 - i[w_2 D R_0(z)]^* u_2 + [|V_0|^{1/2} R_0(z)]^* u_3. \tag{11.2.8}$$

Our next task is to consider the operator

$$Q_0(z) = \left[B(AR_0(\bar{z}))^* \right]. \tag{11.2.9}$$

We carry out the details in the next section.

11.3. The Estimates

The first thing we must show for the operator (11.2.9) is that it is bounded when Im $z \neq 0$. This involves the nine operators

a. $w_2\, D[w_1 R_0(z)]^*$
b. $w_2\, D[w_2\, DR_0(z)]^*$
c. $w_2\, D[|V_0|^{1/2} R_0(z)]^*$,
d. $w_1[w_1 R_0(z)]^*$,
e. $w_1[w_2\, DR_0(z)]^*$,
f. $w_1[|V_0|^{1/2} R_0(z)]^*$,
g. $V_0^{1/2}[w_1 R_0(z)]^*$,
h. $V_0^{1/2}[w_2\, DR_0(z)]^*$,
i. $V_0^{1/2}[|V_0|^{1/2} R_0(z)]^*$.

where we have set $V_0^{1/2} = V_0|V_0|^{-1/2}$. Of these, (d), (f), (g), and (i) can be handled by Lemma 9.14.1. Let us look at (a). By Lemma 9.5.1 and 9.5.2 we see that (a) applied to f equals $\frac{1}{2}i$ times

$$\int_x^\infty e^{i\bar{\kappa}(x-y)} w_2(x) w_1(y) f(y)\, dy - \int_{-\infty}^x e^{i\kappa(y-x)} w_2(x) w_1(y) f(y)\, dy. \tag{11.3.1}$$

If $v \in L^2$, this gives

$$\begin{aligned} 2|(w_2\, D[\, w_1 R_0(z)\,]^* f, v)| &\leq \iint |w_2(x) w_1(y) f(y) v(x)|\, dy\, dx \\ &\leq \|w_2\|\; \|w_1\|\; \|f\|\; \|v\|. \end{aligned}$$

This shows that (a) is bounded. The same reasoning applies to (c). Let us turn to (e). By Lemma 9.5.3, (e) applied to f is $\frac{1}{2}i$ times

$$\int_x^\infty e^{i\bar{\kappa}(x-y)} w_1(x) w_2(y) f(y)\, dy - \int_{-\infty}^x e^{i\bar{\kappa}(y-x)} w_1(x) w_2(y) f(y)\, dy. \tag{11.3.2}$$

The same estimates as above show that (e) is bounded. The same reasoning applies to (h). The only operator that remains is (b). By (9.5.23),

$$\begin{aligned} D[\, w_2\, DR_0(z)\,]^* f(y) &= -w_2(y) f(y) + \tfrac{1}{2} i\bar{\kappa} \int_{-\infty}^\infty e^{-i\bar{\kappa}|y-x|} w_2(x) f(x)\, dx \\ &= -w_2 f + \bar{z}[\, w_2 R_0(z)\,]^* f \end{aligned}$$

(the last equality comes from Lemma 9.5.1). Thus,

$$w_2 D[w_2 DR_0(z)]^* = -w_2^2 + \bar{z}w_2[w_2R_0(z)]^*. \tag{11.3.3}$$

Since $w_2 \in L^2 \cap L^\infty$, the boundedness of (b) follows in the same way as before. Note that all the operators (a)–(i) are densely defined.

Now we turn to compactness. By Lemma 1.14.3, both $w_1R_0(z)$ and $V_0^{1/2}R_0(z)$ are compact operators for Im $z \neq 0$. However, it does not apply to the operator $w_2 DR_0(z)$. For this purpose we shall need Lemma 5.7.2. Let $\{f_n\}$ be a bounded sequence in L^2, and set $v_n = DR_0(z) f_n$. By (9.5.21), the sequence $\{v_n\}$ is also bounded [see (9.5.11)]. Since $(z + D^2)R_0(z) = 1$, we have

$$Dv_n = (1 - zR_0(z))f_n,$$

showing that $\{Dv_n\}$ is also bounded. Now we can apply the lemma mentioned above to conclude that $\{w_2v_n\}$ has a convergent subsequence. This shows that the operator $w_2DR_0(z)$ is compact. Hence the same is true of the operator $AR_0(z)$.

Now we are ready to apply Theorem 9.11.2. By Theorem 9.11.3 and (11.2.5) we note that $G_0(z)$ has a bounded inverse for each nonreal z. As before we take $\Lambda = R - \{0\}$, and we must show that $Q_0(z)$, $Q_0(\bar{z})$ are uniformly continuous in ω_I for each $I \subset\subset \Lambda$. The cases (d), (f), (g), and (i) are covered by Lemma 9.14.4. The cases (a), (c), (e), and (h) can be handled by the same method if we use (11.3.1) and (11.3.2). The remaining case (b) can be reduced to the others by using (11.3.3). Finally, we use Theorem 9.5.1, Lemma 9.5.3 and Theorem 9.12.2 to verify the remaining hypotheses of Theorem 9.11.2. This proves Theorem 11.2.2. □

11.4. A Variation

The theorems of Section 11.2 consider potentials of the form (11.1.2). In this section we shall consider potentials of a slightly different form, namely,

$$V(x) = w_r + V_0(x), \tag{11.4.1}$$

where $r = |x|$ and $w_r = dw/dr$. Clearly,

$$u_r = \begin{cases} u'(x) & \text{when} \quad x > 0 \\ -u'(x) & \text{when} \quad x < 0. \end{cases} \tag{11.4.2}$$

Hence, if we set

$$\tilde{x} = \operatorname{sgn} x = x/r, \tag{11.4.3}$$

then we have

$$u_r = \tilde{x}u'(x), \qquad x \neq 0. \tag{11.4.4}$$

One must be careful when differentiating with respect to r, especially when integration by parts is involved. For instance,

$$(\tilde{x}, v') = \int_0^\infty v'(x)\,dx - \int_{-\infty}^0 v'(x)\,dx = -2v(0). \qquad (11.4.5)$$

Thus,

$$\tilde{x}' = 2\delta(x), \qquad (11.4.6)$$

where $\delta(x)$ is the famous Dirac delta "function" (it is really a functional) defined by

$$\int \delta(x)u(x)\,dx = u(0). \qquad (11.4.7)$$

Thus, we have

$$\begin{aligned}(u_r, v) &= (\tilde{x}u', v) = ([\tilde{x}u]', v) - 2u(0)v(0)^* \\ &= -(\tilde{x}u, v') - 2u(0)v(0)^* \\ &= -(u, v_r) - 2u(0)v(0)^*. \qquad (11.4.8)\end{aligned}$$

Therefore, in place of (11.1.3) we have

$$\begin{aligned}(w_r u, v) = (\tilde{x}w'u, v) &= -(\tilde{x}wu, v') \\ &\quad - (\tilde{x}wu', v) - 2w(0)u(0)v(0)^*. \qquad (11.4.9)\end{aligned}$$

The last term looks like it may cause trouble in carrying out a program similar to that of Sections 11.2 and 11.3. We avoid it by assuming

$$w(0) = 0. \qquad (11.4.10)$$

As we shall see, this assumption will offer little or no impediment in our applications. The counterpart of (11.1.3) is

$$(w_r u, v) = -(\tilde{x}wu, v') - (\tilde{x}wu', v), \qquad (11.4.11)$$

and consequently,

$$(Vu, v) = (Mu, v), \qquad (11.4.12)$$

where

$$M = i\,D\tilde{x}\,w - i\tilde{x}w\,D + V_0. \qquad (11.4.13)$$

We obtain immediately the counterpart of Theorem 11.2.1 in our case. If (11.2.1) holds, then for every $\varepsilon > 0$ there is a K such that (11.2.2) holds. Thus the operator corresponding to the bilinear form (11.2.3) is self-adjoint. We take it to be the Hamiltonian. Our counterpart of Theorem 11.2.2 is

Theorem 11.4.1. *Suppose V is given by* 11.4.1 *with $w \in L^1 \cap L^2$ satisfying* (11.4.10) *and $V_0 \in L^1$. If H is the forms extension of $H_0 + V$, then the wave operators are complete.*

The proof of Theorem 11.4.1 is identical to that of Theorem 11.2.2. We merely replace w by $\tilde{w} = \tilde{x}w$ and note that $\tilde{w} \in L^1 \cap L^2$. Thus all of the steps of the proof of Theorem 11.2.2 go through with w replaced by $\tilde{w}$.

In the next section we give some applications of Theorems 11.2.2 and 11.4.1.

11.5. Examples

Now we consider various examples of potentials to which Theorems 11.2.2 and 11.4.1 can be applied. In every case the wave operators are complete.

1. $V(x) = q(x)e^x \sin e^x$, where $q \in L^1 \cap L^2$ and $q' \in L^1$. To see this we take $w = -q \cos e^x$. Then $w' = -q' \cos e^x + V$. Hence we can take $V_0 = q' \cos e^x$.

2. $V(x) = q(x)e^r \sin e^r$, $r = |x|$, with the same assumptions on q. Set $w = q(\cos 1 - \cos e^r)$. Then $w(0) = 0$ and $w_r = V + q_r(\cos 1 - \cos e^r)$. Thus we can take $V_0 = q_r(\cos e^r - \cos 1)$ and apply Theorem 11.4.1. Note that $w \in L^1 \cap L^2$ and $V_0 \in L^1$.

3. $V(x) = q(x)e^{1/x} \sin e^{1/x}$ with $x^2 q \in L^1 \cap L^2$, $2xq + x^2 q' \in L^1$. Here we take $w = x^2 q \cos e^{1/x}$. Then

$$w' = (2xq + x^2 q') \cos e^{1/x} + V.$$

Thus we may take

$$V_0 = -(2xq + x^2 q') \cos e^{1/x}.$$

Apply Theorem 11.2.2.

4. $V(x) = q(x)e^{1/r} \sin e^{1/r}$ with the same assumptions on q. In this case we take $w = r^2 q \cos e^{1/r}$ and apply Theorem 11.4.1.

5. $V(x) = q(x)x^\beta \sin x^{\beta+1}$, $\beta > 0$. Set $w = -(1 + \beta)^{-1} q \cos x^{\beta+1}$. Then

$$w' = -(1 + \beta)^{-1} q' \cos x^{\beta+1} + V.$$

Thus we can take

$$V_0 = (1 + \beta)^{-1} q' \cos x^{\beta+1}.$$

We see that the wave operators will be complete if $q \in L^1 \cap L^2$ and $q' \in L^1$.

6. $V(x) = q(x)r^\beta \sin r^{\beta+1}$, $\beta \neq 1$, and q satisfying the same assumptions. Here we take $w = (1 + \beta)^{-1} q(\cos 1 - \cos r^{\beta+1})$ and apply Theorem 11.4.1.

7. $V(x) = q(x)xr^{-1}e^r \sin e^r$ with q satisfying the same assumptions. We can take $w = -q \cos e^r$. Then $w' = -q' \cos e^r + V$. We apply Theorem 11.2.2.

8. $V = \delta(x)$, the Dirac delta "function" defined by (11.4.7). To define the Hamiltonian we note that $(1 + r)^{-\alpha}\delta(x) = \delta(x)$. If we take $w = \frac{1}{2}\tilde{x}(1 + r)^{-\alpha}$, $\alpha > 1$, then by (11.4.6)

$$w' = \delta(x) - \tfrac{1}{2}\alpha\tilde{x}(1 + r)^{-\alpha-1}.$$

Thus we can define

$$(Vu, v) = -(wu, v') - (wu', v) + (V_0 u, v),$$

where

$$V_0 = \tfrac{1}{2}\alpha\tilde{x}(1 + r)^{-\alpha-1}.$$

Clearly, w and V_0 are in $L^1 \cap L^2$, so that Theorem 11.2.2 applies and the wave operators are complete.

9. $V(x) = c_0 r^{-\alpha} \sin r^{\sigma}$, $0 < \sigma$, $\alpha < \sigma + 1$. Let us find conditions on α and σ which guarantee that the wave operators are complete. Let $g(r)$ be a continuously differentiable function which equals $c_0\sigma^{-1}r^{1-\alpha-\sigma}$ for $r > 1$, and such that g/r is bounded for $r < 1$. Set $w = g(\cos 1 - \cos r^{\sigma})$. Then

$$w_r = g_r(\cos 1 - \cos r^{\sigma}) + \sigma r^{\sigma-1} g \sin r^{\sigma}.$$

Thus $V = w_r + V_0$, where

$$V_0 = g_r(\cos r^{\sigma} - \cos 1) + \left(c_0 r^{-\alpha} - \sigma r^{\sigma-1} g\right) \sin r^{\sigma}.$$

If $\alpha + \sigma > 2$, then w will be in $L^1 \cap L^2$, and V_0 will be in L^1. Thus we have by Theorem 11.4.1 that a sufficient condition that the wave operators be complete is that $\alpha + \sigma > 2$. This should be compared with the result of Section 7.2, which shows that the wave operators exist everywhere provided $\alpha > \sigma + \frac{1}{2}$ and $\alpha + \sigma > 1$. In particular, if $\alpha = 1$, the results of Chapter 7 show that the wave operators exist everywhere if $0 < \sigma < \frac{1}{2}$. Here we have shown that they are complete (and consequently exist everywhere) if $\sigma > 1$. The only region not covered is $\frac{1}{2} \le \sigma \le 1$.

10. $V(x) = q(x)\varphi'(x) \sin \varphi(x)$, where $q \in L^1 \cap L^2$, $q' \in L^1$ (no assumptions on φ). We take $w = -q \cos \varphi(x)$. Then $w' = -q' \cos \varphi(x) + V$. Apply Theorem 11.2.2.

11. $V(x) = q(x)\varphi_r \sin \varphi(x)$, with the same assumptions on q. We take $w = q(\cos \varphi(0) - \cos \varphi(x))$. Then

$$w_r = q_r(\cos \varphi(0) - \cos \varphi(x)) + V.$$

We apply Theorem 11.4.1.

Exercises

1. Prove (11.2.6).

2. Prove (11.2.8).

3. Show that the domains of the operators (a)–(i) of Section 11.3 are dense in L^2.

4. Show that the operators (a), (b), (c), (e), (h) of Section 11.3 depend continuously on z uniformly in ω_I for any bounded interval I bounded away from the origin.

5. Construct a function $g(r)$ having the properties required in Example 9 of Section 11.4.

12

Eigenfunction Expansions

12.1. The Usefulness

As we mentioned in Section 9.12, the resolvent of the free Hamiltonian satisfies

$$(R_0(z)f)\hat{} = (z - k^2)^{-1}\hat{f}. \tag{12.1.1}$$

As we saw there, this is a convenient formula. One might wonder if there exists a counterpart of (12.1.1) for the perturbed Hamiltonian H. Suppose, for instance, we had a formula

$$(R(z)f)\hat{} = (z - k^2)^{-1}\tilde{f}(k), \qquad \operatorname{Im} z > 0, \tag{12.1.2}$$

where $\tilde{f}(k)$ is a function depending on $f(x)$ similar to the Fourier transform. [We restricted (12.1.2) to the half-plane $\operatorname{Im} z > 0$ because we learned from bitter experience not to be too ambitious.] If $z = s + ia$, we have by (12.1.2)

$$a\int_I (R(z)f, R(z)g)\, ds = \pi \int \int_I \delta_a(s - k^2)\, ds\, \tilde{f}(k)\tilde{g}(k)^*\, dk.$$

In view of Lemma 8.2.1, this implies

$$(\tilde{E}(I)f, g) = \int_{k^2 \in I} \tilde{f}(k)\tilde{g}(k)^*\, dk. \tag{12.1.3}$$

Moreover, if the wave operator W_+ exists, we have by (8.1.10)

$$\begin{aligned}(W_+E_0(I)f, g) &= \lim_{a\to 0} \frac{a}{\pi}\int_I (R_0(z)f, R(z)g)\, ds \\ &= \lim \int \int_I \delta_a(s - k^2)\hat{f}(k)\tilde{g}(k)^*\, ds\, dk = \int_{k^2 \in I} \hat{f}(k)\tilde{g}(k)^*\, dk.\end{aligned}$$

This gives

$$(W_+ f)^{\sim} = \hat{f}. \tag{12.1.4}$$

As we shall see, this is a very useful formula.

There are other useful consequences of (12.1.2). The only problem is that in general a formula of the form (12.1.2) is not true. However, under certain circumstances a formula similar to it is true and consequences such as (12.1.3) and (12.1.4) can be salvaged. The purpose of this chapter is to derive the appropriate formulas and discuss some of the consequences.

12.2. The Problem

In trying to find the function $\tilde{f}(k)$ to satisfy (12.1.2) one might try to imitate the Fourier transform by taking it in the form

$$\tilde{f}(k) = (2\pi)^{-1/2} \int \Phi(k, x)^* f(x)\, dx. \tag{12.2.1}$$

In general we have

$$R(z) - R_0(z) = [R(z) V R_0(z)] \tag{12.2.2}$$

under suitable conditions on V. Thus if we set $X(z) = [1 - VR_0(z)]^{-1}$, we have

$$R(z) = R_0(z) X(z). \tag{12.2.3}$$

This gives

$$(R(z)f)^{\wedge} = (z - k^2)^{-1}(X(z)f)^{\wedge}. \tag{12.2.4}$$

But

$$(X(z)f)^{\wedge} = (2\pi)^{-1/2} \int e^{-ikx} X(z) f(x)\, dx.$$

If we compare this with (12.2.1), we are tempted to take

$$\Phi(k, x) = X(z)^* e^{ikx}. \tag{12.2.5}$$

It is at this point that several difficulties present themselves. First of all, the right-hand side of (12.2.5) depends on z as well as k and x. This suggests that we should really expect $\tilde{f}$ in (12.1.2) to depend not only on k but on z as well. Second, $X(z)$ is an operator acting on L^2 and so is its adjoint. But the function e^{ikx} is not in L^2 and consequently is outside the domain of $X(z)^*$. In other words, (12.2.5) does not make sense. Third, once we are forced to take $\tilde{f}$ to depend on z, we must determine how it behaves as z approaches the real axis in order to derive the desired consequences from (12.1.2). This looks like (and is) a very delicate question. We shall study the details in the next few sections.

12.3. Operators on L^p

We discussed the spaces L^p in Section 10.5. They are not Hilbert spaces unless $p = 2$. However, they have all of the properties of a Hilbert space with the exception of a scalar product. Moreover, they have a norm given by (10.5.3) and (10.5.4) which satisfies

$$\|\alpha f\| = |\alpha| \, \|f\|, \tag{12.3.1}$$

$$\|f + g\| \le \|f\| + \|g\|, \tag{12.3.2}$$

$$\|f\| = 0 \qquad \text{iff} \quad f = 0. \tag{12.3.3}$$

Such spaces are called *Banach spaces*. Not all of the theorems that are true in Hilbert space are true in Banach space, but many of them are.

The function e^{ikx} is not in L^2, but it is in L^∞. Thus in order to deal with the right-hand side of (12.2.5) we shall have to consider operators on L^∞. As we shall see, even L^∞ will not suffice; we shall need to consider a slightly larger space. We begin by defining the operator

$$T(\kappa)f(x) = \int e^{i\kappa|x-y|} f(y)\, dy \tag{12.3.4}$$

for $\kappa = \sigma + i\eta$, $\eta \ge 0$. This operator is defined for $f \in L^1$. In fact we have

Lemma 12.3.1. *$T(\kappa)$ maps L^1 into L^∞ with*

$$\|T(\kappa)f\|_\infty \le \|f\|_1. \tag{12.3.5}$$

PROOF. We have

$$|T(\kappa)f(x)| \le \int e^{-\eta|x-y|} |f(y)|\, dy \le \|f\|_1. \qquad \square \tag{12.3.6}$$

We also have

Lemma 12.3.2. *In $\eta > 0$ the map $T(\kappa)$ depend analytically on κ.*

PROOF. If we differentiate (12.3.4) under the integral sign with respect to κ we get

$$T'(\kappa)f(x) = i\int e^{i\kappa|x-y|} |x-y| f(y)\, dy \tag{12.3.7}$$

and consequently

$$|T'(\kappa)f(x)| \le \int e^{-\eta|x-y|} |x-y| \, |f(y)|\, dy.$$

Now the inequality

$$te^{1-t} \le 1, \qquad 0 \le t \le \infty, \tag{12.3.8}$$

is easily proved. In fact the left-hand side of (12.3.8) is a nonnegative function which vanishes for $t = 0$, ∞, and the only point where its derivative vanishes is at $t = 1$. This gives

$$\|T'(\kappa)f\|_\infty \le \|f\|_1/e\eta. \tag{12.3.9}$$

Thus $T'(\kappa)$ is also a bounded operator from L^1 to L^∞ provided $\eta > 0$. □

In order to carry out the program outlined in Sections 12.1 and 12.2 we shall need $T(\kappa)$ to depend continuously on κ in $\eta \ge 0$. Unfortunately this cannot be done in L^∞. For example, suppose $\kappa = \sigma$, $\kappa' = \sigma'$ are real and $f(y)$ is defined by

$$f(y) = \begin{cases} e^{i\sigma y}, & 0 \le y \le 1 \\ 0, & \text{otherwise.} \end{cases}$$

Then for $x > 1$

$$T(\kappa)f(x) = \int_0^1 e^{i\sigma x}\, dy = e^{i\sigma x}$$

and

$$T(\kappa')f(x) = e^{i\sigma' x}\int_0^1 e^{i\tau y}\, dy,$$

where $\tau = \sigma - \sigma'$. If τ is sufficiently small, then

$$\left| i\int_0^1 e^{i\tau y}\, dy - 1\right| \ge \tfrac{1}{2}.$$

On the other hand, if we take $x = (4k + 1)\pi/2\tau$ for any integer k, we have

$$e^{i\sigma' x} = ie^{i\sigma x}.$$

This gives

$$T(\kappa')f(x) - T(\kappa)f(x) = e^{i\sigma x}\left(i\int_0^1 e^{i\tau y}\, dy - 1\right),$$

which has absolute value $\ge \frac{1}{2}$. Thus,

$$\|[T(\kappa') - T(\kappa)]f\|_\infty \ge \tfrac{1}{2}\|f\|_1$$

no matter how close σ' comes to σ. In order to remedy this situation we shall have to work in a slightly larger space to be described in the next section.

12.4. Weighted L^p-Spaces

For β real we let $L^{p,\beta}$ be the set of functions $f(x)$ such that $f(x)\rho(x)^\beta \in L^p$. Note that $L^{p,0} = L^p$. It is easily checked that $L^{2,\beta}$ is a Hilbert space while $L^{p,\beta}$ is a Banach space for $1 \le p \le \infty$. We shall use the notation

$$\|f\|_{p,\beta} = \|f\rho^\beta\|_p. \tag{12.4.1}$$

If $\beta_1 < \beta_2$ we have $L^{p,\beta_1} \supset L^{p,\beta_2}$ with

$$\|f\|_{p,\beta_1} \leq \|f\|_{p,\beta_2}. \tag{12.4.2}$$

The reason for our use of these spaces is

Lemma 12.4.1. *For each $\theta > 0$, $T(\kappa)$ is a bounded operator from $L^{1,\theta}$ to $L^{\infty,-\theta}$ which depends continuously on κ in $\eta \geq 0$.*

In proving Lemma 12.4.1 we shall make use of the simple

Lemma 12.4.2. *If* $\operatorname{Im} \kappa \geq 0$, $\operatorname{Im} \kappa' \geq 0$, *then*

$$|e^{i\kappa} - e^{i\kappa'}| \leq \min(2, |\kappa - \kappa'|^{\theta}) \tag{12.4.3}$$

for any θ satisfying $0 \leq \theta \leq 1$.

PROOF. Set

$$\kappa(\theta) = (1 - \theta)\kappa' + \theta\kappa.$$

Then $\operatorname{Im} \kappa(\theta) \geq 0$ and

$$e^{i\kappa} - e^{i\kappa'} = e^{i\kappa(1)} - e^{i\kappa(0)} = i[\kappa - \kappa'] \int_0^1 e^{i\kappa(\theta)}\, d\theta.$$

This gives

$$|e^{i\kappa} - e^{i\kappa'}| \leq |\kappa - \kappa'|. \tag{12.4.4}$$

If $|\kappa - \kappa'| \leq 1$, this implies (12.4.3). On the other hand, if $|\kappa - \kappa'| > 1$, then the left-hand side of (12.4.3) is ≤ 2 which equals the right-hand side. □

Using Lemma 12.4.2, we give the

PROOF OF LEMMA 12.4.1. We have

$$|T(\kappa)f - T(\kappa')f| \leq \int |e^{i\kappa|x-y|} - e^{i\kappa'|x-y|}|\, |f(y)|\, dy$$

$$\leq 2|\kappa - \kappa'|^{\theta} \int |x - y|^{\theta} |f(y)|\, dy$$

by Lemma 12.4.2. This in turn is bounded by

$$2|\kappa - \kappa'|^{\theta} \rho(x)^{\theta} \|f\|_{1,\theta}.$$ □

We also have

Lemma 12.4.3. *For any x, x',*

$$|T(\kappa)f(x) - T(\kappa)f(x')| \leq |\kappa|\, |x - x'|\, \|f\|_1. \tag{12.4.5}$$

PROOF. The left-hand side of (12.4.5) is bounded by

$$\int |e^{i\kappa|x-y|} - e^{i\kappa|x'-y|}|\,|f(y)|\,dy \le |\kappa| \int |\,|x-y| - |x'-y|\,|\,|f(y)|\,dy$$

by (12.4.4). This gives (12.4.5). □

Lemma 12.4.4. *For $\eta > 0$ and θ real*

$$\|T(\kappa)f\|_{\infty,\theta} \le C_{\eta,\theta}\|f\|_{1,\theta}. \tag{12.4.6}$$

PROOF. We have

$$|T(\kappa)f(x)| \le \int e^{-\eta|x-y|}|f(y)|\,dy.$$

For $r, \theta \ge 0$ we have

$$e^{-\eta r} \le \theta^\theta e^{\eta-\theta}/\eta^\theta(1+r)^\theta \tag{12.4.7}$$

by (12.3.8). Moreover,

$$\rho(y) \le \rho(x)\rho(x-y), \qquad \rho(x) \le \rho(y)\rho(x-y). \tag{12.4.8}$$

Hence, if we set $r = |x-y|$ in (12.4.7) and apply (12.4.8), we get

$$e^{-\eta|x-y|} \le \theta^\theta e^{\eta-\theta}\eta^{-\theta}\rho(x)^{\pm\theta}\rho(y)^{\mp\theta}$$

This gives (12.4.6). □

Lemma 12.4.5. *If $\eta > 0$, $\beta > \gamma$, and $\beta \ge 0$, then $T(\kappa)$ is a compact operator from $L^{1,\beta}$ to $L^{\infty,\gamma}$.*

PROOF. Suppose $\|f_k\|_{1,\beta} \le C$. Then $\|T(\kappa)f_k\|_{\infty,\beta} \le C'$ by Lemma 12.4.4. Thus,

$$\rho(x)^\gamma|T(\kappa)f_k(x)| \le C'\rho(x)^{\gamma-\beta}, \qquad k = 1, 2, \ldots.$$

Let $\varepsilon > 0$ be given. Take N so large that

$$\rho(x)^\gamma|T(\kappa)f_k(x)| < \varepsilon, \qquad |x| > N, \quad k = 1, 2, \ldots.$$

By Lemmas 12.3.1 and 12.4.3, the sequence $\{T(\kappa)f_k\}$ is bounded and equicontinuous. Thus there is a subsequence (also denoted by $\{f_k\}$) such that $\{T(\kappa)f_k\}$ converges uniformly on any bounded interval. Thus

$$|T(\kappa)f_j(x) - T(\kappa)f_k(x)| < \varepsilon\rho(N)^{-\gamma}, \qquad j, k > M.$$

Hence we have

$$\|T(\kappa)(f_j - f_k)\|_{\infty,\gamma} < 2\varepsilon, \qquad j, k > M,$$

proving the lemma. □

Lemma 12.4.6. *For $\beta > 0$, $T(\sigma)$ is a compact operator from $L^{1,\beta}$ to $L^{\infty,-\beta}$*

PROOF. By Lemma 12.4.5, $T(\sigma + i\eta)$ is compact from $L^{1,\beta}$ to $L^{\infty,-\beta}$ when $\eta > 0$. By Lemma 12.4.1, $T(\sigma + i\eta)$ depends continuously on η when mapping between these spaces. Hence $T(\sigma)$ is the limit in norm of compact operators. It is compact by Theorem 9.7.3. □

Lemma 12.4.7. *If $V \in L^{1,2\theta}$ for some $\theta > 0$, then $T(\kappa)V$ is a compact operator on $L^{\infty,-\theta}$ depending continuously on κ in $\eta \geq 0$.*

PROOF. V maps $L^{\infty,-\theta}$ boundedly into $L^{1,\theta}$. By Lemmas 12.4.1, 12.4.5, and 12.4.6, $T(\kappa)$ is a compact operator from $L^{1,\theta}$ to $L^{\infty,-\theta}$ which depends continuously on κ for $\eta \geq 0$. □

Lemma 12.4.8. *Assume $V \in L^{1,2\theta}$ for some $\theta > 0$. Consider $T(\kappa)V$ as an operator on $L^{\infty,-\theta}$. Then the set of those real σ such that*

$$N(2\sigma i - T(\sigma)V) \neq \{0\} \tag{12.4.9}$$

is a closed set of measure 0. If $\theta > \frac{1}{2}$, then this set consists of at most isolated points having 0 and $\pm\infty$ as the only possible limit points.

PROOF. Define A, B as in Section 8.4. Then $A \in L^{2,\theta}$. Suppose $h \in L^{\infty,-\theta}$ and satisfies

$$(2\sigma i - T(\sigma)V)h = 0. \tag{12.4.10}$$

Set $g = Bh$. Then $g \in L^2$ since

$$\int |Bh|^2\,dx \leq \|B\|_{2,\theta}^2 \|h\|_{\infty,-\theta}^2.$$

Multiplying (12.4.10) by B, we have

$$(2\sigma i - BT(\sigma)A)g = 0.$$

By Theorem 9.5.2, this is equivalent to

$$G_{0\pm}(\sigma^2)g = 0, \tag{12.4.11}$$

where the plus sign is taken when $\sigma > 0$ and the minus sign is taken when $\sigma < 0$ (here we use the definition $G_0(z) = 1 - Q_0(z)$, $Q_0(z) = B[AR_0(\bar{z})]^*$). Hence if (12.4.9) holds, then $N[G_{0\pm}(\sigma^2)] \neq \{0\}$. But this can happen only for σ^2 contained in a closed set of measure 0 (Theorems 9.11.1 and 9.14.1). This proves the first statement. The second statement follows from Theorem 10.5.1 and Lemma 10.3.2. □

12.5. Extended Resolvents

It will be convenient for us to extend the domain of $R(z)$ and $R_0(z)$ from L^2 to a larger class of functions contained in $L^{1,\theta}$. Throughout this section we shall assume that $V \in L^{1,2\theta}$ from some $\theta > 0$. Then we have $A \in L^{2,\theta}$. We want to define $R(z)f$ for $f = Ag$, where $g \in L^2$. We do this by defining

$$R(z)f = (AR(\bar{z}))^*g. \tag{12.5.1}$$

Since A does not vanish, this gives a unique function in L^2. In order that this definition not conflict with our previous usage we must show that (12.5.1) coincides with the old definition when $f \in L^2$. In other words we must show that (12.5.1) holds when $f, g \in L^2$ and $f = Ag$. Let v be any function in L^2. Then we have

$$([AR(\bar{z})]^*g, v) = (g, AR(\bar{z})v) = (f, R(\bar{z})v),$$

showing that (12.5.1) indeed does hold. Next we want to define $R(z)f$ for $f = f_1 + f_2$, where $f_1 \in L^2$ and $f_2 = Ag$, where g is in L^2. This is done easily by setting

$$R(z)f = R(z)f_1 + (AR(\bar{z}))^*g. \tag{12.5.2}$$

When f can be decomposed in this way we write $f \in L^2 + AL^2$. In order that the definition (12.5.2) be meaningful, we must show that it does not depend on the particular f_1, f_2 chosen. So suppose $f = f_3 + f_4$, where $f_3 \in L^2, f_4 = Ah$, and $h \in L^2$. Then

$$f_1 - f_3 = A(h - g).$$

By (12.5.1),

$$R(z)(f_1 - f_3) = (AR(\bar{z}))^*(h - g),$$

which shows that (12.5.2) does not depend on f_1, f_2 but only on f. The same method extends the definition of $R_0(z)f$ to $f \in L^2 + AL^2$.

Now we consider the operators for $\kappa \neq 0$

$$S_1(\kappa) = 1 + \tfrac{1}{2}i\kappa^{-1}VT(\kappa), \tag{12.5.3}$$

$$S_2(\kappa) = 1 + \tfrac{1}{2}i\kappa^{-1}T(\kappa)V, \tag{12.5.4}$$

where $T(\kappa)$ is given by (12.3.4). By Lemmas 12.4.5 and 12.4.6, the operator $S_1(\kappa)$ is a bounded operator on $L^{1,\theta}$ which depends continuously on κ for $\eta \geq 0$ and is such that $S_1(\kappa) - 1$ is compact. Similarly, $S_2(\kappa)$ has the same properties on the space $L^{\infty,-\theta}$ by Lemma 12.4.7. Moreover, we have

Lemma 12.5.1. *The operators $S_i(\kappa)$, $i = 1, 2$, are one-to-one on their respective spaces for $\eta > 0$. Moreover, there is a closed set e of measure 0 on the real line such that the $S_i(\sigma)$ are one-to-one on their respective spaces for $\sigma \notin e$.*

PROOF. Suppose $f \in L^{1,\theta}$ is a solution of $S_1(\kappa)f = 0$. Then $2\kappa if = VT(\kappa)f$. Set $h = BT(\kappa)f$. By Lemma 12.4.1, $h \in L^2$. Since $2\kappa if = Ah$, we see by Theorem 9.5.2 that $G_0(\kappa^2)h = 0$. If $\eta > 0$, this implies that $h = 0$ by Theorem 9.11.3. This in turn implies that $f = 0$. If $\eta = 0$, then $G_{0\pm}(\sigma^2)h = 0$, where the plus sign is taken if $\sigma > 0$ and the minus sign is taken if $\sigma < 0$. Thus $h = 0$ unless σ belongs to a closed set of measure 0 (Theorems 9.11.2 and 9.14.1). This proves the lemma for $S_1(\kappa)$. The proof for $S_2(\kappa)$ is similar and follows as in the proof of Lemma 12.4.8. □

It follows from Lemma 12.5.1 that the operators $S_i(\kappa)$ are one-to-one on their respective spaces for all κ in $\eta \geq 0$ except the points of e on the real axis. If they were operating on Hilbert spaces, it would follow from Theorem 9.7.1 that they have bounded inverses for such κ. Fortunately that theorem holds in Banach space as well (see Appendix C). Thus we have

Theorem 12.5.1. *There is an open subset Γ of the real axis such that $R - \Gamma$ has measure 0 and the $S_i(\kappa)$ have bounded inverses on their respective spaces for κ in the set*

$$\Sigma = \{\eta > 0\} \cup \Gamma. \tag{12.5.5}$$

On this set the inverses depend continuously on κ.

The last statement of Theorem 12.5.1 follows from Lemma 9.6.1 if we note that the proof given there holds in Banach space as well.

Another important property of the operators $S_i(\kappa)$ is given by

Lemma 12.5.2. *If $f \in L^{1,\theta}$ and $g \in L^{\infty,-\theta}$, then*

$$(S_1(\kappa)f, g) = (f, S_2(-\bar{\kappa})g). \tag{12.5.6}$$

PROOF. First we note that the expressions make sense because the integrands are in L^1 even though the various factors are not in L^2. To prove the lemma we note that both sides of (12.5.6) equal (f, g) plus

$$\tfrac{1}{2}i\kappa^{-1}\int\int V(x)e^{i\kappa|x-y|}f(y)g(x)^*\,dx\,dy$$ □

A simple consequence of Lemma 12.5.2 is

Lemma 12.5.3. *If $\kappa \in \Sigma$, then*

$$\big(S_1(\kappa)^{-1}f, g\big) = \big(f, S_2(-\bar{\kappa})^{-1}g\big), \qquad f \in L^{1,\theta}, \quad g \in L^{\infty,-\theta}. \tag{12.5.7}$$

12.6. The Formulas

Now we are ready to derive the formulas we wanted in Sections 12.1 and 12.2. We assume $V \in L^{1,2\theta}$ for some $\theta > 0$. Let $z = s \pm ia$ and take $\kappa = \sigma + i\eta$ such that $\eta > 0$ and $\kappa^2 = z$. Thus

$$s = \sigma^2 - \eta^2, \qquad \pm a = 2\sigma\eta. \tag{12.6.1}$$

By Theorem 9.5.2,

$$R_0(z) = -\tfrac{1}{2}i\kappa^{-1}T(\kappa). \tag{12.6.2}$$

Thus (12.2.2) becomes

$$R(z) - R_0(z) = -\tfrac{1}{2}i\kappa^{-1}(AR(\bar{z}))^*BT(\kappa). \tag{12.6.3}$$

Now suppose $f \in L^2 \cap L^{1,\theta}$. Then $h = BT(\kappa)f/2i\kappa$ is in L^2 and $S_1(\kappa)f = f - Ah$. Hence

$$R(z)S_1(\kappa)f = R(z)f - (AR(\bar{z}))^*h$$

if we use the extended definition of $R(z)$. Comparing this with (12.6.3), we find

$$R(z)S_1(\kappa)f = R_0(z)f. \tag{12.6.4}$$

Similarly, if $f = Ag$, where $g \in L^2$, then $f \in L^{1,\theta}$. The reasoning above then gives (12.6.4) for such f as well. Hence we have

Lemma 12.6.1. *Equation* (12.6.4) *holds for* $f \in L^2 \cap L^{1,\theta} + AL^2$.

Now suppose $f \in L^{1,\theta} \cap L^2$. By Theorem 12.5.1 we know that there is a $w \in L^{1,\theta}$ such that $S_1(\kappa)w = f$. Set $v = BT(\kappa)w/2i\kappa$. Then $v \in L^2$ and $w = f + Av$. Hence $w \in L^2 \cap L^{1,\theta} + AL^2$. Applying (12.6.4) to w, we obtain

$$R(z)f = R_0(z)S_1(\kappa)^{-1}f, \qquad f \in L^2 \cap L^{1,\theta}. \tag{12.6.5}$$

This is the counterpart of (12.2.3) with which we shall work. Immediately we get

$$(R(z)f)\hat{} = (z - k^2)^{-1}\big(S_1(\kappa)^{-1}f\big)\hat{}. \tag{12.6.6}$$

To determine the right-hand side we proceed as follows. Let $v(k)$ be a function in L^2. Then

$$\begin{aligned}\big(S_1(\kappa)^{-1}f, \check{v}\big) &= (2\pi)^{-1/2}\int \big(S_1(\kappa)^{-1}f, e^{ikx}\big)v(k)^*\,dk \\ &= (2\pi)^{-1/2}\int \big(f, S_2(-\bar{\kappa})^{-1}e^{ikx}\big)v(k)^*\,dk\end{aligned}$$

by Lemma 12.5.2. If we set

$$\Phi_\kappa(k, x) = S_2(-\bar{\kappa})^{-1}e^{ikx}, \tag{12.6.7}$$

and

$$f_\kappa(k) = \int f(x)\Phi_\kappa(k, x)^*\,dx, \tag{12.6.8}$$

this gives

$$\big(S_1(\kappa)^{-1}f\big)\hat{} = f_\kappa(k) \tag{12.6.9}$$

by Parseval's identity. Hence

$$(R(z)f)\hat{} = (z - k^2)^{-1}f_\kappa(k), \qquad \operatorname{Im} z \neq 0. \tag{12.6.10}$$

This is the counterpart of (12.1.2) that we shall use. Summarizing, we have

Theorem 12.6.1. *If $V \in L^{1,2\theta}$ for some $\theta > 0$ and $f \in L^2 \cap L^{1,\theta}$, then* (12.6.10) *holds, where $f_\kappa(k)$ is given by* (12.6.8).

Our next step is to try to derive a counterpart of (12.1.3). Let Γ be the set described in Theorem 12.5.1 and suppose $I = (c, d)$ is such that $c > 0$ and $(c^{1/2}, d^{1/2}) \subset\subset \Gamma$. By (12.6.10),

$$a\int_I (R(z)f, R(z)g)\, ds = \pi \int \int_I \delta_a(s - k^2) f_\kappa(k) g_\kappa(k)^*\, ds\, dk. \tag{12.6.11}$$

We want to take the limit as $a \to 0$. Assume first that $z = s + ia$. Thus $\kappa \to s^{1/2}$ as $a \to 0$. By Theorem 12.5.1, $S_2(-\bar{\kappa})^{-1}$ is uniformly continuous on $\omega_I = \{z | s \in I,\ 0 < a < b\}$, b a given positive number. In particular, there is a constant K such that

$$\|S_2(-\bar{\kappa})^{-1}\| \le K, \qquad z \in \omega_I. \tag{12.6.12}$$

By (12.6.7) and (12.6.8), we have

$$\|\Phi_\kappa(k, \cdot)\|_{\infty, -\theta} \le K \tag{12.6.13}$$

and

$$|f_\kappa(k)| \le \|f\|_{1,\theta} \|\Phi_\kappa(k, \cdot)\|_{\infty, -\theta}.$$

Thus

$$|f_\kappa(k)| \le K \|f\|_{1,\theta}, \qquad z \in \omega_I. \tag{12.6.14}$$

We shall show that

$$\int_I \delta_a(s - k^2) f_\kappa(k) g_\kappa(k)^*\, ds \to f_{|k|}(k) g_{|k|}(k)^* \tag{12.6.15}$$

as $a \to 0$ for each k such that $k^2 \in I$. If k^2 is outside I, then the left-hand side of (12.6.15) converges to 0. For any k it is bounded by

$$K^2 \int_I \delta_a(s - k^2)\, ds, \tag{12.6.16}$$

which is an integrable function on $(-\infty, \infty)$. If we now apply (12.6.15) to (12.6.11) and make use of Lemma 8.2.1 and the Lebesgue dominated convergence theorem, we find that

$$(\tilde{E}(I)f, g) = \int_{k^2 \in I} \tilde{f}(k) \tilde{g}(k)^*\, dk, \tag{12.6.17}$$

where

$$\tilde{f}(k) = f_{|k|}(k) \tag{12.6.18}$$

and $\tilde{g}(k)$ is defined similarly. If we define

$$\Phi(k, x) = \Phi_{|k|}(k, x), \tag{12.6.19}$$

then

$$\tilde{f}(k) = \int f(x)\Phi(k, x)^* \, dx. \tag{12.6.20}$$

Thus (12.1.3) does hold with $\tilde{f}(k)$ defined by (12.6.2). On the other hand, if $\bar{I} \subset (-\infty, 0)$, then the left-hand side of (12.6.15) converges to 0 for all k. Thus (12.6.17) is true in this case as well (both sides vanish).

Now suppose $z = s - ia$ with $\bar{I} \subset (0, \infty)$. In this case $\sigma < 0$ and $\kappa \to -s^{1/2}$. Here we need $(-d^{1/2}, -c^{1/2}) \subset\subset \Gamma$. Following the reasoning above, we get

$$(\tilde{E}(I)f, g) = \int_{k^2 \in I} f_{-|k|}(k) g_{-|k|}(k)^* \, dk.$$

If we set

$$\tilde{f}_{\pm}(k) = f_{\pm|k|}(k) \tag{12.6.21}$$

and

$$\Phi_{\pm}(k, x) = \Phi_{\pm|k|}(k, x), \tag{12.6.22}$$

we have

$$\tilde{f}_{\pm}(k) = \int f(x)\Phi_{\pm}(k, x)^* \, dx \tag{12.6.23}$$

and

$$(\tilde{E}(I)f, g) = \int_{k^2 \in I} \tilde{f}_{\pm}(k)\tilde{g}_{\pm}(k)^* \, dk, \tag{12.6.24}$$

provided the set $k^2 \in I$ is contained in Γ. Summarizing, we have

Theorem 12.6.2. *If $V \in L^{1,2\theta}$ for some $\theta > 0$, then there are an open set $\Gamma \subset R$ such that $C\Gamma = R - \Gamma$ has measure 0 and functions $\Phi_{\pm}(k, x)$ defined for $k \in \Gamma$ and $x \in R$ such that* (a) *$\Phi_{\pm}(k, \cdot) \in L^{-\infty,-\theta}$ for each $k \in \Gamma$ and is continuous in this norm with respect to k.* (b) *Equation* (12.6.24) *holds for all intervals I such that $k^2 \in I$ implies $k \in \Gamma$, where $\tilde{f}_{\pm}(k)$ is given by* (12.6.23) *and $\tilde{g}_{\pm}(k)$ is defined similarly.*

It remains to prove (12.6.15). For this purpose we use

Lemma 12.6.2. *If $h(s)$ is continuous in $\bar{I}$, then as $a \to 0$*

$$\begin{aligned} \int_I \delta_a(s - t)h(s)\, ds &\to h(t), \qquad && t \in I, \\ &\to \tfrac{1}{2}h(t), && t \in \partial I, \\ &\to 0, && t \notin \bar{I}. \end{aligned} \tag{12.6.25}$$

PROOF. It is easily checked that

$$h(t)\int_I \delta_a(s-t)\,ds$$

converges to these limits as $a \to 0$ [see (8.2.10)]. Thus it suffices to show that

$$\int_I \delta_a(s-t)[h(s)-h(t)]\,ds \to 0. \tag{12.6.26}$$

Let $\varepsilon > 0$ be given, and take $\delta > 0$ so small that $|h(s)-h(t)| < \varepsilon$ for $|s-t| < \delta$. Then

$$\int_{|s-t|<\delta} \delta_a(s-t)|h(s)-h(t)|\,ds < \varepsilon$$

On the other hand,

$$\int_{|s-t|>\delta} \delta_a(s-t)\,ds \to 0 \tag{12.6.27}$$

as is easily verified. This gives (12.6.26). □

Once Lemma 12.6.2 is known, the proof of (12.6.15) is straightforward.

12.7. Some Consequences

We describe some consequences of Theorem 12.6.2. First we note that (12.6.24) implies

$$(E(\Lambda)f, g) = \int_\Gamma \tilde{f}_\pm(k)\tilde{g}_\pm(k)^*\,dk, \qquad f, g \in L^2 \cap L^{1,\theta}, \tag{12.7.1}$$

where $\Lambda = \{k^2 | k \in \Gamma\}$. Since $C\Gamma$ has measure 0, we can integrate over the whole real line. In particular, we have

$$\|E(\Lambda)f\| = \|\tilde{f}_\pm\|, \qquad f \in L^2 \cap L^{1,\theta}. \tag{12.7.2}$$

This allows us to define $\tilde{f}_\pm$ for every $f \in L^2$ as the limits in L^2 of $\tilde{f}_\pm$ defined for f in the dense set $L^{1,\theta} \cap L^2$. These limits are unique by (12.7.2), and they satisfy (12.7.1).

Next we note that

$$(E(\Lambda)f, g) = \int \int_\Gamma \tilde{f}_\pm(k)\Phi_\pm(k, x)g(x)^*\,dk\,dx$$

by (12.7.1) and (12.6.23). This gives

$$E(\Lambda)f = \int_\Gamma \tilde{f}_\pm(k)\Phi_\pm(k, x)\,dk. \tag{12.7.3}$$

Similarly, we have by (8.1.10) and (12.6.10) that

$$(W_\pm E_0(I)f, g) = \lim_{a\to 0} \int \int_I \delta_a(s - k^2)\hat{f}(k)g_\kappa(k)^*\, ds\, dk$$

$$= \int_{k^2 \in I} \hat{f}(k)\tilde{g}_\pm(k)^*\, dk, \qquad I \subset\subset \Lambda. \tag{12.7.4}$$

In view of (12.6.24) this suggests that

$$(W_\pm f)^\sim_\pm = \hat{f}. \tag{12.7.5}$$

This is indeed true, but we must be careful. First we should note that (12.6.24) and (12.7.4) imply that

$$\int v(k)\tilde{g}_\pm(k)^*\, dk = 0, \qquad g \in L^2,$$

where $v = \hat{f} - (W_\pm f)^\sim_\pm$. Since $v \in L^2$, $h = \check{v} \in L^2$, and (12.7.4) gives

$$(W_\pm E_0(I)h, g) = 0, \qquad g \in L^2.$$

This means that

$$W_\pm E_0(I)h = 0, \qquad I \subset\subset \Lambda,$$

and consequently,

$$W_\pm E_0(\Lambda)h = 0.$$

Since the complement of Λ in $(0, \infty)$ has measure 0 and $\mathcal{H}_{ac}(H_0) = L^2$ (Theorem 9.12.3), we see that

$$E_0(\Lambda) = 1. \tag{12.7.6}$$

Thus,

$$W_\pm h = 0,$$

which implies that $h = 0$. Consequently, $v = 0$, and (12.7.5) holds. Combining (12.7.4) and (12.7.6), we obtain

$$(W_\pm f, g) = \int \hat{f}(k)\tilde{g}_\pm(k)^*\, dk. \tag{12.7.7}$$

Next let $\varphi(\lambda)$ be a continuous function in $\bar{I} \subset\subset \Lambda$ and vanish outside. Then by definition,

$$(\varphi(H)f, g) = \int_I \varphi(s)\, d(E(s)f, g)$$

$$= \lim \sum_j \varphi(s_j)(E(I_j)f, g)$$

$$= \lim \sum_j \varphi(s_j) \int_{k^2 \in I_j} \tilde{f}_\pm(k)\tilde{g}_\pm(k)^*\, dk$$

$$= \int_I \varphi(k^2)\tilde{f}_\pm(k)\tilde{g}_\pm(k)^*\, dk,$$

where $\bar{I} = \cup \bar{I}_j$ is a partition of $\bar{I}$ into intervals whose lengths tend to 0, and $s_j \in \bar{I}_j$. This implies

$$[\varphi(H)f]^{\sim}_{\pm} = \varphi(k^2)\tilde{f}_{\pm} \tag{12.7.8}$$

by the argument used above.

Another result is

Theorem 12.7.1. *If $k \in \Gamma$ and $f \in D(H)$, then*

$$(Hf)^{\sim}_{\pm}(k) = k^2\tilde{f}_{\pm}(k). \tag{12.7.9}$$

PROOF. Let $\sigma \in \Gamma$ be given and let $\varphi(k)$ be any function in C_0^∞ such that $\varphi(\sigma) = 1$. Let g_n be such that $\hat{g}_n(k) = j_n(k - \sigma)\varphi(k)$ (see Section 7.8). Then by (12.7.7),

$$(W_{\pm}\, g_n, h) = \int j_n(k - \sigma)\varphi(k)\tilde{h}_{\pm}(k)^*\, dk \to \tilde{h}_{\pm}(\sigma)^*.$$

On the other hand,

$$\begin{aligned}(W_{\pm}\, g_n, [\sigma^2 - H]f) &= (W_{\pm}[\sigma^2 - H_0]\, g_n, f)\\ &= \int j_n(k - \sigma)\varphi(k)(\sigma^2 - k^2)\tilde{f}_{\pm}(k)^*\, dk \to 0\end{aligned}$$

by the intertwining relations (see Lemma 9.13.1). This gives (12.7.9). □

Another way of writing (12.7.9) is

$$(\Phi_{\pm}(k, \cdot), [k^2 - H]f) = 0. \tag{12.7.10}$$

If $\Phi_{\pm}(k, x)$ were in L^2, this would mean that it is an eigenfunction of H. This is why (12.6.24) is known as an eigenfunction expansion for H. On the other hand, (12.6.24) shows us that $E(I)f \to 0$ as $|I| \to 0$ for $f \in L^2 \cap L^1$ and $I \subset\subset \Lambda = \{k^2 | k \in \Gamma\}$. This shows that H has no eigenvalues in Λ. Hence $\Phi_{\pm}(k, x)$ cannot be in L^2 for $k \in \Gamma$. But it does satisfy

$$(k^2 - H)\Phi_{\pm}(k, \cdot) = 0$$

in the following sense.

Theorem 12.7.2. *At each point x where $V(x)$ is continuous, $\Phi_{\pm}(k, x)$ has continuous second derivatives and satisfies*

$$(D^2 + V(x) - k^2)\Phi_{\pm}(k, x) = 0, \qquad D = d/i\,dx. \tag{12.7.11}$$

In proving the theorem we shall make use of

Lemma 12.7.1. *If $w \in L^1$ is continuous at x, then $T(k)w$ has continuous second derivatives at x and satisfies*

$$(D^2 - k^2)T(k)w(x) = -2ikw(x) \tag{12.7.12}$$

for each k.

PROOF. Since

$$T(k)w(x) = \int_{-\infty}^{x} e^{ik(x-y)}w(y)\,dy + \int_{x}^{\infty} e^{ik(y-x)}w(y)\,dy$$
$$= T_1(k)w + T_2(k)w, \qquad (12.7.13)$$

we can differentiate under the integral sign to obtain

$$DT(k)w(x) = kT_1(k)w(x) - iw(x) + kT_2(k)w(x) + iw(x) \qquad (12.7.14)$$

and

$$D^2T(k)\,w(x) = k^2T(k)w(x) - 2ikw(x), \qquad (12.7.15)$$

which is just (12.7.12). □

PROOF OF THEOREM 12.7.2. By (12.6.7), $\Phi_\pm(k, x)$ is a solution of

$$\left(1 \mp i(2|k|)^{-1}T(\mp|k|)V\right)\Phi_\pm(k, x) = e^{ikx}. \qquad (12.7.16)$$

Since $\Phi_\pm(k, \cdot) \in L^{\infty, -\theta}$ and $V \in L^{1, 2\theta}$, the product is in $L^{1,\theta} \subset L^1$. Lemma 12.4.3 shows us that

$$T(\mp|k|)V\Phi_\pm(k, x) \qquad (12.7.17)$$

is continuous in x for all x. Thus the same is true of $\Phi_\pm(k, x)$. This means that $V(x)\Phi_\pm(k, x)$ is continuous at each x where $V(x)$ is continuous. By Lemma 12.7.1,

$$\left[D^2 - k^2 \mp i(2|k|)^{-1}(\pm 2i|k|)V\right]\Phi_\pm(k, x) = (D^2 - k^2)e^{ikx} = 0.$$

This gives (12.7.11). □

12.8. Summary

We gather all of the information obtained in the preceding sections into

Theorem 12.8.1. *Assume that $V \in L^{1, 2\theta}$ for some $\theta > 0$. Then there is an open subset Γ of R such that $C\Gamma$ has measure 0 and functions $\Phi_\pm(k, x)$ defined for $k \in \Gamma$ and $x \in R$ such that*

a. *$\Phi_\pm(k, x) \in L^\infty$ for each $k \in \Gamma$,*
b. *$\Phi_\pm(k, x)$ is continuous in k with respect to the $L^{\infty, -\theta}$ norm,*
c. *for each $f \in L^2$ we can define functions $\tilde{f}_\pm \in L^2$ by (12.6.23) such that*

$$(E(\Lambda)f, g) = \int\!\!\int \tilde{f}_\pm(k)\tilde{g}_\pm(k)^*\,dk, \qquad f, g \in L^2, \qquad (12.8.1)$$

where $\Lambda = \{k^2 | k \in \Gamma\}$,

d. *the mappings $f \to \tilde{f}_\pm$ are onto,*
e. *(12.7.3), (12.7.5), (12.7.7), (12.7.8), (12.7.9) hold for all $f, g \in L^2$,*
f. *$\Phi_\pm(k, x)$ is Lipschitz continuous in x uniformly in k for $k \in I \subset\subset \Gamma$,*

g. *at each point* x *where* $V(x)$ *is continuous,* $\Phi_{\pm}(k, x)$ *has continuous second derivatives and satisfies* (12.7.11).

PROOF. The only item that was not mentioned so far is (a). This follows from (12.7.16) and the fact that the expression (12.7.17) is in L^{∞} for each k. □

Exercises

1. Assume (12.12) and prove (12.1.3) and (12.1.4).

2. Give sufficient conditions that (12.2.2) hold.

3. Show that (12.2.1) and (12.2.4) imply (12.2.5).

4. Show that L^p is not a Hilbert space for $p \neq 2$.

5. Prove (12.3.7).

6. Show that $L^{p,\beta}$ is a Banach space for $1 \leq p \leq \infty$, β real, and that $L^{2,\beta}$ is a Hilbert space.

7. Prove Lemma 12.5.3.

8. Prove (12.6.4) for $f = Ag$, $g \in L^2$.

9. Show that the function (12.6.16) is in L^1.

10. Prove (12.6.26).

11. Prove (12.6.15) using Lemma 12.6.2.

12. Prove (12.7.1).

13. Prove (12.7.14) and (12.7.15).

13
Restricted Particles

13.1. A Particle between Walls

Until now we have studied only particles which theoretically can be anywhere on the real line. We have placed no restrictions on the wave functions ψ, and by suitably choosing ψ we can make the probability that the particle is in any interval I differ from zero. In this chapter we shall discuss circumstances in which the particle is so restricted that it cannot be situated on certain parts of the line at any time. Suppose, for instance, that the particle is contained between two impenetrable walls located at $x = \pm b$. Then we must have $\psi(x, t) = 0$ for $|x| \geq b$ for all times t.

At first glance one might question why there is need for any special consideration of such cases. Why do we not merely consider the Hamiltonian H for the whole line restricted to those $\psi \in D(H)$ which vanish for $|x| \geq b$. The answer is simple. Such an operator is not self-adjoint. For instance, a function $u \in L^2$ which equals $Ax + B$ in $J = (-b, b)$ satisfies

$$(u, H_0\psi) = 0 \tag{13.1.1}$$

for all $\psi \in D(H_0)$ which vanish outside J. Thus in every case of restricted particles we must start from scratch and define the Hamiltonian all over again. We shall find that several important adjustments have to be made.

Until now we have always assumed that the potential $V(x)$ is locally integrable. This insured us that the bilinear form used to define the Hamiltonian was defined on C_0^∞, which is dense in L^2. At this point we want to try to relax this assumption. Our motives will become clear in the next chapter. The bilinear form which we shall use to determine our Hamiltonian should obviously be

$$h_J(u, v) = (u', v')_J + (Vu, v)_J, \tag{13.1.2}$$

but it is not quite clear what its domain should be. Of course they should be functions in L^2 which vanish outside J. To get a clearer picture of what other restrictions should be imposed, let us consider the unperturbed (or free) Hamiltonian. The corresponding bilinear form is

$$h_{0J}(u, v) = (u', v')_J. \tag{13.1.3}$$

Let $C_0^\infty(J)$ be the set of those functions in C_0^∞ which vanish outside J. We would want to include such functions in $D(h_{0J})$. To obtain a self-adjoint operator associated with (13.1.3) we need it to be closed (see Section 5.6). The set $C_0^\infty(J)$ will not quite suffice for this, but it is not difficult to find one that will. If we set

$$D(h_{0J}) = \{u \in L^2(J) | u' \in L^2(J) \text{ and } u(\pm b) = 0\}, \tag{13.1.4}$$

then h_{0J} will be closed. We should explain a little more precisely what we mean by $u' \in L^2(J)$ since we do not stipulate that $u \in C'$. What is meant is as follows. There is a $w \in L^2(J)$ such that

$$(w, \varphi) = -(u, \varphi'), \qquad \varphi \in C_0^\infty(J). \tag{13.1.5}$$

If such a w exists, we define it to be u'. There can be only one such w, for if there were another, the difference between the two would satisfy

$$(v, \varphi) = 0, \qquad \varphi \in C_0^\infty(J).$$

That this implies $v = 0$ follows from

Lemma 13.1.1. *$C_0^\infty(J)$ is dense in $L^2(J)$.*

The simple proof of Lemma 13.1.1 will be given at the end of this section. Now let us verify that h_{0J} is closed. Suppose $\{u_n\} \subset D(h_{0J})$ and that $u_n \to u$ in $L^2(J)$ and $h_{0J}(u_n - u_m) \to 0$. Then $\{u_n'\}$ is a Cauchy sequence in $L^2(J)$. Consequently there is a $w \in L^2(J)$ such that $u_n' \to w$. Thus if $\varphi \in C_0^\infty(J)$, we have

$$(w, \varphi) \leftarrow (u_n', \varphi) = (u_n, \varphi') \to -(u, \varphi').$$

Thus $w = u'$ is in $L^2(J)$. Moreover, by (2.7.6),

$$|u(\pm b) - u_n(\pm b)| \le C(\|u - u_n\|_J + \|u' - u_n'\|_J) \to 0.$$

Since $u_n(\pm b) = 0$ for each n, this implies $u(\pm b) = 0$. Thus $u \in D(h_{0J})$. Moreover,

$$h_{0J}(u_n - u) = \|u_n' - u'\|_J^2 \to 0.$$

This shows that h_{0J} is closed. Also, $D(h_{0J})$ is dense in $L^2(J)$ since it contains $C_0^\infty(J)$ (Lemma 13.1.1). Finally, we note that

$$h_{0J}(u) = \|u'\|_J^2 \ge 0.$$

Now we can apply Theorem 5.6.2 to conclude that the operator H_{0J}

associated with h_{0J} is self-adjoint and

$$\sigma(H_{0J}) \subset [0, \infty). \tag{13.1.6}$$

We take H_{0J} as the Hamiltonian for the free particle in J.

To find the Hamiltonian for $V \neq 0$, we shall need h_J to be closed, densely defined, and bounded from below. Set

$$\rho(x) = V_+(x)^{1/2}, \qquad \sigma(x) = V_-(x)^{1/2}. \tag{13.1.7}$$

If we assume

$$\|\sigma u\|_J^2 \leq \theta h_{0J}(u) + K\|u\|_J^2 \tag{13.1.8}$$

for some $\theta < 1$, then h_J will be bounded from below. For we have

$$\begin{aligned} h_J(u) &= h_{0J}(u) + \|\rho u\|_J^2 - \|\sigma u\|_J^2 \\ &\geq (1-\theta)h_{0J}(u) + \|\rho u\|_J^2 - K\|u\|_J^2. \end{aligned} \tag{13.1.9}$$

If we define

$$D(h_J) = D(h_{0J}) \cap D(\rho) \cap D(\sigma), \tag{13.1.10}$$

where

$$D(\rho) = \{u \in L^2 | \rho u \in L^2\}, \tag{13.1.11}$$

then h_J will be closed. For if $\{u_n\} \subset D(h_J)$, $u_n \to u$ in $L^2(J)$ and $h_J(u_n - u_m) \to 0$, then (13.1.8) and (13.1.9) imply

$$h_{0J}(u_n - u_m) \to 0, \qquad \rho(u_n - u_m) \to 0, \qquad \sigma(u_n - u_m) \to 0.$$

Since h_{0J} is closed, this implies $u \in D(h_{0J})$ and $h_{0J}(u_n - u) \to 0$. Now there is a subsequence (also denoted by $\{u_n\}$) such that $u_n \to u$ almost everywhere in J. Thus $\rho u_n \to \rho u$ and $\sigma u_n \to \sigma u$ a.e. But these sequences converge in $L^2(J)$. Hence ρu and σu are in $L^2(J)$. Consequently, $u \in D(h_J)$ and $h_J(u_n - u) \to 0$. Hence h_J is closed. The following theorem is now a simple consequence of Theorem 5.6.2.

Theorem 13.1.1. *Assume that* (13.1.8) *holds and that the set* (13.1.10) *is dense in $L^2(J)$. Then the operator H_J associated with h_J is self-adjoint. Thus the operator $H_0 + V$ has a self-adjoint forms extension.*

Now we turn our attention to describing sufficient conditions for the hypotheses of Theorem 13.1.1 to hold. First we turn our attention to (13.1.10). We shall say that ρ is *almost locally* in $L^p(J)$ if for every $\varepsilon > 0$ there is an open set $Q \subset J$ such that $|J - Q| < \varepsilon$ and $\rho \in L^p(Q)$. We have

Theorem 13.1.2. *If ρ, σ are almost locally in $L^2(J)$, then the set* (13.1.10) *is dense in $L^2(J)$.*

PROOF. Let $\varepsilon > 0$ be given and let u be any function in $L^2(J)$. Then there is a $\varphi \in C_0^\infty(J)$ such that $\|u - \varphi\| < \varepsilon$ (Lemma 13.1.1). Let $M = \max|\varphi(x)|$. By hypothesis, there is an open set $Q \subset J$ such that $|J - Q| < \varepsilon^2$ and $\rho, \sigma \in L^2(Q)$. Now I claim that there is a closed set $F \subset Q$ such that $|Q - F| < \varepsilon^2$. To see this, suppose

$$Q = \bigcup_1^\infty I_j,$$

where the I_j are nonoverlapping intervals. Then there is an N such that

$$\sum_N^\infty |I_j| < \tfrac{1}{2}\varepsilon^2$$

and there are closed intervals $F_j \subset\subset I_j$ such that

$$\sum_1^N |I_j - F_j| < \tfrac{1}{2}\varepsilon^2.$$

If we take

$$F = \bigcup_1^N F_j,$$

then F has the desired properties. By Theorem 7.8.3, there is a $\psi \in C_0^\infty$ such that $0 \le \psi \le 1$, $\psi = 1$ on $\bar{J} - \bar{Q}$, and $\psi = 0$ on F. The function $\varphi(1 - \psi)$ is in $C_0^\infty(J)$ and vanishes on $\bar{J} - Q$. Hence it is in $D(\rho) \cap D(\sigma)$. Moreover, we have

$$\begin{aligned}\|u - \varphi(1 - \psi)\| &\le \|u - \varphi\| + \|\varphi\psi\| \\ &< \varepsilon + \left(\int_{J-F} |\varphi(x)|^2\, dx\right)^{1/2} < (1 + 2M)\varepsilon\end{aligned}$$

since $|J - F| < 2\varepsilon^2$ and $|\varphi| \le M$. Hence the set (13.1.10) is dense in $L^2(J)$. □

Next we turn our attention to (13.1.8). For this purpose we shall use

Theorem 13.1.3. *If*

$$\int_J |\sigma(x)|^2(b^2 - x^2)\, dx < \infty, \tag{13.1.12}$$

then for each $\varepsilon > 0$ there is a constant K_ε such that

$$\|\sigma u\|_J^2 \le \varepsilon h_{0J}(u) + K_\varepsilon \|u\|_J^2, \qquad u \in D(h_{0J}). \tag{13.1.13}$$

PROOF. Let $\varepsilon > 0$ be given and take $0 < c < b$ so close to b that

$$\int_c^b |\sigma(x)|^2(b - x)\, dx < \varepsilon.$$

This can be done because $b \le b + x$ for $c \le x \le b$. Now,

$$u(x) = -\int_x^b u'(y)\,dy, \tag{13.1.14}$$

and consequently,

$$|u(x)|^2 \le (b - x)\int_x^b |u'(y)|^2\,dy \tag{13.1.15}$$

by Schwarz's inequality. Hence

$$\int_c^b |\sigma(x)u(x)|^2\,dx \le \int_c^b \int_x^b (b - x)|\sigma(x)|^2|u'(y)|^2\,dy\,dx$$
$$= \int_c^b \int_c^y (b - x)|\sigma(x)|^2\,dx|u'(y)|^2\,dy \le \varepsilon\int_c^b |u'(y)|^2\,dy. \tag{13.1.16}$$

If we take c such that

$$\int_{-b}^{-c} |\sigma(x)|^2(b + x)\,dx < \varepsilon,$$

we obtain

$$\int_{-b}^{-c} |\sigma(x)u(x)|^2\,dx < \varepsilon\int_{-b}^{-c} |u'(y)|^2\,dy \tag{13.1.17}$$

as well. Moreover, the inequality

$$\int_{-c}^{c} |\sigma(x)u(x)|^2\,dx \le \varepsilon\int_{-c}^{c} |u'(y)|^2\,dy + K\int_{-c}^{c} |u(y)|^2\,dy \tag{13.1.18}$$

follows from inequality (1.8.6) and the fact that

$$\int_{-c}^{c} |\sigma(x)|^2\,dx < \infty.$$

Combining (13.1.16)–(13.1.18), we obtain (13.1.13). □

From Theorems 13.1.1–13.1.3, we obtain

Theorem 13.1.4. *If*

$$\int_J (b^2 - x^2)V_-(x)\,dx < \infty \tag{13.1.19}$$

and $V(x)$ is almost locally in $L^1(J)$, then the operator $H_0 + V$ on $C_0^\infty(J) \cap D(V)$ has a self-adjoint forms extension H_J on $L^2(J)$.

As we have done in the past, we take H_J as the Hamiltonian operator for the case of a particle between walls. We shall investigate its properties in the next few sections. We conclude this section by giving the

PROOF OF LEMMA 13.1.1. Let u be any function in $L^2(J)$. Define $\tilde{u} = u$ in J, $\tilde{u} = 0$ outside J. Then $\tilde{u} \in L^2$. Let $\varepsilon > 0$ be given and take $I \subset\subset J$ such that

$$\int_{J-I} |u(x)|^2 \, dx < \varepsilon^2. \tag{13.1.20}$$

Set $w = u$ in I, $w = 0$ outside I, and $w_n = J_n w$ (see Section 6.8). For n sufficiently large, $w_n \in C_0^\infty(J)$ and

$$\|w_n - w\| < \varepsilon$$

(Theorem 6.8.1). Moreover, (13.1.20) says

$$\|\tilde{u} - w\| < \varepsilon.$$

Combining these inequalities, we obtain the desired results. □

13.2. The Energy Levels

When we consider the spectrum of the Hamiltonian for restricted particles, we find that the situation can be completely different than in the unrestricted case. To illustrate this we consider a free particle contained within walls. Thus $V = 0$ in the Hamiltonian H. We designate this Hamiltonian by H_{0J}.

Let us first describe the spectrum of H_{0J}. Now,

$$(H_{0J}\psi, \psi) = -(\psi'', \psi) = \|\psi'\|^2 \geq 0 \tag{13.2.1}$$

since every state function in $D(H_{0J})$ must vanish at $\pm b$. Hence $(-\infty, 0)$ is contained in $\rho(H_{0J})$ (Theorem 2.2.1). On the other hand, if k^2 is an eigenvalue with corresponding eigenfunction $u(x)$, then

$$k^2 u + u'' = 0, \qquad u(\pm b) = 0. \tag{13.2.2}$$

The most general solution of (13.2.2) is

$$u = Ae^{ikx} + Be^{-ikx} \tag{13.2.3}$$

with

$$Ae^{ikb} + Be^{-ikb} = 0, \qquad Ae^{-ikb} + Be^{ikb} = 0. \tag{13.2.4}$$

Thus the only way we can have $u \not\equiv 0$ is when

$$e^{4ikb} = 1. \tag{13.2.5}$$

Thus, we must have

$$k = \pi n/2b \tag{13.2.6}$$

with n an integer. We can exclude $n = 0$, for then $u = A + B$, a constant which must vanish at $\pm b$. On the other hand, each integer $n \neq 0$ gives a nonvanishing solution (13.2.3) of (13.2.2) with k given by (13.2.6). Note that (13.2.4) and (13.2.6) imply

$$B = (-1)^{n+1} A. \tag{13.2.7}$$

Consequently, a change in the sign of n merely changes the sign of u. Thus, we see that H_{0J} has eigenvalues

$$\lambda_{0n} = \pi^2 n^2/4b^2, \qquad n = 1, 2, 3, \ldots, \tag{13.2.8}$$

and there are no others.

Are there any other points in $\sigma(H_{0J})$? We shall show in the next section that $\sigma(H_J)$ consists only of eigenvalues $\lambda_n \to \infty$. Moreover, if u_n are the corresponding eigenfunctions, then they form a *complete orthonormal sequence* in $L^2(J)$. This means that every $f \in L^2(J)$ can be written in the form

$$f = \Sigma(f, u_n)u_n, \tag{13.2.9}$$

where the convergence is in $L^2(J)$. Since

$$(z - H_J)u_n = (z - \lambda_n)u_n,$$

we have

$$R_J(z)u_n = (z - \lambda_n)^{-1}u_n,$$

where

$$R_J(z) = (z - H_J)^{-1}.$$

Consequently, by (13.2.9),

$$\begin{aligned} R_J(z)f &= \Sigma(f, u_n)(z - \lambda_n)^{-1}u_n \\ &= \Sigma \int_J (z - \lambda_n)^{-1}u_n(x)u_n(y)^* f(y)\, dy. \end{aligned} \tag{13.2.10}$$

Set

$$w_n(x, y) = (z - \lambda_n)^{-1}u_n(x)u_n(y)^*. \tag{13.2.11}$$

Then

$$\begin{aligned} \int_J\int_J \left| \sum_j^k w_n(x, y) \right|^2 dx\, dy &= \sum_{m,\, n=j}^k \int_J\int_J (z - \lambda_n)^{-1}(z - \lambda_m)^{-1} \\ &\quad \times u_n(x)u_n(y)^* u_m(x)^* u_m(y)\, dx\, dy \\ &= \sum_j^k |z - \lambda_n|^{-2}. \end{aligned}$$

Thus, if

$$\sum_1^\infty |z - \lambda_n|^{-2} < \infty, \tag{13.2.12}$$

then

$$\sum_1^N w_n(x, y)$$

converges in $L^2(J \times J)$ to a function $K_z(x, y)$. Consequently,

$$(R_J(z)f, v) = \Sigma \int_J \int_J w_n(x, y) f(y) v(x)^* \, dx \, dy$$
$$= \int_J \int_J K_z(x, y) f(y) v(x)^* \, dx \, dy.$$

Thus

$$R_J(z)f = \int_J K_z(x, y) f(y) \, dy. \tag{13.2.13}$$

Since (13.2.12) holds for $\lambda_n = \lambda_{0n}$, we see that (13.2.13) is true in the case of H_{0J}. It is not difficult to compute the function $K_z(x, y)$ in this case.

13.3. Compact Resolvents

In this section we shall prove

Theorem 13.3.1. *Under the hypotheses of Theorem* 13.1.2, $\sigma(H_J)$ *consists only of isolated eigenvalues* $\lambda_n \to \infty$.

In proving the theorem we shall make use of two lemmas.

Lemma 13.3.1. *If A is a self-adjoint operator and $\sigma(A) \subset [-N, N]$, then A is bounded and defined everywhere and $\|A\| \le N$.*

PROOF. By Theorem 1.10.1(g),

$$\|Au\|^2 = \int_{-N}^{N} \lambda^2 \, d\|E(\lambda)u\|^2 \le N^2 \int_{-\infty}^{\infty} d\|E(\lambda)u\|^2 = N^2\|u\|^2.$$

This shows that A is bounded. Since it is densely defined, it must be defined everywhere. □

Lemma 13.3.2. *Let A be a compact self-adjoint operator with eigenvalues μ_n and corresponding eigenfunctions u_n. Then*

$$Af = \Sigma \, \mu_n(f, u_n)u_n. \tag{13.3.1}$$

PROOF. Order the eigenvalues so that $|\mu_1| \ge |\mu_2| \ge \cdots \ge |\mu_{n-1}| \ge |\mu_n| \ge \cdots$, and let H_n be the set of those elements orthogonal to $u_1, \ldots, u_{n-1}$. Clearly, H_n is a Hilbert space. Moreover, A maps H_n into itself. In fact, if $v \perp u_j$, then

$$(Av, u_j) = (v, Au_j) = \mu_j(v, u_j) = 0.$$

Let A_n be the restriction of A to H_n. It is compact and self-adjoint. Also, $\mu_1, \ldots, \mu_{n-1}$ are not eigenvalues of A_n, but all of the other μj are. Thus $\sigma(A_n) \subset [-|\mu_n|, |\mu_n|]$. By Lemma 13.3.1, this implies that

$$\|A_n\| \le |\mu_n|.$$

Set

$$f_n = f - \sum_1^{n-1} (f, u_j)u_j.$$

Then

$$(f_n, u_k) = 0, \qquad 1 \le k < n.$$

Thus, $f_n \in H_n$. Consequently,

$$\|Af_n\| = \|A_n f_n\| \le |\mu_n| \, \|f_n\|.$$

But

$$\|f_n\|^2 = \|f\|^2 - 2\sum_1^{n-1} |(f, u_j)|^2 + \sum_1^{n-1} |(f, u_j)|^2 \le \|f\|^2.$$

Hence

$$\|Af_n\| \le |\mu_n| \, \|f\| \to 0 \qquad \text{as} \quad n \to \infty$$

by Theorem 9.7.1. This gives (13.3.1). □

To prove (13.2.9) we have

Corollary 13.3.1. *If A is also one-to-one, then every element can be written in the form*

$$f = \Sigma(f, u_n)u_n.$$

PROOF. Since $\Sigma_1^\infty |(f, u_n)|^2 \le \|f\|^2$,

$$\left\| \sum_M^N (f, u_n)u_n \right\|^2 = \sum_M^N |(f, u_n)|^2 \to 0 \qquad \text{as} \quad M, N \to \infty.$$

Hence

$$f_0 = \sum_1^\infty (f, u_n)u_n$$

is a convergent series. But

$$(f - f_0, u_n) = (f, u_n) - (f, u_n) = 0$$

for each n. Thus $A(f - f_0) = 0$ by (13.3.1). Since A is one-to-one, $f = f_0$. This gives the desired result. □

Now we can give the

PROOF OF THEOREM 13.3.1. By (13.1.9), there are constants K, $c > 0$ such that

$$c(\|u'\|^2 + \|u\|^2) \le ([H_J + K]u, u), \qquad u \in D(H_J). \tag{13.3.2}$$

Hence it is one-to-one and onto (2.2.1). Set $L = (H_J + K)^{-1}$. I claim that L is a compact operator. To see this suppose $\|v_n\| \le c$. If $u_n = Lv_n$, then

$$c|u_n|_{1,b}^2 \le (v_n, u_n) \le C'$$

by (13.3.2) [see (4.7.6)]. Thus $\{u_n\}$ has a subsequence which converges uniformly on J (Lemma 4.7.2). Hence L is a compact operator. Now

$$(H_J + K)(\mu - L) = \mu(H_J + K - \mu^{-1}), \qquad (13.3.3)$$

$$\mu - L = \mu(H_J + K - \mu^{-1})L. \qquad (13.3.4)$$

This shows that for $\mu \neq 0$, $\mu - L$ is one-to-one and onto iff $H_J + K - \mu^{-1}$ is. Hence $\mu \neq 0$ is in $\sigma(L)$ iff $\mu^{-1} - K$ is in $\sigma(H_J)$. Now we know that $\sigma(L)$ consists only of isolated eigenvalues $\{\mu_n\}$ having 0 as the only possible limit point. By (13.3.2),

$$0 \leq (Lv, v).$$

Hence the eigenvalues of L are nonnegative. If there were only a finite number of them, we would have

$$Lf = \sum_1^N \mu_n(f, u_n)u_n \qquad (13.3.5)$$

by Lemma 13.3.2, where the u_n are the corresponding eigenfunctions. This would imply that L is not one-to-one, contradicting the fact that L is the inverse of $H_J + K$. Hence the sequence $\{\mu_n\}$ is infinite and consists of positive numbers. Since 0 is the only possible limit point, we have $0 < \mu_n \to 0$. Thus the only points in $\sigma(H_J)$ are $\lambda_n = \mu_n^{-1} - K \to \infty$. □

To prove (13.2.9) we note that by (13.3.3) the eigenfunctions of H_J are precisely those of L (the eigenvalues differ, however). We apply Corollary 13.3.1 to obtain the desired conclusion. Under the hypothesis of Theorem 13.1.1 there is not much of a scattering theory to talk of for H_{0J} and H_J. Neither of them has continuous spectrum, and unless they share an eigenvalue and its corresponding eigenfunction, we have $M_\pm = \{0\}$ (see Theorem 6.3.2).

13.4. One Opaque Wall

Next let us consider the case of a particle restricted to the interval $x \geq 0$ by an impenetrable wall at $x = 0$. Here we consider the bilinear form

$$h(u, v) = (u', v') + (Vu, v), \qquad u, v \in C_0^\infty(0, \infty). \qquad (13.4.1)$$

Here we shall assume that V is locally integrable so that $h(u, v)$ is defined on the whole of $C_0^\infty(0, \infty)$. The theory of Section 4.2 will give a self-adjoint operator associated with $h(u, v)$, provided we have

$$c(\|u'\|^2 + \|u\|^2) \leq h(u) + K\|u\|^2, \qquad u \in C_0^\infty(0, \infty), \qquad (13.4.2)$$

for some constants K, $c > 0$. A sufficient condition for this can be obtained from

Theorem 13.4.1. *If*

$$\int_a^{a+1} |\sigma(x)|^2 \min(x, 1)\, dx \leq C_0, \qquad a \geq 0, \qquad (13.4.3)$$

then for any $\varepsilon > 0$ *there is a constant* K_ε *such that* (13.1.13) *holds for all* $u \in C_0^\infty(0, \infty)$.

PROOF. Take $\delta > 0$ so small that

$$\int_0^\delta x|\sigma(x)|^2\,dx < \varepsilon.$$

If $u \in C_0^\infty(0, \infty)$, we have

$$u(x) = \int_0^x u'(y)\,dy, \tag{13.4.4}$$

and consequently,

$$|u(x)|^2 \le x\int_0^x |u'(y)|^2\,dy.$$

Thus,

$$\begin{aligned}\int_0^\delta |\sigma(x)u(x)|^2\,dx &\le \int_0^\delta\int_0^x x|\sigma(x)|^2|u'(y)|^2\,dy\,dx\\ &= \int_0^\delta\int_y^\delta x|\sigma(x)|^2\,dx\,|u'(y)|^2\,dy \le \varepsilon\int_0^\delta |u'(y)|^2\,dy.\end{aligned} \tag{13.4.5}$$

For the interval $[\delta, \infty)$ we use inequality (2.7.7). □

As a consequence of Theorem 13.4.1 we have

Theorem 13.4.2. *If* V *is locally integrable on* $(0, \infty)$ *and*

$$\int_a^{a+1} V_-(x)\min(x, 1)\,dx \le C_0, \qquad a \ge 0, \tag{13.4.6}$$

then $H_0 + V$ *on* $C_0^\infty(0, \infty)$ *has a self-adjoint forms extension.*

We take this forms extension as the Hamiltonian corresponding to a particle restricted to the nonnegative real axis. The proof of Theorem 13.4.2 is similar to that of Theorem 13.1.4 and is omitted.

Let H denote the Hamiltonian for a particle on a half-line and let H_0 be the corresponding Hamiltonian for a "free" particle on the half-line (i.e., for the case $V \equiv 0$). First we determine the spectrum of H_0. Since

$$(H_0\psi, \psi) = -(\psi'', \psi) = \|\psi'\|^2 \ge 0, \qquad \psi \in C_0^\infty(0, \infty), \tag{13.4.7}$$

we see that $(-\infty, 0) \subset \rho(H_0)$. To see whether H_0 has any positive eigenvalues, suppose

$$k^2u + u'' = 0, \qquad x > 0, \qquad u(0) = 0. \tag{13.4.8}$$

The most general solution of (13.4.8) is of the form (13.2.3). Since $u(0) = 0$, we must have $A + B = 0$. Consequently, we must have

$$u(x) = 2iA\sin kx. \tag{13.4.9}$$

This cannot be in $L^2(0, \infty)$ unless $A = 0$. Thus H_0 has no eigenvalues. However, the functions (13.4.9) just miss being in L^2. This makes us suspect that all nonnegative numbers λ are in $\sigma(H_0)$. For $\lambda > 0$ this can be proved as in Section 2.5. For $\lambda = 0$ it follows from the fact that the spectrum is a closed set. Thus we have

Theorem 13.4.3. *For the Hamiltonian H_0 corresponding to a free particle in a half-line, $\sigma(H_0) = [0, \infty)$.*

Concerning the essential spectrum of the total Hamiltonian H, all of the theorems of Chapter 5 apply. For instance, the following counterpart of Theorem 5.8.1 holds.

Theorem 13.4.4. *Assume that V is locally integrable in* $(0, \infty)$ *and that* (13.4.3) *holds. If*

$$\limsup_{t\to 0} \limsup_{x\to\infty} \int_x^{x+t} |V(y)|\, dy = 0 \tag{13.4.10}$$

and

$$\lim_{t\to 0} \lim_{x\to\infty} t^{-1} \int_x^{x+t} V(y)\, dy = \mu, \tag{13.4.11}$$

then

$$\sigma_e(H) = [\mu, \infty). \tag{13.4.12}$$

13.5. Scattering on a Half-Line

For the case of a half-line the scattering theories of Chapters 7–12 hold with little change. The most important difference is the form of the resolvent operator $R_0(z) = (z - H_0)^{-1}$ for this case. If

$$u(x) = R_0(z)f, \tag{13.5.1}$$

then $u \in L^2(0, \infty)$ and it satisfies

$$zu + u'' = f, \qquad x > 0, \qquad u(0) = 0. \tag{13.5.2}$$

There can be only one solution of (13.5.2) in $L^2(0, \infty)$. For the difference of any two solutions must satisfy

$$zw + w'' = 0, \qquad x > 0, \qquad w(0) = 0.$$

This implies

$$z\|w\|^2 = \|w'\|^2.$$

In particular,

$$(\operatorname{Im} z)\|w\|^2 = 0,$$

and consequently, $w = 0$. Hence we are looking for a solution of (13.5.2) which is in $L^2(0, \infty)$. Fortunately, it is not difficult to find. In fact, we have

Theorem 13.5.1. *If* $\kappa^2 = z$ *and* $\operatorname{Im} \kappa > 0$, *then*

$$R_0(z)f(x) = \frac{1}{2\kappa i}\int_0^\infty \left[e^{i\kappa|x-y|} - e^{i\kappa|x+y|}\right] f(y)\,dy \tag{13.5.3}$$

for the free particle on the half-line $x \geq 0$.

The proof of this theorem is similar to that of Theorem 9.5.2 and is omitted. If one sets

$$f_1(x) = \begin{cases} f(x), & x \geq 0 \\ -f(-x), & x < 0, \end{cases}$$

then (13.5.3) can be put in the form

$$u(x) = \frac{1}{2\kappa i}\int_{-\infty}^\infty e^{i\kappa|x-y|} f_1(y)\,dy. \tag{13.5.4}$$

This shows us immediately that u is a solution of (13.5.2).

Once we have a formula for $R_0(z)$ we can follow the theory of Chapter 9 to prove completeness. For instance we have the following theorem.

Theorem 13.5.2. *If*

$$\int_0^\infty |V(x)| \min(1, x)\,dx < \infty, \tag{13.5.5}$$

then the wave operators are complete.

In proving the theorem we shall use several lemmas similar to those of Section 9.14. In particular, we shall use

Lemma 13.5.1. *If*

$$\int_0^\infty |A(x)|^2 \min(1, x)\,dx \int_0^\infty |B(x)|^2 \min(1, x)\,dx \leq C_0^2, \tag{13.5.6}$$

then

$$\|B[AR_0(z)]^*\| \leq C_0/\left(1 + |z|^{-1/2}\right) \tag{13.5.7}$$

PROOF. By Lemma 12.4.2, we have

$$\begin{aligned} |e^{i\kappa|x-y|} - e^{i\kappa|x+y|}| &\leq 2\min(1, |\kappa y|, |\kappa x|) \\ &\leq 2(1 + |\kappa|)\min(1, |x|, |y|). \end{aligned} \tag{13.5.8}$$

Hence

$$|(B[AR_0(z)]^*u, v| \leq \left(|z|^{-1/2} + 1)\right)\int_0^\infty\int_0^\infty \min(1, |x|, |y|)$$

$$|B(x)A(y)u(y)v(x)|\,dx\,dy.$$

The double integral is bounded by

$$\left(\int_0^\infty \int_0^\infty \min(1, x)|B(x)u(y)|^2 \, dx \, dy\right)^{1/2}$$

$$\times \left(\int_0^\infty \int_0^\infty \min(1, y)|A(y)v(x)|^2 \, dx \, dy\right)^{1/2} \le C_0\|u\| \, \|v\|.$$

This gives (13.5.7). □

Lemma 13.5.2. *If*

$$C_1 = \int_0^\infty \rho(x)^2(|A(x)|^2 + |B(x)|^2) \, dx < \infty, \tag{13.5.9}$$

where $\rho(x) = 1 + |x|$, *then* $Q_0(z) = [B(AR_0(z))^*]$ *is uniformly continuous in* ω_I *for any bounded interval* I *away from the origin* [*for the definition of* ω_I *see* (9.6.5)].

PROOF. Note that by (12.4.4),

$$|e^{i\kappa|x-y|} - e^{i\kappa'|x-y|}| \le |\kappa - \kappa'| \, |x - y|,$$

with a similar expression for $x + y$. Hence

$$\left|\frac{1}{2\kappa i}\left[e^{i\kappa|x-y|} - e^{i\kappa|x+y|}\right] - \frac{1}{2\kappa' i}\left[e^{i\kappa'|x-y|} - e^{i\kappa'|x+y|}\right]\right|$$

$$\le \tfrac{1}{2}|\kappa|^{-1}|\kappa - \kappa'|(|x - y| + |x + y|) + \left|\frac{1}{\kappa} - \frac{1}{\kappa'}\right|.$$

Thus

$$|R_0(z)f - R_0(z')f| \le |\kappa|^{-1}(1 + |\kappa'|^{-1})|\kappa - \kappa'|\int_0^\infty \rho(x)\rho(y)|f(y)| \, dy.$$

Consequently,

$$|([Q_0(z) - Q_0(z')]u, v)| \le |\kappa|^{-1}(1 + |\kappa'|^{-1})|\kappa - \kappa'|$$

$$\times \int_0^\infty \int_0^\infty \rho(x)\rho(y)|B(x)A(y)u(y)v(x)| \, dx \, dy.$$

The integral is bounded by

$$\left(\int_0^\infty \int_0^\infty \rho(x)^2|B(x)u(y)|^2 \, dx \, dy\right)^{1/2}\left(\iint \rho(y)^2|A(y)v(x)|^2 \, dx \, dy\right)^{1/2}$$

$$\le \left(\int_0^\infty \rho(x)^2|B(x)|^2 \, dx \int_0^\infty \rho(y)^2|A(y)|^2 \, dy\right)^{1/2}\|u\| \, \|v\|.$$

This gives

$$\|Q_0(z) - Q_0(z')\| \le |\kappa|^{-1}(1 + |\kappa'|^{-1})|\kappa - \kappa'|C_1. \ \square \tag{13.5.10}$$

Lemma 13.5.3. *If A and B satisfy* (13.5.6), *then $Q_0(z)$ is uniformly continuous in ω_I for any bounded I a positive distance from the origin.*

PROOF. For each n set

$$A_n(x) = A(x), \qquad B_n(x) = B(x), \qquad n^{-1} \le x \le n,$$
$$A_n(x) = B_n(x) = 0 \qquad \text{elsewhere},$$

and

$$Q_n(z) = \left[B_n(A_n R_0(\bar{z}))^* \right].$$

Since

$$Q_n(z) - Q_0(z) = \left[(B_n - B)(A_n R_0(\bar{z}))^* \right] + \left[B(\{A_n - A\} R_0(\bar{z}))^* \right],$$

we have by Lemma 13.5.1,

$$\begin{aligned} &\|Q_n(z) - Q_0(z)\|(1 + |z|^{-1/2}) \\ &\quad \le \left(\int_0^\infty |B(x) - B_n(x)|^2 \min(1, x)\, dx \int_0^\infty |A_n(x)|^2 \min(1, x)\, dx \right)^{1/2} \\ &\quad + \left(\int_0^\infty |B(x)|^2 \min(1, x)\, dx \int_0^\infty |A(x) - A_n(x)|^2 \min(1, x)\, dx \right)^{1/2}. \end{aligned}$$

This converges to 0 uniformly in ω_I as long as I is bounded away from the origin. Now,

$$\int_0^\infty \rho(x)^2 |A_n(x)|^2\, dx \le n\rho(n)^2 \int_0^\infty |A(x)|^2 \min(1, x)\, dx. \tag{13.5.11}$$

Hence $Q_n(z)$ is uniformly continuous in ω_I for each n (Lemma 13.5.2). Since the $Q_n(z)$ converge uniformly to $Q_0(z)$ in ω_I, we see that $Q_0(z)$ is uniformly continuous there as well. □

Lemma 13.5.4. *If $u(x)$ is given by* (13.5.1), *then*

$$|u(x)|^2 \le (3/2)|z|^{-1}\eta^{-1}\|f\|^2, \tag{13.5.12}$$

$$\|u\| \le (3/4)|z|^{-1/2}\eta^{-1}\|f\|, \tag{13.5.13}$$

and

$$\|u'\| \le (3/4)\eta^{-1}\|f\|, \qquad |u'(x)|^2 \le \frac{3}{2}\eta^{-1}\|f\|^2, \tag{13.5.14}$$

where $\eta = \operatorname{Im} \kappa > 0$.

PROOF. By (13.5.3), we have

$$|u(x)| \le (1/2)|\kappa|^{-1} \int_0^\infty (e^{-\eta|x-y|} + e^{-\eta|x+y|})|f(y)|\, dy. \tag{13.5.15}$$

Hence

$$\begin{aligned} |u(x)|^2 \le (1/4)|z|^{-1} &\int_0^\infty (e^{-\eta|x-y|} + e^{-\eta|x+y|})\, dy \\ \times &\int_0^\infty (e^{-\eta|x-y|} + e^{-\eta|x+y|})|f(y)|^2\, dy \end{aligned}$$

Now,

$$\int_0^\infty e^{-\eta|x-y|}\,dy \le 2/\eta \tag{13.5.16}$$

and

$$\int_0^\infty e^{-\eta|x+y|}\,dy \le 1/\eta. \tag{13.5.17}$$

Hence

$$|u(x)|^2 \le (3/4)|z|^{-\frac{1}{2}}\eta^{-\frac{1}{2}}\eta^{-1}\int_0^\infty (e^{-\eta|x-y|} + e^{-\eta|x+y|})|f(y)|^2\,dy. \tag{13.5.18}$$

This gives (13.5.13). Integrating with respect to x we get

$$\int_0^\infty |u(x)|^2\,dx \le (3/4)|z|^{-\frac{1}{2}}\eta^{-1}\int_0^\infty\int_0^\infty (e^{-\eta|x-y|} + e^{-\eta|x+y|})\,dx|f(y)|^2\,dy$$
$$\le (9/16)|z|^{-\frac{1}{2}}\eta^{-2}\|f\|^2.$$

This gives (13.5.13). Next we note that

$$2u'(x) = \int_0^x e^{i\kappa(x-y)}f(y)\,dy - \int_x^\infty e^{i\kappa(y-x)}f(y)\,dy - \int_0^\infty e^{i\kappa(x+y)}f(y)\,dy. \tag{13.5.19}$$

Hence

$$|u'(x)| \le (1/2)\int_0^\infty (e^{-\eta|x-y|} + e^{-\eta|x+y|})|f(y)|\,dy. \tag{13.5.20}$$

Calculating as above, we obtain (13.5.14). □

Lemma 13.5.5. *If* Im $z \ne 0$ *and*

$$\int_0^\infty |A(x)|^2 \min(1, x)\,dx < \infty,$$

then $AR_0(z)$ is a compact operator on $L^2(0, \infty)$.

PROOF. Let $\varepsilon > 0$ be given, and take N so large that

$$\int_N^\infty |A(x)|^2\,dx < \varepsilon.$$

Then, if u satisfies (13.5.1), we have by (13.5.12)

$$\int_N^\infty |A(x)u(x)|^2\,dx \le (3/2)\varepsilon|z|^{-\frac{1}{2}}\eta^{-1}\|f\|^2.$$

By Theorem 13.4.1, there is a constant K_ε such that

$$\int_0^N |A(x)u(x)|^2\,dx \le \varepsilon\|u'\|^2 + K_\varepsilon\int_0^N |u(x)|^2\,dx.$$

Now let $\{f_n\}$ be a sequence in $L^2(0, \infty)$ such that $\|f_n\| \le C_2$. Set $u_n =$

$R_0(z)f_n$. Then

$$\int_N^\infty |A(x)u_n(x)|^2\,dx \le (3/2)\varepsilon|z|^{-1}\eta^{-1}C_2^2.$$

By Lemma 13.5.4,

$$\|u_n\|^2 + \|u_n'\|^2 \le (9/16)\eta^{-2}C_2^2(1+|z|^{-1}).$$

We can apply Lemma 4.7.2 to show that there is a subsequence (also denoted by $\{u_n\}$) which converges uniformly on $[0, N]$. Hence

$$\int_0^N |A(x)[u_n(x) - u_m(x)]|^2\,dx \le (9/16)\varepsilon\eta^{-2}C_2^2$$

$$+ K_\varepsilon \int_0^N |u_n(x) - u_m(x)|^2\,dx.$$

If we take M so large that

$$|u_n(x) - u_m(x)|^2 \le \varepsilon/NK_\varepsilon, \qquad n, m > M,$$

we find that

$$\|A[u_n - u_m]\|^2 \le \left(6|z|^{-1}\eta^{-1}C_2^2 + \eta^{-2}C_2^2 + 1\right)\varepsilon.$$

Hence Au_n converges in $L^2(0,\infty)$. This shows that $AR_0(z)$ is compact. □

We are well on our way to proving Theorem 13.5.1. However, we shall need a bit more. In particular, we shall need to find $E_0(I)$ for an interval I. This will be done in the next section, together with the rest of the proof.

13.6. The Spectral Resolution for the Free Particle on a Half-Line

In order to apply the theory of Chapter 9 we must know something about $\mathcal{H}_{ac}(H_0)$. In order to determine this we compute $E_0(I)$ for an arbitrary interval. Let u, f satisfy (13.5.1), and define them both to vanish for $x < 0$. Thus,

$$(u')^\wedge = (2\pi)^{-1/2}\int_0^\infty e^{-ikx}u'(x)\,dx = ik\hat{u}. \tag{13.6.1}$$

There are no boundary terms since $u(0) = 0$ and $u(\infty) = 0$ by (13.5.18). Also,

$$(u'')^\wedge = (2\pi)^{-1/2}\int_0^\infty e^{-ikx}u''(x)\,dx = ik(u')^\wedge - (2\pi)^{-1/2}u'(0)$$

$$= -k^2\hat{u} - (2\pi)^{-1/2}u'(0) \tag{13.6.2}$$

since $u'(\infty) = 0$ [see (13.5.20)]. By (13.5.19),

$$u'(0) = -\int_0^\infty e^{i\kappa y}f(y)\,dy. \tag{13.6.3}$$

Hence

$$(u'')\hat{} = -k^2\hat{u} - f_\kappa, \tag{13.6.4}$$

where

$$f_\kappa = (2\pi)^{-1/2}\int_0^\infty e^{i\kappa y}f(y)\,dy. \tag{13.6.5}$$

Thus we have by (13.5.2)

$$(z - k^2)\hat{u} = \hat{f} + f_\kappa = \tilde{f}(\kappa, k). \tag{13.6.6}$$

Therefore, if $z = s + ia$,

$$\frac{a}{\pi}\int_I (R_0(z)f, R_0(z)g)\,ds = \int_I\int_0^\infty \delta_a(s - k^2)\tilde{f}\tilde{g}^*\,dk\,ds. \tag{13.6.7}$$

Let $I \subset\subset (0, \infty)$. Then $\kappa \to s^{1/2}$ as $a \to 0$. If we take the limit in (13.6.7) as $a \to 0$, we have by Lemma 8.2.1

$$(E_0(I)f, g) = \int_{k^2\in I} \tilde{f}(|k|, k)\tilde{g}(|k|, k)^*\,dk \tag{13.6.8}$$

[note that $\tilde{E}_0(I) = E_0(I)$ for each I since H_0 has no eigenvalues]. This shows that

$$\Sigma(E_0(I_k)f, g) \to 0 \qquad \text{as} \quad \Sigma|I_k| \to 0.$$

Hence we have proved

Theorem 13.6.1. $\mathcal{H}_{ac}(H_0) = L^2(0, \infty)$.

Now we are ready for the

PROOF OF THEOREM 13.5.2. We apply Theorems 9.11.2 and 9.11.3. We take

$$A = \begin{cases} |V(x)|^{1/2}, & V(x) \neq 0 \\ e^{-x^2}, & V(x) = 0 \end{cases}$$

and $B = V/A$. Then (9.11.8) holds with $\mathcal{H} = \mathcal{K} = L^2(0, \infty)$. The fact that functions in $D(H)$ or $D(H_0)$ vanish at $x = 0$ is used here. Next we note that

$$\|AR_0(z)f\|^2 + \|BR_0(z)f\|^2 \le 3C_0(1 + |z|^{-1})\|f\|^2/\eta \tag{13.6.9}$$

To see this we use (13.5.12) to obtain

$$\int_1^\infty |A(x)u(x)|^2\,dx \le (3/2)|z|^{-1}\eta^{-1}\|f\|^2\int_1^\infty |A(x)|^2\,dx$$

and (13.4.4) to get

$$\int_0^1 |A(x)u(x)|^2\,dx \le \int_0^1 x|A(x)|^2\,dx\int_0^1 |u'(y)|^2\,dy$$

$$\le (3/2)\eta^{-1}\|f\|^2\int_0^1 x|A(x)|^2\,dx$$

by (13.5.14). This gives (13.6.9). Since a/η is bounded in any I, we see that (9.11.6) holds. Next we note that by Theorem 13.4.1,

$$\|Au\|^2 + \|Bu\|^2 \le \varepsilon\|u'\|^2 + K\|u\|^2$$

for all $u \in D(H_0^{1/2})$. Thus $D(|H|^{1/2}) = D(H_0^{1/2})$ by Theorem 8.4.2. This gives $D(H_0) \cup D(H) \subset D(A) \cap D(B)$. $Q_0(z)$ is bounded, everywhere defined, and uniformly continuous in each ω_I by Lemmas 13.5.1 and 13.5.3, and $AR_0(z)$ is a compact operator on $L^2(0, \infty)$ for each z by Lemma 13.5.5. $Q(z)$ is bounded and everywhere defined since

$$B(AR(\bar{z}))^* = B|R(i)|^{1/2}(A|i - H|^{1/2}R(\bar{z}))^*, \qquad \operatorname{Im} z \neq 0.$$

Moreover, $\mathcal{H}_{ac}(H_0) = L^2(0, \infty)$ by Theorem 13.6.1. All of the hypotheses of Theorems 9.11.2 and 9.11.3 are satisfied. The conclusion now follows. □

Exercises

1. Prove (13.1.1) under the circumstances stated.
2. If $u \in L^2(J)$ and $(u,\varphi') = 0$ for all $\varphi \in C_0^\infty(J)$, show that u is a constant a.e.
3. Prove (13.1.14).
4. Prove (13.1.17) and (13.1.18).
5. Show that w_n given in Section 13.1 is in $C_0^\infty(J)$ for n sufficiently large.
6. Calculate the function $K_z(x, y)$ in (13.2.13) for the case $V(x) \equiv 0$.
7. For $\mu \neq 0$ show that $\mu \in \sigma(L)$ iff $\mu^{-1} - K \in \sigma(H_J)$ [see (13.3.3) and (13.3.4)].
8. Why cannot (13.3.5) hold?
9. Prove $M_\pm = \{0\}$ for H_{0J} and H_J.
10. Use Theorem 13.4.1 to prove (13.4.2).
11. Show that $u(x)$ given by (13.4.8) cannot be in $L(0, \infty)$ unless $A = 0$.
12. Prove Theorem 13.4.3.
13. Prove Theorem 13.4.4.
14. Prove Theorem 13.5.1.
15. Prove (13.5.8).
16. Verify (13.5.16) and (13.5.17).

17. Prove (13.6.8).

18. Prove Theorem 13.6.1.

19. Show that (13.2.12) implies (13.2.13).

14
Hard-Core Potentials

14.1. Local Absorption

In proving completeness for the wave operators we have assumed until now that the potential $V(x)$ is locally integrable. On the other hand, the behavior of $V(x)$ at ∞ seemed to be the deciding factor in determining whether or not the wave operators were complete. This raises the question whether local singularities can affect completeness. This question is more than academic because there are potentials of importance in physics which contain strong local singularities. In particular, one may want to know the following: if a potential is not locally integrable in the neighborhood of a point but otherwise satisfies all of the requirements for completeness in Chapter 9, then will completeness be violated? The answer is that this indeed can happen, although examples are not easily found. When this happens, the particle can be "captured" and remain in a bounded region as $t \to \infty$.

The purpose of this chapter is to give conditions on $V(x)$ which allow strong local singularities and yet guarantee completeness. As before, we shall use the bilinear form

$$h(u, v) = (u', v') + (Vu, v) \tag{14.1.1}$$

to define the Hamiltonian H. But now that we do not require V to be locally integrable, $h(u, v)$ is not defined on the whole of C_0^∞. However, we shall have to define h on a dense set. An assumption we will want to make is that there exist constants $K, c > 0$ such that

$$c(\|u'\|^2 + \|u\|^2) \le h(u) + K\|u\|^2, \qquad u \in D(h). \tag{14.1.2}$$

It will be used to show that the operator associated with h is self-adjoint. To see this, let ρ, σ be defined by (13.1.7). Then we have

Theorem 14.1.1. *Assume* (A) *there is a constant C such that*

$$\|\sigma u\|^2 \le C(\|H_0^{1/2}u\|^2 + \|u\|^2), \qquad u \in D(H_0^{1/2}) \cap D(\sigma) \quad (14.1.3)$$

and (B)

$$D(h) = D(H_0^{1/2}) \cap D(\rho) \cap D(\sigma) \textit{ and } (14.1.2) \textit{ holds.}$$

Then h is closed.

PROOF. Suppose $u_n \in D(h)$, $u_n \to u$ in L^2, and $h(u_n - u_m) \to 0$. Then (14.1.2) implies $H_0^{1/2}(u_n - u_m) \to 0$, while (14.1.3) implies $\sigma(u_n - u_m) \to 0$. This in turn implies $\rho(u_n - u_m) \to 0$. Now there is a subsequence (also denoted by $\{u_n\}$) such that $u_n \to u$ almost everywhere. Hence $\sigma u_n \to \sigma u$ and $\rho u_n \to \rho u$ a.e. Since they converge in L^2 as well, they must converge to the same limits there. Hence σu and ρu are in L^2. Since $H_0^{1/2}$ is a closed operator, $u \in D(H_0^{1/2})$ as well. Hence $u \in D(h)$. Moreover,

$$h(u_n - u) = \|H_0^{1/2}(u_n - u)\|^2 + \|\rho(u_n - u)\|^2 - \|\sigma(u_n - u)\|^2 \to 0 \quad \text{as} \quad n \to \infty.$$

Hence h is closed. □

In addition we have

Corollary 14.1.1. *Under hypotheses* (A), (B), *and* (C) *$D(h)$ is dense in L^2 the operator H associated with h is self-adjoint.*

PROOF. We apply Theorem 5.6.2. □

We take H as the Hamiltonian operator for our case. To obtain a sufficient condition for condition (C) to hold, we shall say that ρ is almost locally in L^p if it is almost locally in $L^p(J)$ for each bounded interval J. The following is an immediate consequence of Theorem 13.1.2.

Theorem 14.1.2. *If ρ is almost locally in L^2, then* (C) *holds.*

PROOF. Let u be any function in L^2, and let $\varepsilon > 0$ be given. Then there is an N such that

$$\int_{|x|>N} |u(x)|^2\, dx < \tfrac{1}{2}\varepsilon^2.$$

Since ρ is almost locally in $L^2(J)$, where $J = [-N, N]$, there is a $\varphi \in C_0^\infty(J) \cap D(\rho)$ such that

$$\int_{|x|<N} |u(x) - \varphi(x)|^2\, dx < \tfrac{1}{2}\varepsilon^2.$$

This gives $\|u - \varphi\| < \varepsilon$. Hence (C) holds. □

Potentials $V(x)$ having strong (not locally integrable) singularities at finite points are called *hard-core* potentials. The purpose of this chapter is to show that one can have a scattering theory for some of them. We shall assume that (14.1.3) holds and that V is locally integrable outside a finite interval J. Then we shall compare our Hamiltonian with that for a particle that cannot enter or leave J. For the latter situation the singularities inside J do not affect scattering theory since the spectrum for a particle inside J is discrete (consists only of isolated eigenvalues). We carry out the program in the next few sections.

A sufficient condition for (A) and (B) to hold is easily obtained. In fact we have

Theorem 14.1.3. *If*

$$\sup_a \int_a^{a+1} |\sigma(x)|^2 \, dx < \infty, \tag{14.1.4}$$

then $D(H_0^{1/2}) \subset D(\sigma)$, and for each $\varepsilon > 0$ there is a constant K_ε such that

$$\|\sigma u\|^2 \leq \varepsilon \|u'\|^2 + K\|u\|^2, \qquad u \in D\big(H_0^{1/2}\big). \tag{14.1.5}$$

Moreover, (A) *and* (B) *hold.*

PROOF. Inequality (14.1.5) follows from (2.7.7). Since $D(H_0^{1/2}) = D(h_0)$ by Theorem 8.4.1, we see that $D(H_0^{1/2}) \subset D(\sigma)$. Now (14.1.5) implies (14.1.3). It also implies (14.1.2), for we have

$$h(u) = \|u'\|^2 + \|\rho u\|^2 - \|\sigma u\|^2 \geq (1 - \varepsilon)\|u'\|^2 - K\|u\|^2.$$

This gives (B). □

14.2. The Modified Hamiltonian

We want to develop a scattering theory for hard-core potentials. We shall assume the following:

I. There is a positive number c such that

$$\int_{|x|>c} |V(x)| \, dx < \infty. \tag{14.2.1}$$

II.

$$\int_{-c}^{c} |V_-(x)| \, dx < \infty. \tag{14.2.2}$$

III. $V(x)$ is almost locally in L^1.

By Theorems 14.1.2 and 14.1.3 and Corollary 14.1.1 the Hamiltonian H for this particle is well defined. We shall prove

Theorem 14.2.1. *Under hypotheses* I–III *the wave operators are complete.*

The proof of Theorem 14.2.1 will be given in Section 14.9. In proving the theorem we shall construct the Hamiltonian for a particle that cannot enter or leave the interval $J = (-b, b)$, $b \geq c$. We do this as follows. Put $J_1 = J$, $J_2 = (b, \infty)$, and $J_3 = (-\infty, -b)$. We consider the bilinear forms

$$h_i(u, v) = (u', v')_{J_i} + (Vu, v)_{J_i}, \qquad i = 1, 2, 3. \tag{14.2.3}$$

We set

$$D(h_i) = \{u \in L^2(J_i) | u', \rho u, \sigma u \in L^2(J_i), u(\pm b) = 0\}. \tag{14.2.4}$$

By Theorems 13.1.4 and 13.4.2, there are self-adjoint operators H_i on $L^2(J_i)$ associated with these bilinear forms. We define the modified Hamiltonian as follows. For any function u defined on the real line, let u_{J_i} be the restriction of u to J_i.

We shall say that $u \in L^2$ is in $D(\tilde{H})$ and $\tilde{H}u = f$ if $u_{J_i} \in D(H_i)$ and $H_i u_{J_i} = f_{J_i}$. We must show that $\tilde{H}$ is a self-adjoint operator. This is not difficult. In fact, suppose $u, f \in L^2$ and

$$(u, \tilde{H}v) = (f, v), \qquad v \in D(\tilde{H}).$$

If $w \in D(H_i)$ and v is given by

$$v = \begin{cases} w & \text{in } J_i \\ 0 & \text{in } J_k, \quad k \neq i, \end{cases}$$

then $v \in D(\tilde{H})$. Thus,

$$(u, H_i w)_{J_i} = (f, w)_{J_i}, \qquad w \in D(H_i).$$

Since H_i is self-adjoint, $u_{J_i} \in D(H_i)$ and $H_i u_{J_i} = f_{J_i}$. Thus $u \in D(\tilde{H})$ and $\tilde{H}u = f$. Thus $\tilde{H}$ is self-adjoint. We are also going to use a special case of $\tilde{H}$. Let

$$V_1(x) = \begin{cases} V(x) & \text{for } |x| \leq c \\ 0 & \text{for } |x| > c. \end{cases} \tag{14.2.5}$$

We let $\tilde{H}_1$ be the operator just constructed corresponding to $H_0 + V_1$ instead of $H_0 + V$. Thus the bilinear forms h_2, h_3 for this case are

$$\tilde{h}_i(u, w) = (u', v')_{J_i}, \qquad i = 2, 3, \tag{14.2.6}$$

in place of (14.2.3), but the form h_1 remains the same. We shall use $\tilde{H}_1$ in our scattering theory. In fact we shall first consider the wave operators $W_{\pm}(\tilde{H}_1, H_0)$. In so doing we shall make use of the fact that we can construct the resolvent operator for $\tilde{H}_1$ explicitly. This will be carried out in the next section.

14.3. The Resolvent Operator for $\tilde{H}_1$

Now we shall construct $\tilde{R}_1(z) = (z - \tilde{H}_1)^{-1}$ for use in our scattering theory. It will be fairly easy to do so if we make use of the results of

Chapter 13. For we can actually split up the resolvent over the intervals J_i. To see this, suppose

$$u = \tilde{R}(z)f = (z - \tilde{H})^{-1}f, \tag{14.3.1}$$

or

$$(z - \tilde{H})u = f.$$

By definition,

$$(z - H_i)u_{J_i} = f_{J_i},$$

and thus

$$u_{J_i} = R_i(z)f_{J_i},$$

where

$$R_i(z) = (z - H_i)^{-1}.$$

Thus,

$$(\tilde{R}(z)f)_{J_i} = R_i(z)f_{J_i}. \tag{14.3.2}$$

Therefore, to construct $\tilde{R}(z)$ it suffices to construct each $R_i(z)$. For the case of $\tilde{H}_1$, H_2 and H_3 correspond to a free particle. Let us denote these operators by H_{20}, H_{30}, respectively. Both of these operators represent a free particle on a half-line. The resolvent of such an operator has already been constructed in Section 13.5. Using the method described there we obtain

Theorem 14.3.1. *If* $z = \kappa^2$, Im $\kappa > 0$, *and*

$$R_{i0}(z) = (z - H_{i0})^{-1}, \qquad i = 2, 3,$$

then

$$R_{20}(z)f(x) = \frac{1}{2\kappa i}\int_b^\infty \left[e^{i\kappa|x-y|} - e^{i\kappa|x+y-2b|}\right]f(y)\,dy, \qquad x > b \tag{14.3.3}$$

and

$$R_{30}(z)f(x) = \frac{1}{2\kappa i}\int_{-\infty}^{-b}\left[e^{i\kappa|x-y|} - e^{i\kappa|x+y+2b|}\right]f(y)\,dy, \qquad x < -b. \tag{14.3.4}$$

The proof of Theorem 14.3.1 is simple and is omitted. We shall not compute $R_1(z)$ explicitly, but we shall note its form, which was given by (13.2.10).

Theorem 14.3.2. *There are eigenvalues $\lambda_n \to \infty$ of H_1 and corresponding eigenfunctions $\{u_n\}$ such that*

$$R_1(z)f(x) = \sum \int_J (z - \lambda_n)^{-1} u_n(x) u_n(y)^* f(y)\, dy, \qquad x \in J. \tag{14.3.5}$$

Hence we have

Corollary 14.3.1. *$\tilde{R}_1(z)f(x)$ is given by* (14.3.3) *in J_2,* (14.3.4) *in J_3, and* (14.3.5) *in J.*

Let us describe this result in a slightly different way. Suppose

$$u = \tilde{R}_1(z)f. \tag{14.3.6}$$

Then $u = u_1 + u_2 + u_3$, where u_1 is given by (14.3.5) in J and vanishes in $J_2 \cup J_3$; u_2 is given by (14.3.3) in J_2 and vanishes in $J_1 \cup J_3$; and u_3 is given by (14.3.4) in J_3 and vanishes in $J_1 \cup J_2$. We wish to compute the Fourier transforms $\hat{u}_2$ and $\hat{u}_3$. Let us first consider $\hat{u}_2$. Assume $f \in L^1$ and set $w = u_2$. Then

$$zw + w'' = \begin{cases} f, & x > b \\ 0, & x < b, \end{cases}$$

and $w(b) = 0$; we have by integration by parts

$$\begin{aligned} (w'')\hat{} &= (2\pi)^{-1/2} \int_b^\infty e^{-ixk} w''(x)\, dx \\ &= -(-ik)(w')\hat{} + (2\pi)^{-1/2} e^{-ixk} w'(x) \big]_b^\infty \\ &= ik(w')\hat{} - (2\pi)^{-1/2} e^{ibk} w'(b), \end{aligned} \tag{14.3.7}$$

where we made use of the fact that $w'(\infty) = 0$. To see this note that

$$\begin{aligned} 2\kappa i w(x) &= \int_b^x e^{i\kappa(x-y)} f(y)\, dy \\ &\quad + \int_x^\infty e^{i\kappa(y-x)} f(y)\, dy - \int_b^\infty e^{i\kappa(x+y-2b)} f(y)\, dy. \end{aligned} \tag{14.3.8}$$

Thus,

$$\begin{aligned} 2w'(x) &= \int_b^x e^{i\kappa(x-y)} f(y)\, dy \\ &\quad - \int_x^\infty e^{i\kappa(y-x)} f(y)\, dy - \int_b^\infty e^{i\kappa(x+y-2b)} f(y)\, dy. \end{aligned} \tag{14.3.9}$$

Now

$$|e^{i\kappa|x-y|} f(y)| \le e^{-\eta|x-y|} |f(y)| \le |f(y)|,$$

where $\eta = \operatorname{Im} \kappa > 0$. Thus it is majorized by a function in L^1 and it converges a.e. to 0 as $x \to \infty$. This shows that $w(x)$ and $w'(x) \to 0$ as $x \to \infty$. Hence (14.3.7) holds. Also we have

$$(w')\hat{} = (2\pi)^{-1/2}\int_b^\infty e^{-ixk}w'(x)\,dx = -(-ik)\hat{w} = ik\hat{w}.$$

This time there are no boundary terms because $w(b) = 0$. Hence we have

$$(w'')\hat{} = -k^2\hat{w} - (2\pi)^{-1/2}e^{-ibk}w'(b).$$

Now by (14.3.9),

$$w'(b) = -\int_b^\infty e^{i\kappa(y-b)}f(y)\,dy. \tag{14.3.10}$$

Hence we have

$$(z - k^2)\hat{w} = (f_{J_2})\hat{} - f_{2\kappa}(k), \tag{14.3.11}$$

where

$$f_{2\kappa}(k) = (2\pi)^{-1/2}e^{ib(k+\kappa)}\int_b^\infty e^{i\kappa y}f(y)\,dy. \tag{14.3.12}$$

Similarly, if $v = u_3$, then

$$zv + v'' = \begin{matrix} f, & x < -b \\ 0, & x > -b \end{matrix}$$

and $v(-b) = 0$. Thus

$$\begin{aligned}(v'')\hat{} &= (2\pi)^{-1/2}\int_{-\infty}^{-b} e^{-ixk}v''(x)\,dx \\ &= ik(v')\hat{} + (2\pi)^{-1/2}e^{-ixb}v'(x)\big]_{-\infty}^{-b} \\ &= ik(v')\hat{} + (2\pi)^{-1/2}e^{ibk}v'(-b).\end{aligned}$$

Now,

$$\begin{aligned}2\kappa i v(x) = &\int_{-\infty}^{x} e^{i\kappa(x-y)}f(y)\,dy + \int_x^{-b} e^{i\kappa(y-x)}f(y)\,dy \\ &- \int_{-\infty}^{-b} e^{-i\kappa(x+y+2b)}f(y)\,dy.\end{aligned} \tag{14.3.13}$$

Hence

$$\begin{aligned}2v'(x) = &\int_{-\infty}^{x} e^{i\kappa(x-y)}f(y)\,dy - \int_x^{-b} e^{i\kappa(y-x)}f(y)\,dy \\ &+ \int_{-\infty}^{-b} e^{i\kappa(x+y+2b)}f(y)\,dy\end{aligned} \tag{14.3.14}$$

and consequently

$$v'(-b) = \int_{-\infty}^{b} e^{i\kappa(y+b)}f(y)\,dy. \tag{14.3.15}$$

Moreover, $(v')\hat{} = ik\hat{v}$ since $v(-b) = 0$. Hence we have

$$(z - k^2)\hat{v} = (f_{J_3})\hat{} - f_{3\kappa}(k), \tag{14.3.16}$$

where

$$f_{3\kappa}(k) = (2\pi)^{-1/2} e^{ib(k-\kappa)} \int_{-\infty}^{-b} e^{-i\kappa y} f(y)\, dy. \tag{14.3.17}$$

Combining these formulas we get

Theorem 14.3.3. *If $f_J = 0$, then*

$$(z - k^2)(\tilde{R}_1(z)f)\hat{} = \hat{f} - f_\kappa(k), \tag{14.3.18}$$

where

$$f_\kappa(k) = A(\kappa)e^{ikb} + B(\kappa)e^{-ikb} \tag{14.3.19}$$

and

$$A(\kappa) = (2\pi)^{-1/2} \int_{-\infty}^{-b} e^{-i\kappa(y+b)} f(y)\, dy,$$

$$B(\kappa) = (2\pi)^{-1/2} \int_{b}^{\infty} e^{i\kappa(y-b)} f(y)\, dy. \tag{14.3.20}$$

Our proof of Theorem 14.3.3 has assumed that $f \in L^1$. This is all that we shall need. But it is easily shown that it is also true for $f \in L^2$ (see Section 12.6).

14.4. The Wave Operators $W_\pm(\tilde{H}_1, H_0)$

In this section we discuss the existence of the wave operators $W_\pm(\tilde{H}_1, H_0)$. We are going to use the criterion of Theorem 8.2.1. In particular, we want to compute

$$\lim_{a\to 0} a \int_I \left(R_0(z)f, \left[\tilde{R}_1(z) - R_0(z)\right] g\right) ds \tag{14.4.1}$$

for each interval I. First we note that

$$[f, u_n]_I = 0, \qquad n = 1, 2, \ldots,$$

where the u_n are the eigenfunctions of H_J defined to vanish outside J and

$$[f, g]_I = \lim \frac{a}{\pi} \int_I \left(R_0(z)f, \tilde{R}_1(z)g\right) ds \tag{14.4.2}$$

when the limit exists. To see this note that the right-hand side of (14.4.2) for $g = u_n$ equals

$$\frac{a}{\pi} \int_I \int (z - k^2)^{-1} (\bar{z} - \lambda_n)^{-1} \hat{f}(k) \hat{u}_n(k)^*\, dk\, ds, \tag{14.4.3}$$

and this converges to 0 as $a \to 0$. In fact the integral

$$a\int_I (z - k^2)^{-1}(\bar{z} - \lambda_n)^{-1}\, ds \tag{14.4.4}$$

is uniformly bounded in a as long as the endpoints of I do not coincide with λ_n. Moreover, (14.4.4) tends to 0 as long as $k^2 \neq \lambda_n$, that is, for almost all k. Thus,

$$[f, g]_I = 0 \tag{14.4.5}$$

whenever g is a linear combination of the u_n. If g is any function vanishing outside J, then

$$g = \sum (g, u_n) u_n \tag{14.4.6}$$

by (13.2.9). Since

$$|[f, g]_I| \leq \|f\|\, \|g\| \tag{14.4.7}$$

by (8.2.17), we have

$$\Big[f, \sum (g, u_n) u_n\Big]_I = \sum (u_n, g)[f, u_n]_I = 0.$$

Hence (14.4.5) holds for all g which vanish outside J. Now by Theorem 14.3.3, for any g

$$(z - k^2)(\tilde{R}_1(z) g_{CJ})^{\wedge} = (g_{CJ})^{\wedge} - g_\kappa(k). \tag{14.4.8}$$

Hence

$$\begin{aligned}[f, g]_I = [f, g_{CJ}]_I &= \lim \frac{a}{\pi} \int_I (R_0(z) f, \tilde{R}_1(z) g_{CJ})\, ds \\ &= \lim \int_I \int \delta_a(s - k^2) \hat{f}(k)[\hat{g}_{CJ}(k) - g_\kappa(k)]^*\, dk\, ds \\ &= \int_{k^2 \in I} \hat{f}(k)[\hat{g}_{CJ}(k) - g_{|k|}(k)]^*\, dk. \end{aligned} \tag{14.4.9}$$

Thus the limit (14.4.1) exists and equals

$$\tilde{J}_I^+(f, g) = -\int_{k^2 \in I} \hat{f}(k)[\hat{g}_J(k) + g_{|k|}(k)]^*\, dk. \tag{14.4.10}$$

This is the first step in proving the existence of the wave operators. The next step is to verify (8.2.4). This will be done in the next section.

14.5. Propagation

In proving the existence of the wave operators we shall make use of an important theorem which has many applications.

Theorem 14.5.1. *Let $\varphi(k)$ be piecewise twice continuously differentiable on $(-\infty, \infty)$ and such that $\varphi'(k) > 0$ except at a finite number of points. Then for each $g \in L^2$ and each $a > -\infty$*

$$\|g\|^2 \geq \int_a^\infty \left| \int e^{ixk + it\varphi(k)} g(k)\, dk \right|^2 dx \to 0 \quad \text{as} \quad t \to \infty. \tag{14.5.1}$$

By piecewise twice continuously differentiable we mean that the real line can be divided up into a finite number of intervals, in the closure of each of which φ is twice continuously differentiable. The proof of Theorem 14.5.1 will be given in the next section. Here we shall draw some consequences and make some observations. The following are simple consequences of (14.5.1). In each, a is assumed finite.

$$\int_a^\infty \left| \int_0^\infty e^{ixk + itk^2} g(k)\, dk \right|^2 dx \to 0 \quad \text{as} \quad t \to \infty. \tag{14.5.2}$$

$$\int_a^\infty \left| \int_0^\infty e^{-ixk - itk^2} g(k)\, dk \right|^2 dx \to 0 \quad \text{as} \quad t \to \infty. \tag{14.5.3}$$

$$\int_{-\infty}^a \left| \int_0^\infty e^{ixk - itk^2} g(k)\, dk \right|^2 dx \to 0 \quad \text{as} \quad t \to \infty. \tag{14.5.4}$$

$$\int_{-\infty}^a \left| \int_0^\infty e^{-ixk + itk^2} g(k)\, dk \right|^2 dx \to 0 \quad \text{as} \quad t \to \infty. \tag{14.5.5}$$

$$\int_{-\infty}^a \left| \int_{-\infty}^0 e^{ixk + itk^2} g(k)\, dk \right|^2 dx \to 0 \quad \text{as} \quad t \to \infty. \tag{14.5.6}$$

$$\int_a^\infty \left| \int_{-\infty}^0 e^{ixk - itk^2} g(k)\, dk \right|^2 dx \to 0 \quad \text{as} \quad t \to \infty. \tag{14.5.7}$$

To prove (14.5.2) we define $g(k) = 0$ and $\varphi(k) = -k^2$ for $k < 0$. Then all of the hypotheses of Theorem 14.5.1 are fulfilled. To prove (14.5.3) we take the complex conjugate of the inner integral and use $g(k)^*$ in place $g(k)$. In (14.5.4) the substitution $y = -x$ transforms the expression into

$$\int_{-a}^\infty \left| \int e^{-iyk - itk^2} g(k)\, dk \right|^2 dy$$

which converges to 0 by (14.5.3). The same trick works for (14.5.5). To obtain (14.5.6) we set $l = -k$ in the inner integral to obtain

$$\int_{-\infty}^a \left| \int_0^\infty e^{-ixl + itl^2} g(-l)\, dl \right|^2 dx,$$

which reduces to (14.5.5) and so on. There are other combinations which we have not bothered to write down. One interpretation of (14.5.1) is that the wave function

$$\psi(x, t) = \int e^{ixk + it\varphi(k)} g(k)\, dk \tag{14.5.8}$$

propagates toward $-\infty$ as $t \to \infty$, so that eventually very little is left in any interval of the form (a, ∞). Similar interpretations hold for (14.5.2)–(14.5.7). Another consequence of Theorem 14.5.1 is that for each bounded interval I and $g \in L^2$

$$\int_I \left| \int e^{ixk+itk^2} g(k)\, dk \right|^2 dx \to 0 \qquad \text{as} \quad t \to \infty. \tag{14.5.9}$$

To see this we split the inner integral up into an integral over $(-\infty, 0)$ and one over $(0, \infty)$ and apply (14.5.6) and (14.5.2), respectively. Of course this makes use of the fact that both endpoints of I are finite. Thus the wave function

$$\psi(x, t) = \int e^{ixk+tk^2} g(k)\, dk \tag{14.5.10}$$

is the sum of two wave functions, one propagating to ∞ and the other to $-\infty$ as $t \to \infty$.

To complete the proof that the wave operators $W_\pm(\tilde{H}_1, H_0)$ exist, we show that

$$\tilde{J}_I^+(f_t, f_t) \to 0 \qquad \text{as} \quad t \to \infty, \tag{14.5.11}$$

where $\tilde{J}_I^+$ is given by (14.4.10) and

$$f_t = e^{-itH_0} f, \qquad \hat{f}_t = e^{-itk^2} \hat{f}. \tag{14.5.12}$$

Now,

$$\left| \int_{k^2 \in I} \hat{f}(k) \hat{g}_J(k)^*\, dk \right| \le \|\hat{f}\|\, \|\hat{g}_J\| = \|f\|\, \|g_J\|.$$

Thus this term will be taken care of if we can show that

$$\|(f_t)_J\| \to 0 \qquad \text{as} \quad t \to \infty.$$

But

$$\|(f_t)_J\|^2 = \int_J \left| \int e^{ixk-itk^2} \hat{f}(k)\, dk \right|^2 dx \to 0$$

by (14.5.9). Hence we can concentrate on the term

$$\begin{aligned} &\int_{k^2 \in I} \hat{f}(k) g_{|k|}(k)^*\, dk \\ &\quad = \int_{k^2 \in I} \hat{f}(k) \left[\int_{-\infty}^{-b} e^{-ikb+i|k|(y+b)} g(y)^*\, dy \right. \\ &\qquad\qquad \left. + \int_b^\infty e^{ikb-i|k|(y-b)} g(y)^*\, dy \right] dk. \end{aligned} \tag{14.5.13}$$

If $I = (c^2, d^2)$, $0 < c < d$, set $I_1 = (c, d)$ and $I_2 = (-d, -c)$. The right-hand side of (14.5.13) becomes

$$\int_{I_1} \hat{f}(k)\left[\int_{-\infty}^{-b} e^{iky} g(y)^* \, dy + \int_b^{\infty} e^{2ikb - iky} g(y)^* \, dy\right] dk$$
$$+ \int_{I_2} \hat{f}(k)\left[\int_{-\infty}^{-b} e^{-2ikb - iky} g(y)^* \, dy + \int_b^{\infty} e^{iky} g(y)^* \, dy\right] dk.$$

Estimates for these terms are easily found. For instance,

$$\left|\int_{-\infty}^{-b} g(y)^* \int_{I_1} e^{iky}\hat{f}(k)\, dk\, dy\right|^2 \le \|g\|^2 \int_{-\infty}^{-b} \left|\int_{I_1} e^{iky}\hat{f}(k)\, dk\right|^2 dy.$$

Now, if we set $f = g = f_t$, then this converges to 0 by (14.5.4). Similarly,

$$\left|\int_b^{\infty} g(y)^* \int_{I_1} e^{ik(2b-y)}\hat{f}(k)\, dk\, dy\right|^2 \le \|g\|^2 \int_{-b}^{\infty} \left|\int_{I_1} e^{-ikx}\hat{f}(k)\, dk\right|^2 dx,$$

where we made the substitution $x = y - 2b$. Again, if we set $f = g = f_t$, then this converges to 0 by (14.5.3). Now,

$$\int_{-\infty}^{-b} g(y)^* \int_{I_2} e^{-ik(2b+y)}\hat{f}(k)\, dk\, dy = \int_{-\infty}^{b} g(x - 2b)^* \int_{I_1} e^{ikx}\hat{f}(-k)\, dk\, dx.$$

Thus the absolute value squared of this term is bounded by

$$\|g\|^2 \int_{-\infty}^{b} \left|\int_{I_1} e^{ikx}\hat{f}(-k)\, dk\right|^2 dx.$$

Thus, if $f = g = f_t$, this term converges to 0 as well by (14.5.4). Finally, we note that

$$\left|\int_b^{\infty} g(y)^* \int_{I_2} e^{iky}\hat{f}(k)\, dk\, dy\right|^2 \le \|g\|^2 \int_b^{\infty} \left|\int_{I_2} e^{iky}\hat{f}(k)\, dk\right|^2 dy,$$

and if we set $f = g = f_t$, then this converges to 0 by (14.5.7). Thus we have

Theorem 14.5.2. $M_{\pm}(\tilde{H}_1, H_0) = L^2$.

PROOF. We have just shown that (8.2.4) holds. Thus, $M_+(\tilde{H}_1, H_0) = L^2$ by Theorem 8.2.1. If we substituted $-\tilde{H}_1$, $-H_0$ for $\tilde{H}_1$, H_0, the same proof goes through. □

The next step is the proof of completeness. This will be carried out in Section 14.7 after we prove Theorem 14.5.1 in the next section.

14.6. Proof of Theorem 14.5.1

A *step function* is one that is constant in each of a finite number of bounded intervals and vanishes elsewhere. We shall use

Theorem 14.6.1. *The step functions are dense in* L^2.

We shall prove this theorem at the end of the section. Now we shall give the

PROOF OF THEOREM 14.5.1. First we note that if the first integration on the right-hand side of (14.5.1) were from $-\infty$ to ∞ instead of a to ∞, then we would have equality in (14.5.1). For then the right-hand side would equal the L^2-norm of $e^{it\varphi(k)}g(k)$ by Parseval's identity. Next we note that it suffices to prove the limit in (14.5.1) for step functions. Let g be any function in L^2 and denote the right-hand side of (14.5.1) by $\|g\|_t^2$. Let $\varepsilon > 0$ be given, and let v be a step function such that $\|g - v\| < \varepsilon$. Then take t so large that $\|v\|_t < \varepsilon$. Then we have $\|g\|_t \le \|g - v\| + \|v\|_t < 2\varepsilon$ for t sufficiently large. Thus the limit in (14.5.1) holds for g. Next we note that if the limit in (14.5.1) holds for each of two functions, it holds for their sum. Thus it suffices to prove the limit in (14.5.1) for a function which is constant in a bounded interval I and vanishes outside I. For the same reason we may take the interval I such that $\varphi'(k)$ does not vanish in its interior. Moreover, if $\varphi'(k)$ vanishes at an endpoint of I we shrink I slightly to avoid this. Our error will be at most the constant value of g times the length cut off by the inequality in (14.5.1) (actually this amount squared). For any $\varepsilon > 0$ we can make this error $< \varepsilon$ and proceed as above. Thus we may assume that $g = \alpha$ in I and vanishes outside, and that $\varphi'(k) \ge c > 0$ in I.

The inner integral in (14.5.1) equals

$$- i\alpha \int_I \tau(k)\, d(e^{ixk + it\varphi(k)}), \tag{14.6.1}$$

where $\tau(k) = (x + t\varphi'(k))^{-1}$ is positive as long as $t > |a|/c$. Moreover, it will satisfy

$$\tau(k) \le (x + tc)^{-1} \le 2/tc, \qquad x \ge a, \tag{14.6.2}$$

as long as $t > 2|a|/c$. Integrating (14.6.1) by parts we get $-i\alpha$ times

$$\tau(k)e^{ixk + it\varphi(k)}\Big]_c^d - \int_I e^{ixk + it\varphi(k)}\tau'(k)\, dk. \tag{14.6.3}$$

Now

$$\tau'(k) = -(x + t\varphi')^{-2}t\varphi'' = -\tau(k)^2 t\varphi''.$$

Thus, if $|\varphi''| \le M$ in I, we have

$$\int_I |\tau'(k)|\, dk \le tM \int_I \tau(k)^2\, dk \le tM|I|(x + tc)^{-2}.$$

Hence (14.6.3) is bounded in absolute value by

$$(2(x + tc) + tM|I|)/(x + tc)^2 \le 2(1 + M|I|c^{-1})/(x + tc).$$

Thus the right-hand side of (14.5.1) is bounded by

$$2|\alpha|^2(1 + M|I|c^{-1})^2 \int_a^\infty (x + tc)^{-2}\,dx$$

$$= 2|\alpha|^2(1 + M|I|c^{-1})^2/(a + tc) \to 0 \qquad \text{as} \quad t \to \infty.$$

This completes the proof. □

Now we turn to the

PROOF OF THEOREM 14.6.1. Let $\varepsilon > 0$ be given and let u be any function in L^2. Then there is a $v \in C_0^\infty$ such that $\|u - v\| < \varepsilon$ (Theorem 7.8.2). Let I be a bounded interval containing the support of v (i.e., v vanishes outside I). Break up I into subintervals $I_1, \ldots, I_N$ such that $v(x)$ does not vary more than $\varepsilon/|I|^{1/2}$ in each I_n. Pick a point $x_n \in I_n$ and define the step function w to equal $v(x_n)$ in I_n and to vanish outside I. Then $|v - w| < \varepsilon/|I|^{1/2}$ in I and vanishes outside I. Hence

$$\int |v - w|^2\,dx \leq |I|\varepsilon^2/|I| = \varepsilon^2.$$

Thus,

$$\|u - w\| \leq \|u - v\| + \|v - w\| < 2\varepsilon.$$

□

14.7. Completeness of the Wave Operators $W_\pm(\tilde{H}_1, H_0)$

Now that we have shown the existence of the wave operators $W_\pm(\tilde{H}_1, H_0)$ on the whole of L^2, we must discuss their completeness. This turns out to be easier than expected because of the following results.

Theorem 14.7.1. *Set*

$$\tilde{f}_\pm(k) = \hat{f}_{CJ}(k) - f_{\pm|k|}(k), \tag{14.7.1}$$

where $f_\kappa(k)$ is defined by (14.3.18), (14.3.19). *Then for any bounded function $\varphi(\lambda)$ we have*

$$(\varphi(\tilde{H}_1)f, g) = \int \varphi(k^2)\tilde{f}_\pm(k)\tilde{g}_\pm(k)^*\,dk \tag{14.7.2}$$

when $f_J = g_J = 0$.

PROOF. Set $z = s \pm ia$. Then by Theorem 14.3.3

$$a\int_I (\tilde{R}_1(z)f, \tilde{R}_1(z)g)\,ds = \pi \int_I \int \delta_a(s - k^2)\tilde{f}_\kappa(k)\tilde{g}_\kappa(k)^*\,dk\,ds, \tag{14.7.3}$$

where $\tilde{f}_\kappa = \hat{f}_{CJ} - f_\kappa$ with f_κ given by (14.3.19) and with corresponding formula for $\tilde{g}_\kappa$. Taking the limit as $a \to 0$, we get $\kappa \to \pm |k|$ and

$$(E_1(I)f, g) = \int_{k^2 \in I} \tilde{f}_\pm(k)\tilde{g}_\pm(k)^* \, dk. \tag{14.7.4}$$

In particular, this implies

$$(E_1(\lambda)f, g) = \begin{cases} \int_{-\lambda^{1/2}}^{\lambda^{1/2}} \tilde{f}_\pm(k)\tilde{g}_\pm(k)^* \, dk, & \lambda \geq 0 \\ 0, & \lambda < 0. \end{cases} \tag{14.7.5}$$

Thus,

$$d(E_1(\lambda)f, g)/d\lambda = \tfrac{1}{2}\lambda^{-1/2}\left[\tilde{f}_\pm(\lambda^{1/2})\tilde{g}_\pm(\lambda^{1/2})^* + \tilde{f}_\pm(-\lambda^{1/2})\tilde{g}_\pm(-\lambda^{1/2})^*\right], \tag{14.7.6}$$

and consequently,

$$\begin{aligned}(\varphi(\tilde{H}_1)f, g) &= \int_0^\infty \varphi(\lambda) \, d(E_1(\lambda)f, g) \\ &= \frac{1}{2}\int_0^\infty \varphi(\lambda)\lambda^{-1/2}\left[\tilde{f}_\pm(\lambda^{1/2})\tilde{g}_\pm(\lambda^{1/2})^* + \tilde{f}_\pm(-\lambda^{1/2})\tilde{g}_\pm(-\lambda^{1/2})^*\right] d\lambda \\ &= \int_{-\infty}^\infty \varphi(k^2)\tilde{f}_\pm(k)\tilde{g}_\pm(k)^* \, dk. \end{aligned} \tag{14.7.7}$$

This gives (14.7.2). □

Theorem 14.7.2. *If*

$$\tilde{W}_\pm f = W_\pm(\tilde{H}_1, H_0)f, \tag{14.7.8}$$

then

$$(\tilde{W}_\pm f, g) = \int \hat{f}(k)\tilde{g}_\pm(k)^* \, dk. \tag{14.7.9}$$

PROOF. We know that the limit (14.7.8) exists for each $f \in L^2$ (Theorem 14.5.2). Now by (8.1.10) and (8.1.14), if $g_J = 0$, then

$$\begin{aligned}(\tilde{W}_\pm f, g) &= \lim \frac{a}{\pi} \int (R_0(z)f, \tilde{R}_1(z)g) \, ds \\ &= \lim \int\int \delta_a(s - k^2)\hat{f}(k)\tilde{g}_\kappa(k)^* \, dk \, ds.\end{aligned}$$

This gives (14.7.9). On the other hand, if g vanishes outside J, then $(\tilde{W}_\pm f, g) = 0$ by (14.4.5). Thus, (14.7.9) holds for any g. □

Our main result is

Theorem 14.7.3. *The wave operators $W_\pm(\tilde{H}_1, H_0)$ are strongly complete. In fact, we have*

$$R_\pm(\tilde{H}_1, H_0) = \mathcal{H}_{ac}(\tilde{H}_1) = \mathcal{H}_c(\tilde{H}_1) = \{f \in L^2 | f_J = 0\}. \tag{14.7.10}$$

PROOF. By (13.2.9), every $f \in L^2$ which vanishes outside J is in the subspace spanned by the eigenfunctions of $\tilde{H}_1$ (note that any eigenfunction of H_J when defined so as to vanish outside J becomes an eigenfunction of $\tilde{H}_1$). Thus

$$\mathcal{H}_c(\tilde{H}_1) \subset \{f \in L^2 | f_J = 0\}. \tag{14.7.11}$$

On the other hand, (14.7.4) shows that

$$\{f \in L^2 | f_J = 0\} \subset \mathcal{H}_{ac}(\tilde{H}_1). \tag{14.7.12}$$

This together with (14.7.11) gives all but the first equality in (14.7.10). Now, if $g_{CJ} = 0$, then $(\tilde{W}_{\pm} f, g) = 0$ for each f by (14.4.5). Thus, $g \perp R(\tilde{W}_{\pm})$. On the other hand, if g is any function orthogonal to $R(\tilde{W}_{\pm})$, then $\tilde{g}_{\pm}(k) = 0$ by (14.7.9). But by (14.7.2),

$$\|g_{CJ}\| = \|\tilde{g}_{\pm}\| = 0.$$

Hence we have

$$R(W_{\pm})^{\perp} = \{g \in L^2 | g_{CJ} = 0\},$$

which is the same as

$$R(W_{\pm}) = \{g \in L^2 | g_J = 0\}.$$

This gives the theorem. □

14.8. The Wave Operators $W_{\pm}(H, \tilde{H}_1)$

Next we compare the operators H and $\tilde{H}_1$. In this case we can apply the theory of Chapter 9. It will make things much easier if we take $b = c + 1$ in the definition of J, where c is the constant in (14.2.1) and (14.2.3). The reason for this will soon be apparent.

The difficulty in comparing the operators H and $\tilde{H}_1$ is that the domains of their corresponding bilinear forms $h, \tilde{h}_1$ are different. Thus, $D(h)$ is given by (B) of Section 14.1, while $D(\tilde{h}_1)$ equals

$$D(\tilde{h}_1) = \{u \in L^2 | u_{J_i} \in D(h_i)\} \tag{14.8.1}$$

with each $D(h_i)$ given by (14.2.4). However, since

$$\tilde{h}_1(u, v) = (u', v') + (V_1 u, v), \tag{14.8.2}$$

we can extend this form so as to be defined for functions in $D(h)$. Let us denote the extended form by $\tilde{h}$. We must be careful in using this form since it no longer corresponds to the operator $\tilde{H}_1$. However, this is where the assumption that $V_1(x)$ vanishes for $|x| > b - 1$ becomes useful. It allows us to prove

Lemma 14.8.1. *If $u \in D(\tilde{H}_1)$ and $v \in D(h)$, then*

$$\tilde{h}(u, v) = (\tilde{H}_1 u, v) + \langle u, v \rangle, \tag{14.8.3}$$

where

$$\langle u, v\rangle = [u'(b-0) - u'(b+0)]v(b)^* + [u'(-b-0) - u'(-b+0)]v(-b)^*. \quad (14.8.4)$$

Here we use the notation

$$w(x \pm 0) = \lim_{0<\delta\to 0} w(x \pm \delta). \quad (14.8.5)$$

A proof of Lemma 14.8.1 will be given in Section 14.10. Here we shall show how it can be used to study the wave operators $W_{\pm}(H, \tilde{H}_1)$. Set $V_2 = V - V_1$. Then by hypotheses I and II, we have

$$\int |V_2(x)|\, dx < \infty. \quad (14.8.6)$$

Now by Lemma 14.8.1, if $u \in D(\tilde{H}_1)$ and $v \in D(H)$, we have

$$h(u, v) = \tilde{h}(u, v) + (V_2 u, v) = (\tilde{H}_1 u, v) + (Au, Bv)_{\mathcal{K}}, \quad (14.8.7)$$

where

$$Au = \{u'(b-0) - u'(b+0), u'(-b-0) - u'(-b+0), A_2 u\},$$

$$Bu = \{v(b-0), v(-b+0), B_2 u\},$$

$$A_2(x) = \begin{cases} |V_2(x)|^{1/2}, & V_2(x) \neq 0 \\ e^{-x^2}, & V_2(x) = 0, \end{cases}$$

$$B_2(x) = V_2(x)/A_2(x),$$

$$\mathcal{K} = E^2 \oplus L^2(-\infty, \infty).$$

Note that elements of $\mathcal{K}$ consist of vectors of the form $\{\alpha_1, \alpha_2, w\}$, where α_1, α_2 are scalars and $w \in L^2$. The scalar product that makes $\mathcal{K}$ into a Hilbert space is obvious. Note that functions in $D(H)$ are continuous. Thus we could have used $v(b)$ and $v(-b)$ in place of $v(b-0)$ and $v(-b+0)$ in the definition of B. The definition we gave is more convenient in that it allows B to have a larger domain.

The next step is to show that $A\tilde{R}_1(z)$ and $B\tilde{R}(z)$ are bounded operators from L^2 to $\mathcal{K}$ for $\operatorname{Im} z \neq 0$. Let φ be a function in C_0^∞ which vanishes outside $b-1 < x < b+1$ and such that $\varphi(b) = 1$. Clearly, $\varphi \in D(H)$. By (14.8.2) and (14.8.3),

$$(u', \varphi') = \tilde{h}(u, \varphi) = (\tilde{H}_1 u, \varphi) + u'(b-0) - u'(b+0) \quad (14.8.8)$$

since V_1 vanishes on the support of φ. Hence

$$|u'(b-0) - u'(b+0)| \le |(\tilde{H}_1 u, \varphi)| + |(u', \varphi')|. \quad (14.8.9)$$

Now, if $u = \tilde{R}_1(z)f$, the right-hand side is bounded by

$$\|zu - f\|\, \|\varphi\| + \|u'\|\, \|\varphi'\|. \quad (14.8.10)$$

Also, by Lemma 13.5.4,

$$\|u_{CJ}\| \le c\|f_{CJ}\|/\eta|z|^{1/2} \tag{14.8.11}$$

and

$$\|u'_{CJ}\| \le c\|f_{CJ}\|/\eta. \tag{14.8.12}$$

Moreover, by (13.2.10) and (13.1.9), we have

$$\|u_J\|^2 = \sum |z - \lambda_n|^{-2}|(f, u_n)|^2 \le \|f_J\|^2/d^2, \tag{14.8.13}$$

$$\|u'_J\|^2 \le C_1 h_J(u_J) + C_2\|u_J\|^2, \tag{14.8.14}$$

where d is the distance from z to $\sigma(H_J)$. Since $h_J(u_J) = (H_J u_J, u_J)$, (14.8.14) gives

$$\|u'_J\| \le C(|z| + d^{-1} + 1)\|f_J\|. \tag{14.8.15}$$

Combining (14.8.10)–(14.8.15), we obtain

$$|u'(b - 0) - u'(b + 0)| \le C(|z| + 1)(\eta^{-1} + d^{-1} + 1)\|f\|. \tag{14.8.16}$$

A similar estimate holds for $|u'(-b - 0) - u'(-b + 0)|$. Also, by (14.8.12) and (14.8.15),

$$\|u'\| + \|u\| \le C(|z| + |z|^{-1/2} + \eta^{-1} + d^{-1})\|f\|. \tag{14.8.17}$$

This implies that $|u|$ is bounded by this constant as well [see (2.7.6)]. Since A_2 and B_2 are in L^2, we see that $A\tilde{R}_1(z)$ and $B\tilde{R}_1(z)$ are bounded operators from L^2 to $\mathcal{K}$ and

$$\|A\tilde{R}_1(z)\| + \|B\tilde{R}_1(z)\| \le C(|z| + |z|^{-1/2} + 1)(\eta^{-1} + d^{-1} + 1). \tag{14.8.18}$$

Also, we note that

$$B\tilde{R}_1(z)\sum_1^N (f, u_n)u_n = B\sum_1^N (z - \lambda_n)^{-1}(f, u_n)u_n$$

$$= \left\{0, 0, B_2\sum_1^N (z - \lambda_n)^{-1}(f, u_n)u_n\right\}.$$

Since $B\tilde{R}_1(z)$ is bounded, this gives

$$B\tilde{R}_1(z)f = \{0, 0, B_2\tilde{R}_1(z)f\}. \tag{14.8.19}$$

It is easily checked that this is a compact operator. On the intervals J_2, J_3 this follows as shown in Lemma 9.14.3. In the interval J we note that

$$\|u_J\| + \|u'_J\| \le C\|f_J\| \tag{14.8.20}$$

by (14.8.13) and (14.8.15). Thus, if $\{f_n\}$ is a bounded sequence in $L^2(J)$, then $\{\tilde{R}_1(z)f_n\}$ has a subsequence converging uniformly on J (Lemma 4.7.2). Hence $\{B_2\tilde{R}_1(z)f_n\}$ has a subsequence converging in $L^2(J)$.

Next we note by (14.3.9) and (14.3.14) that

$$A\tilde{R}_1(z)f = \Big\{ \sum (z - \lambda_n)^{-1}(f, u_n)u_n'(b) + \int_b^\infty e^{i\kappa(y-b)}f(y)\,dy,$$
$$\sum (z - \lambda_n)^{-1}(f, u_n)u_n'(-b) - \int_{-\infty}^{-b} e^{-i\kappa(y+b)}f(y)\,dy,\ A_2\tilde{R}_1(z)f \Big\}. \tag{14.8.21}$$

From this we can calculate $[AR_1(z)]^*$. If $W = \{\beta_1, \beta_2, v\}$, then

$$\begin{aligned}[A\tilde{R}_1(z)]^*W(y) &= \beta_1\chi_{J_1}e^{-i\kappa(y-b)} + \beta_2\chi_{J_2}e^{i\kappa(y+b)} \\ &\quad + \frac{i}{2\kappa}\chi_{J_1}\int_b^\infty \left[e^{-i\kappa|x-y|} - e^{-i\kappa|x+y-2b|}\right]A_2(x)v(x)\,dx \\ &\quad + \frac{i}{2\kappa}\chi_{J_2}\int_{-\infty}^{-b} \left[e^{-i\kappa|x-y|} - e^{-i\kappa|x+y+2b|}\right]A_2(x)v(x)\,dx \\ &\quad + \sum (z-\lambda_n)^{-1}\left[(A_2v, u_n) + \beta_1u_n'(b) + \beta_2u_n'(-b)\right]u_n(y).\end{aligned} \tag{14.8.22}$$

Thus,

$$B[A\tilde{R}_1(z)]^*W = \{0, 0, B_2[A\tilde{R}_1(z)]^*W\}. \tag{14.8.23}$$

Before we can proceed we must show that these operators are bounded. We shall prove this and finish the proof of Theorem 14.2.1.

14.9. A Regularity Theorem

In our proof of Theorem 14.2.1, and especially in our proof of Lemma 14.8.1, we make use of the fact that $V_1(x)$ vanishes in $c = b - 1 < |x| < b$. This gives us a clear picture of the nature of the u_n on this set. To obtain this we shall use the following regularity theorem, which will be proved in the next section.

Theorem 14.9.1. *If $u \in L^2(I)$ satisfies*

$$(u, \varphi'') = \lambda(u, \varphi), \qquad \varphi \in C_0^\infty(I), \tag{14.9.1}$$

then $u \in C^\infty(I)$ and $u'' = \lambda u$.

We should remark that the statement of Theorem 14.9.1 is not absolutely precise. For if we change u on a set of measure 0, hypothesis (14.9.1) is not altered. So we cannot possibly claim that u is infinitely differentiable. The only thing we can claim is that it is possible to change u on a set of measure 0 so that the resulting function is infinitely differentiable. Another way of stating it is to say that u is almost everywhere equal to a function in $C^{\infty}(I)$.

As a consequence of Theorem 14.9.1 we have

Theorem 14.9.2. *Suppose* $\lambda_n > 0$ *for* $n > N$, *and set* $k_n = \lambda_n^{1/2}$. *Then*

$$\begin{aligned} u_n(x) &= C_n \sin k_n(x - b), \qquad c < x < b \\ &= D_n \sin k_n(x + b), \qquad -b < x < -c. \end{aligned} \tag{14.9.2}$$

PROOF. Set $I = (c, b)$. Then by definition,

$$(u_n', \varphi') = \tilde{h}_1(u_n, \varphi) = \lambda_n(u_n, \varphi), \qquad \varphi \in C_0^{\infty}(I).$$

By Theorem 14.9.2, $u_n \in C^{\infty}(I)$ and $u_n'' = \lambda_n u_n$. The most general solution of this equation is

$$u_n(x) = C_n \sin k_n(x - \alpha).$$

Since $u_n(b) = 0$, we see that $\alpha = b$. A similar argument works in the interval $(-b, -c)$. □

Theorem 14.9.3

$$\sum_N^{\infty} (C_n^2 + D_n^2)/\lambda_n < \infty. \tag{14.9.3}$$

PROOF. Set

$$u(x) = \begin{cases} x - c, & c < x < b \\ 0, & -b < x < c. \end{cases}$$

Then

$$(u, u_n) = C_n \int_c^b (x - c) \sin k_n(x - b)\, dx.$$

The integral equals

$$\begin{aligned} \int_0^1 (y - 1) \sin k_n y\, dy &= k_n^{-1} \int_0^1 (y - 1)\, d \cos k_n y \\ &= (1 - k_n^{-1} \sin k_n)/k_n \end{aligned}$$

Thus for n sufficiently large,

$$|(u, u_n)|^2 \geq C_n^2/2\lambda_n \tag{14.9.4}$$

Since $\lambda_n \to \infty$ by Theorem 13.3.1. Since $u \in L^2(J)$ and the u_n are orthonormal, we have

$$\sum |(u, u_n)|^2 \leq \|u\|^2.$$

This gives (14.9.3) for the C_n. A similar argument works for the D_n. □

We set $\Lambda = R - \{0\} - \sigma(H_J)$ and let χ_c be the characteristic function of the set $|x| > c$.

Theorem 14.9.4. *If* $u = \tilde{R}_1(z)\chi_c f$, *then*

$$\|u\|_\infty \le C\Big(|z|^{-1/2} + \sum (C_n^2 + D_n^2)|z - \lambda_n|^{-1}\Big)\|f\|_1 \tag{14.9.5}$$

and

$$|u'(b-0) - u'(b+0)| + |u'(-b+0) - u'(-b-0)|$$
$$\le C\Big(|z|^{-1/2} + |z|^{1/2} + (1 + |z|)\sum (C_n^2 + D_n^2)|z - \lambda_n|^{-1}\Big)\|f\|_1. \tag{14.9.6}$$

PROOF. Assume f vanishes in $|x| \le c$. Now,

$$u_J = \sum (z - \lambda_n)^{-1}(f, u_n)u_n.$$

Hence

$$|u_J| \le C\sum |z - \lambda_n|^{-1}(C_n^2 + D_n^2)\|f\|_1.$$

Also,

$$|u_{CJ}| \le \|f\|_1/2|\kappa|$$

by (13.5.15). This gives (13.9.5). To obtain (13.9.6), let φ be a function in C_0^∞ which vanishes outside $c < x < b + 1$ and such that $\varphi(b) = 1$. Then by (14.8.9),

$$\begin{aligned}|u'(b-0) - u'(b+0)| &\le |(zu - f, \varphi)| + |(u', \varphi')| \\ &\le |z|\,\|u\|_\infty\|\varphi\|_1 + \|f\|_1\|\varphi\|_\infty + |(u', \varphi')|.\end{aligned} \tag{14.9.7}$$

Now,

$$(u', \varphi') = -(u, \varphi'')$$

since $u(b) = \varphi'(c) = 0$. Hence

$$|(u', \varphi')| \le \|u\|_\infty\|\varphi''\|_1.$$

Hence the left-hand side of (14.9.7) is bounded by

$$C_1(|z| + 1)\|u\|_\infty + C_2\|f\|_1.$$

If we use (14.9.5), this gives the first part of (14.9.6). The second part is obtained in a similar way. □

Corollary 14.9.1. *The operator* $A\tilde{R}_1(z)\chi_c$ *is bounded from* L^1 *to* $\mathcal{K}$. *It is uniformly bounded in* ω_I *for any* $I \subset\subset \Lambda$.

PROOF. This follows from inequalities (14.9.5) and (14.9.6) and the fact that $A_2 \in L^2$. Note that

$$\sum_1^\infty (C_n^2 + D_n^2)|z - \lambda_n|^{-1}$$

is uniformly bounded in ω_I for $I \subset\subset \Lambda$. □

Corollary 14.9.2. *The operator $B[A\tilde{R}_1(z)]^*$ is bounded on $\mathcal{H}$ uniformly in ω_I for $I \subset\subset \Lambda$.*

PROOF. We have by Corollary 14.9.1

$$|(B_2[A\tilde{R}_1(z)]^*w, h)| = |(w, A\tilde{R}_1(z)\chi_c B_2 h)|$$
$$\leq \|w\|_{\mathcal{H}} C(z)\|B_2 h\|_1 \leq C(z)\|B_2\| \; \|w\|_{\mathcal{H}}\|h\|,$$

where $C(z)$ is uniformly bounded in ω_I. The result now follows by (14.8.23). □

Theorem 14.9.5. *The operator $G(z) = 1 - B[A\tilde{R}_1(z)]^*$ has a bounded inverse for all $z \in \rho(H_1)$.*

PROOF. The equation

$$G(z)\{\beta_1, \beta_2, v\} = \{\gamma_1, \gamma_2, w\} \tag{14.9.8}$$

is equivalent to $\beta_i = \gamma_i$ and

$$(1 - [B_2\tilde{R}_1(z)A_2])v = w + \gamma_1 B_2\chi_{J_2}e^{-i\kappa(x-b)} + \gamma_2 B_2\chi_{J_3}e^{i\kappa(x+b)}$$
$$+ B_2\sum(z-\lambda_n)^{-1}[\gamma_1 u_n'(b) + \gamma_2 u_n'(-b)]u_n. \tag{14.9.9}$$

Note that the right-hand side of (14.9.9) is in L^2 by Corollary 14.9.2. Now, the operator on the left-hand side has a bounded inverse. In fact, we have

$$1 + [B_2\tilde{R}(z)A_2] = (1 - [B_2\tilde{R}_1(z)A_2])^{-1} \tag{14.9.10}$$

(see Theorem 9.11.3). This gives the theorem. □

Finally we are ready for the

PROOF OF THEOREM 14.2.1. We have shown that all of the hypotheses of Theorem 9.11.2 hold. By Theorem 14.7.3, $E_1(\Lambda)$ is the projection onto $\mathcal{H}_{ac}(\tilde{H}_1) = \mathcal{H}_c(\tilde{H}_1) = \{f \in L^2 | f_J = 0\}$. $B\tilde{R}_1(z)$ was shown to be compact for $\operatorname{Im} z \neq 0$. $G(z)$ is uniformly bounded in ω_I for $I \subset\subset \Lambda$ and it has a bounded inverse for $\operatorname{Im} z \neq 0$. We show that $G(z)$ is uniformly continuous in ω_I as in the proof of Lemma 9.14.4. [Note that $R_J(z)$ is continuous on Λ.] We have

$$a\|A\tilde{R}_1(z)\|^2 + a\|B\tilde{R}_1(z)\|^2 \leq C_I, \qquad z, \bar{z}, \in \omega_I, \tag{14.9.11}$$

for $I \subset\subset \Lambda$ by (14.8.14). Hence the wave operators are complete. □

14.10. A Family of Spaces

Now we turn our attention to the proof of Theorem 14.9.1. For this purpose we introduce a family of Hilbert spaces as follows. For each real s we let H^s be the set of functions $u(x)$ such that $(1 + k^2)^{s/2}\hat{u}(k)$ is in L^2.

The norm of H^s is given by

$$|u|_s^2 = \int (1 + k^2)^s |\hat{u}(k)|^2 \, dk. \tag{14.10.1}$$

Clearly, we have

Lemma 14.10.1. *$C_0^\infty \subset H^s$ for any s.*

Another simple consequence of the definition is

Lemma 14.10.2. $|u'|_s \leq |u|_{s+1}$.

It is also easily checked that when s is a nonnegative integer we have

$$|u|_s^2 = \sum_{j \leq s} \binom{s}{j} \int |D^j u(x)|^2 \, dx. \tag{14.10.2}$$

Another important statement is

$$|(u, v)| \leq |u|_s |v|_{-s}. \tag{14.10.3}$$

We also have

$$|u|_s = \sup_{v \in H^{-s}} |(u, v)|/|v|_{-s}. \tag{14.10.4}$$

To see this note that the right-hand side of (14.10.4) is less than or equal to the left-hand side. Now if $u \in H^{2s}$, let v be such that $\hat{v} = (1 + k^2)^s \hat{u}$, that is, take v to be

$$v(x) = (2\pi)^{-1/2} \int e^{ixk} (1 + k^2)^s \hat{u}(k) \, dk.$$

Then clearly $v \in H^{-s}$. Moreover, $(u, v) = |u|_s^2$ and $|v|_{-s} = |u|_s$. This gives (14.10.4). Now if u is only known to be in H^s, we let $\{\varphi_n\}$ be a sequence of functions in C_0^∞ converging to $(1 + k^2)^{s/2} \hat{u}$ in L^2 (Theorem 7.8.2), and set $\hat{v}_n = (1 + k^2)^{s/2} \varphi_n$. Then

$$(u, v_n) = \left(\hat{u}, (1 + k^2)^{s/2} \varphi_n\right) \to |u|_s^2$$

and

$$|v_n|_{-s} = \|\varphi_n\| \to |u|_s.$$

Hence

$$|(u, v_n)|/|v_n|_{-s} \to |u|_s.$$

Another important fact is

Theorem 14.10.1. *If $\varphi \in C_0^\infty$ and s is an integer, then there is a constant C depending only on s and φ such that*

$$|\varphi u|_s \leq C |u|_s, \qquad u \in H^s. \tag{14.10.5}$$

PROOF. For s nonnegative, this follows easily from (14.10.2) and the rule for differentiating a product. If s is negative, we have

$$|(\varphi u, v)| \le |u|_s |\varphi v|_{-s} \le C|u|_s |v|_{-s}$$

and we can apply (14.10.4). □

Theorem 14.10.2. *If $u \in H^s$ and $s > \frac{1}{2}$, then u is bounded and continuous. It satisfies*

$$2\pi \|u\|_\infty^2 \le \int (1 + k^2)^{-s}\, dk\, |u|_s^2. \tag{14.10.6}$$

PROOF. Since

$$u(x) = (2\pi)^{-1/2} \int e^{ixk} (1 + k^2)^{-s/2} (1 + k^2)^{-s/2} \hat{u}(k)\, dk,$$

(14.10.6) follows from Schwarz's inequality. We also have

$$|u(x) - u(y)|^2 \le \int |e^{ixk} - e^{iyk}| (1 + k^2)^{-s}\, dk\, |u|_s^2. \tag{14.10.7}$$

This shows that u is continuous. □

Theorem 14.10.3. *For each real s, C_0^∞ is dense in H^s.*

PROOF. It suffices to prove this for s a positive integer. Suppose $D^j u \in L^2$ for $j \le s$. Let $\varphi(x)$ be a function in C_0^∞ which equals 1 when $|x| < 1$, and set $\varphi_n(x) = \varphi(x/n)$. Then $D^j \varphi_n \to 0$ as $n \to \infty$ uniformly when $j > 0$. Set $u_n = \varphi_n J_n u$, where J_n is the mollifier described in Section 7.8. Then $u_n \in C_0^\infty$ and

$$D^j u_n = \sum \binom{j}{i} D^{j-i} \varphi_n\, J_n\, D^i u. \tag{14.10.8}$$

It is easily checked that this converges to $D^j u$ in L^2 as $n \to \infty$. The result now follows from (14.10.2). □

Theorem 14.10.4. *If $0 \le j < s - \frac{1}{2}$ and $u \in H^s$, then u has continuous derivatives up to order j.*

PROOF. The case $j = 0$ is covered by Theorem 14.10.2. Assume $j \ge 0$ and set

$$w_m(x) = (2\pi)^{-1/2} \int e^{ixk} k^m \hat{u}(k)\, dk, \qquad m \le j. \tag{14.10.9}$$

Then each w_m is a bounded continuous function by Theorem 14.10.2. Moreover,

$$(J_n w_m, v) = (w_m, J_n v) = (u, D^m J_n v) = (D^m J_n u, v) \tag{14.10.10}$$

for any $v \in C_0^\infty$. Thus

$$D^m J_n u = J_n w_m \to w_m \qquad \text{as} \quad n \to \infty \tag{14.10.11}$$

uniformly on any bounded interval. This shows that u has continuous derivatives up to order j. □

Theorem 14.10.5. *If $u, f \in H^s$, $g \in H^{s+1}$, and*

$$(u, \varphi'') = (f, \varphi) + (g, \varphi'), \qquad \varphi \in C_0^\infty, \tag{14.10.12}$$

then $u \in H^{s+2}$.

PROOF. Hypothesis (14.10.12) implies

$$\left(k^2\hat{u} + \hat{f} + ik\hat{g}, \hat{\varphi}\right) = 0, \qquad \varphi \in C_0^\infty. \tag{14.10.13}$$

Also by hypothesis, $(1 + k^2)^{(s-1)/2}(k^2\hat{u} + \hat{f} + ik\hat{g})$ is in L^2. Since C_0^∞ is dense in each H^t (Theorem 14.10.3), (14.10.13) implies that

$$k^2\hat{u} + \hat{f} + ik\hat{g} = 0. \tag{14.10.14}$$

Thus

$$(1 + k^2)^{s/2}k^2\hat{u} = -(1 + k^2)^{s/2}\left(\hat{f} + ik\hat{g}\right) \in L^2.$$

Since

$$(1 + k^2)^{s/2}\hat{u} \in L^2$$

by hypothesis, we have

$$(1 + k^2)^{(s+2)/2}\hat{u} \in L^2.$$

This means that $u \in H^{s+2}$. □

Now we are ready for the

PROOF OF THEOREM 14.9.1. We want to prove that $u \in C^j(I)$ for each positive integer j. This will be true if we can show that $\varphi u \in C^j$ for any $\varphi \in C_0^\infty(I)$. For we can always take $\varphi \equiv 1$ in the neighborhood of any particular point. Since continuity and differentiability are local properties, this will give $u \in C^j(I)$. Now by Theorem 14.10.4 we will have $\varphi u \in C^j$ if we have $\varphi u \in H^s$ for s sufficiently large. Thus it suffices to show that $\varphi u \in H^s$ for every $\varphi \in C_0^\infty(I)$ and every s.

We proceed by induction. We know that $\varphi u \in H^0 = L^2$ for all $\varphi \in C_0^\infty(I)$. Assume $\varphi u \in H^s$ for some $s \geq 0$ and all $\varphi \in C_0^\infty(I)$. If $\psi \in C_0^\infty$, we have

$$\begin{aligned}(\varphi u, \psi'') &= (u, (\varphi\psi)'') - 2(u, \varphi'\psi') - (u, \varphi''\psi) \\ &= \lambda(u, \varphi\psi) - 2(\varphi' u, \psi') - (\varphi'' u, \psi) \\ &= (\lambda\varphi u - \varphi'' u, \psi) - 2(\varphi' u, \psi').\end{aligned}$$

Now $\varphi, \varphi', \varphi''$ are all in $C_0^\infty(I)$. Hence $\lambda\varphi u - \varphi'' u$ and $\varphi' u$ are in H^s by hypothesis. Applying Theorem 14.10.5, we see that $\varphi u \in H^{s+1}$ for any $\varphi \in C_0^\infty(I)$. This completes the proof. □

Finally we turn to the

PROOF OF LEMMA 14.8.1. If $u \in D(\tilde{H}_1)$, then in J

$$u = \sum (u, u_n)u_n \tag{14.10.15}$$

and

$$\tilde{H}_1 u = \sum \lambda_n (u, u_n)u_n \tag{14.10.16}$$

[see (13.2.9)]. Since $\tilde{H}_1 u \in L^2$, we have

$$\sum \lambda_n^2 |(u, u_n)|^2 < \infty. \tag{14.10.17}$$

Now by (14.10.15),

$$u' = \sum (u, u_n)u_n',$$

and consequently, in $c \leq |x| \leq b$

$$|u'(x)| \leq \sum |(u, u_n)| k_n \max(|C_n|, |D_n|)$$

by Theorem 14.9.2. This implies

$$|u'(x)|^2 \leq \sum \lambda_n^2 |(u, u_n)|^2 \sum |\lambda_n|^{-1}(C_n^2 + D_n^2). \tag{14.10.18}$$

Thus u' is uniformly bounded by Theorem 14.9.3. Moreover, if we set

$$v_N = \sum_1^N (u, u_n)u_n,$$

then we see that u' is the uniform limit of v_N', which is continuous in $c \leq |x| \leq b$. Thus u' is continuous on this set.

Next set $f = \tilde{H}_1 u$ and $I = (-b, -c) \cup (c, b)$. Then for $\varphi \in C_0^\infty(I)$ and $\psi \in C_0^\infty$ we have

$$\begin{aligned} (\varphi u, \psi'') &= -((\varphi u)', \psi') = -(u', (\varphi\psi)') + (u', \varphi'\psi) - (\varphi' u, \psi') \\ &= -\tilde{h}_1(u, \varphi\psi) + (\varphi' u', \psi) - (\varphi' u, \psi') \\ &= (\varphi' u' - \varphi f, \psi) - (\varphi' u, \psi'). \end{aligned}$$

Since $\varphi' u' - \varphi f \in L^2$ and $\varphi' u \in H^1$, we see that $\varphi u \in H^2$ for each $\varphi \in C_0^\infty(I)$ (Theorem 14.10.5). Moreover,

$$(u, \varphi'') = -h_1(u, \varphi) = -(f, \varphi), \qquad \varphi \in C_0^\infty(I).$$

Hence

$$u'' = -f \qquad \text{in} \quad I. \tag{14.10.19}$$

Now let v be any function in $D(h)$ and let φ be a function in C_0^∞ which vanishes outside $c < |x| < b + 1$ and satisfies $\varphi(\pm b) = v(\pm b)$. Then

$$\tilde{h}(u, v - \varphi) = \tilde{h}_1(u, v - \varphi) = (f, v - \varphi). \tag{14.10.20}$$

But

$$\tilde{h}(u, \varphi) = -(u'', \varphi) + \langle u, \varphi \rangle, \tag{14.10.21}$$

where we used integration by parts and made use of the fact that u' is continuous in $\bar{I}$. But

$$\langle u, \varphi \rangle = \langle u, v \rangle. \tag{14.10.22}$$

The identity (14.8.3) follows from (14.10.19)–(14.10.22). □

Exercises

1. Prove Theorem 14.3.1.

2. Prove (14.3.7) for $f \in L^2$.

3. Show that the expressions (14.4.3) and (14.4.4) tend to 0 as $a \to 0$.

4. Prove (14.6.2).

5. Show that (14.7.4) and (14.7.7) imply (14.7.12).

6. Prove (14.7.5) and (14.7.6).

7. Verify (14.8.7).

8. Verify (14.8.11)–(14.8.16).

9. Prove (14.8.19) and show that $B\tilde{R}_1(z)$ is a compact operator from L^2 to $\mathcal{H}$.

10. Prove (14.8.22).

11. Show that (14.9.8) is equivalent to the statements $\beta_i = \gamma_i$, $i = 1, 2$, and (14.9.9).

12. Show that the right-hand side of (14.9.9) is in L^2.

13. Prove (14.9.10).

14. Show that (14.9.11) follows from (14.8.18).

15. Prove Lemmas 14.10.1 and 14.10.2.

16. Prove (14.10.2) and (14.10.3).

17. Show how (14.10.7) implies that u is continuous.

18. Why does it suffice to prove Theorem 14.10.3 for s a positive integer?

19. Prove (14.10.8).

20. Show that (14.10.9) implies that u has continuous derivatives up to order j.

21. If u is continuous, show that $J_n u$ converges to u uniformly on any bounded interval.

22. Show that (14.10.11) implies (14.10.12).

23. Prove (14.10.19).

15
The Invariance Principle

15.1. Introduction

Let H_0, H be self-adjoint operators on a Hilbert space $\mathcal{H}$. The wave operators for this pair are the limits as $t \to \pm\infty$ of

$$W(t)u = e^{itH}e^{-itH_0}u \tag{15.1.1}$$

defined on those $u \in \mathcal{H}$ for which the limits exist (see Section 6.2). The existence of the wave operators implies certain relationships between H_0 and H (see Section 6.3). A useful and surprising fact is that under suitable conditions the wave operators $W_\pm(\varphi(H), \varphi(H_0))$ exist for a fairly large class of functions $\varphi(s)$ and coincide with the $W_\pm(H, H_0)$. This implies that the same relationships that hold between H_0 and H hold as well between $\varphi(H_0)$ and $\varphi(H)$. This is of both practical and theoretical interest. In this chapter we shall explore conditions under which

$$W_\pm(\varphi(H), \varphi(H_0)) = W_\pm(H, H_0). \tag{15.1.2}$$

15.2. A Simple Result

Our first theorem concerning the invariance principle is

Theorem 15.2.1. *Let $\varphi(s)$ be a real-valued function such that*

$$\int_0^\infty \left| \int_I e^{-isk - it\varphi(k)}\, dk \right|^2 ds \to 0 \qquad \textit{as} \quad t \to \infty \tag{15.2.1}$$

for each bounded interval I, and

$$\int e^{-it\varphi(k)} w(k)\, dk \to 0 \qquad \textit{as} \quad t \to \infty, \quad w \in L^1. \tag{15.2.2}$$

Let u be an element in $\mathcal{H}_{ac}(H_0) \cap M_+$ such that $d(E_0(s)u, u)/ds$ is bounded. Assume that

$$\int_0^\infty \|[W_+ - W(s)]u\|^2 \, ds < \infty. \tag{15.2.3}$$

Then $u \in M_+(\varphi(H), \varphi(H_0))$ and

$$W_+(\varphi(H), \varphi(H_0))u = W_+u. \tag{15.2.4}$$

Before we prove the theorem, we want to point out that sufficient conditions for (15.2.1) to hold have already been given (Theorem 14.5.1). We shall show that the same conditions are sufficient for (15.2.2) to hold (see Section 15.3). In proving Theorem 15.2.1 we shall make use of

Lemma 15.2.1. *If $u \in \mathcal{H}_{ac}(H)$ and*

$$d(E(s)u, u)/ds \le m^2, \tag{15.2.5}$$

then

$$\int |(e^{-itH}u, v)|^2 \, dt \le 2\pi m^2 \|v\|^2 \tag{15.2.6}$$

Lemma 15.2.2. *If $u \in \mathcal{H}_{ac}(H)$, $v \in \mathcal{H}$, then $(E(s)u, v)$ is absolutely continuous and*

$$|d(E(s)u, v)/ds|^2 \le \frac{d}{ds}(E(s)u, u)\frac{d}{ds}(E(s)v_0, v_0) \quad a.e.,$$

where v_0 is the projection of v onto $\mathcal{H}_{ac}(H)$.

Lemmas 15.2.1 and 15.2.2 will be proved in the next section. Meanwhile we give the

PROOF OF THEOREM 15.2.1. Let $\varepsilon > 0$ be given and take I so large that $\|E_0(CI)u\| < \varepsilon$. Set

$$f(t) = [W_+ - 1]e^{-itH_0}.$$

Then

$$(f(t)u, v) = \int e^{-its}h(s) \, ds, \tag{15.2.7}$$

where

$$h(s) = d(E_0(s)u, [W_+^* - 1]v)/ds. \tag{15.2.8}$$

Define

$$(Z(t)u, v) = \int_I e^{-it\varphi(s)}h(s) \, ds. \tag{15.2.9}$$

Now, by (15.2.7), $(f(t)u, v)$ is $(2\pi)^{1/2}$ times the Fourier transform of $h(s)$. Thus

$$h(s) = (2\pi)^{-1}\int e^{iks}(f(k)u, v) \, dk. \tag{15.2.10}$$

If we make use of Parseval's identity, (15.2.9) becomes

$$(Z(t)u, v) = \int g_t(k)(f(k)u, v)\, dk, \tag{15.2.11}$$

where

$$g_t(k) = (2\pi)^{-1} \int_I e^{isk - it\varphi(s)}\, ds. \tag{15.2.12}$$

Thus,

$$Z(t)u = \int g_t(k) f(k)u\, dk.$$

Set

$$Z_1(t)u = \int_0^\infty g_t(k) f(k)u\, dk$$

and

$$Z_2(t)u = \int_{-\infty}^0 g_t(k) f(k)u\, dk.$$

Now,

$$|(Z_2(t)u, v)|^2 \le \int_{-\infty}^0 |g_t(k)|^2\, dk \int_{-\infty}^0 |(f(k)u, v)|^2\, dk.$$

By Lemma 15.2.1, the last integral is bounded by $8\pi m^2 \|v\|^2$. Thus

$$\|Z_2(t)u\|^2 \le 8\pi m^2 \int_{-\infty}^0 |g_t(k)|^2\, dk. \tag{15.2.13}$$

If we change k to $-k$, we will get precisely the expression on the right of (15.2.1). Since I is a bounded interval, it follows from (15.2.1) that

$$Z_2(t)u \to 0 \quad \text{as} \quad t \to \infty. \tag{15.2.14}$$

Next we note that

$$Z_1(t)u = \int_I e^{-it\varphi(s)} w(s)\, ds, \tag{15.2.15}$$

where

$$w(s) = (2\pi)^{-1/2} \int_0^\infty e^{isk} f(k)u\, dk. \tag{15.2.16}$$

Now, $\|f(k)u\| = \|[W_+ - W(s)]u\|$. Hence $f(k)u$ is in $L^2(0, \infty)$ by (15.2.3). Thus $w \in L^2$. In particular, $w\chi_I \in L^1$. Consequently,

$$Z_1(t)u \to 0 \quad \text{as} \quad t \to \infty \tag{15.2.17}$$

by (15.2.2). Combining (15.2.14) and (15.2.17), we get

$$Z(t)u \to 0 \quad \text{as} \quad t \to \infty.$$

Next we note by Lemma 15.2.2

$$\int_{CI} |h(s)|\,ds \le \int_{CI} \left(\frac{d}{ds} \|E_0(s)u\|^2 \right)^{1/2}$$

$$\times \left(\frac{d}{ds} \|E_0(s)([W_+^* - 1]v)_0\|^2 \right)^{1/2} ds \le \left(\int_{CI} \frac{d}{ds} \|E_0(s)u\|^2\, ds \right)^{1/2}$$

$$\times \left(\int_{CI} \frac{d}{ds} \|E_0(s)([W_+^* - 1]v)_0\|^2\, ds \right)^{1/2} = \|E_0(CI)u\| \, \|([W_+^* - 1]v)_0\|$$

$$\le 2\|E_0(CI)u\| \; \|v\|.$$

Thus, if we define

$$(Y(t)u, v) = \int_{CI} e^{-it\varphi(s)} h(s)\, ds,$$

we have

$$\|Y(t)u\| \le 2\|E_0(CI)u\| < 2\varepsilon.$$

Now,

$$\left([W_+ - 1]e^{-it\varphi(H_0)}u, v\right) = \int e^{-it\varphi(s)} h(s)\, ds. \tag{15.2.18}$$

Hence

$$[W_+ - 1]e^{-it\varphi(H_0)}u = Y(t)u + Z(t)u.$$

Since ε was arbitrary, we have

$$[W_+ - 1]e^{-it\varphi(H_0)}u \to 0 \qquad \text{as} \quad t \to \infty.$$

But

$$\left\|[W_+ - e^{it\varphi(H)}e^{-it\varphi(H_0)}]u\right\| = \left\|[W_+ - 1]e^{-it\varphi(H_0)}u\right\|.$$

This gives (15.2.4). □

Let us apply Theorem 15.2.1 to our Hamiltonian. A sufficient condition for (15.2.3) to hold is

$$\left\|[W_+ - W(t)]u\right\| = O(t^{-\alpha}) \qquad \text{as} \quad t \to \infty \tag{15.2.19}$$

for some $\alpha > 1/2$. This condition is not readily applicable. However, if we note that

$$[W_+ - W(s)]u = \int_s^\infty W'(t)u\, dt,$$

we see that (15.2.19) will hold if

$$\|W'(t)u\| = O(t^{-\beta}) \qquad \text{as} \quad t \to \infty \tag{15.2.20}$$

for some $\beta > 3/2$. In fact, (15.2.20) implies (15.2.19) with $\alpha = \beta - 1$.

When $H = H_0 + V$, we have

$$W'(t)u = e^{itH}Ve^{-itH_0}u. \tag{15.2.21}$$

Thus, (15.2.2) is equivalent to

$$\|Ve^{-itH_0}u\| = O(t^{-\beta}). \tag{15.2.22}$$

We have considered such estimates before. In fact, if u is the Fourier transform of the function ψ_s given by (6.4.11), then (15.2.22) is implied by

$$\int_{-\infty}^{\infty} |V(x)(x-s)|^2 \exp\left\{-\frac{(x-s)^2}{2(1+t^2)}\right\} dx = O(t^{-\varepsilon}) \tag{15.2.23}$$

for some $\varepsilon > 0$ [see (6.4.16)]. If we use inequality (6.4.9), we see that (15.2.23) is implied by

$$\int |V(x)|^2|x-s|^\gamma \, dx < \infty \tag{15.2.24}$$

for some $\gamma > 2$. This in turn is implied by V locally in L^2 and satisfying

$$V(x) = O(|x|^{-\beta}) \quad \text{as} \quad |x| \to \infty \tag{15.2.25}$$

for some $\beta > 3/2$.

In order to use such functions, we must verify that they satisfy the other hypotheses of Theorem 15.2.1. By Theorem 9.12.1, $\mathcal{H}_{ac}(H_0)$ is the whole of L^2. Moreover, by (9.12.13),

$$(E_0(s)u, u) = \int_{-s^{1/2}}^{s^{1/2}} |\hat{u}(k)|^2 \, dk = \int_{-s^{1/2}}^{s^{1/2}} k^2 e^{-2k^2} \, dk. \tag{15.2.26}$$

Hence

$$d(E_0(s)u, u)/ds = \begin{matrix} s^{1/2}e^{-2s}, & s \geq 0 \\ 0, & s < 0. \end{matrix} \tag{15.2.27}$$

This is clearly bounded. Finally, we note that if V satisfies (15.2.23), then $M = L^2$ by Theorem 6.4.2. Hence we have

Theorem 15.2.2. *Suppose V satisfies either* (15.2.23), (15.2.24), *or* (15.2.25), *and that φ satisfies both* (15.2.1) *and* (15.2.2). *Then*

$$M_\pm(\varphi(H), \varphi(H_0)) = L^2 \tag{15.2.28}$$

and

$$W_\pm(\varphi(H), \varphi(H_0))u = W_\pm u, \qquad u \in L^2. \tag{15.2.29}$$

PROOF. We have shown that each u which is the Fourier transform of a function ψ_s is in $M_+(\varphi(H), \varphi(H_0))$ and that (15.2.4) holds. The same reasoning applies to M_- and W_- as well. Now all we need note is that such functions are dense in L^2 (Lemma 6.4.1). □

We shall show in Section 15.4 how Theorems 15.2.1 and 15.2.2 can be strengthened to yield sharper results.

15.3. The Estimates

Now we give the proofs of Lemmas 15.2.1 and 15.2.2. First we look at Lemma 15.2.2.

PROOF OF LEMMA 15.2.2. Since $E(s)$ maps $\mathcal{H}_{ac}(H)$ into itself, we have

$$(E(s)u, v) = (E(s)u, v_0).$$

Thus we may assume that $v = v_0$. Next we note that

$$\begin{aligned}(E(s)u, v) = {} & (E(s)(u+v), u+v) - (E(s)(u-v), u-v) \\ & + i(E(s)(u+iv), u+iv) - i(E(s)(u-iv), u-iv).\end{aligned}$$

This shows that $(E(s)u, v)$ is absolutely continuous. Next we note that

$$|(E(I)u, v)| = |(E(I)u, E(I)v)| \le \|E(I)u\| \, \|E(I)v\|.$$

Thus,

$$\frac{|([E(s) - E(s_0)]u, v)|^2}{s - s_0} \le \frac{([E(s) - E(s_0)]u, u)}{s - s_0} \frac{([E(s) - E(s_0)]v, v)}{s - s_0}.$$

Taking the limit as $s \to s_0$, we obtain the lemma. □

PROOF OF LEMMA 15.2.1. We note that

$$(e^{-itH}u, v) = \int e^{-its} \frac{d}{ds}(E(s)u, v)\, ds.$$

Thus $(e^{-itH}u, v)$ is $(2\pi)^{1/2}$ times the Fourier transform of $d(E(s)u, v)/ds$. Thus, to prove (15.2.6) it suffices to show that $d(E(s)u, v)/ds$ is in L^2 and its L^2-norm is bounded by $m\|v\|$. By (15.2.5) and Lemma 15.2.2 we have

$$|d(E(s)u, v)/ds|^2 \le m^2 \frac{d}{ds}(E(s)v_0, v_0).$$

Hence

$$\int \left| \frac{d}{ds}(E(s)u, v) \right|^2 ds \le m^2 \|v_0\|^2 \le m^2 \|v\|^2.$$

This is precisely what we need. □

15.4. An Extension

We shall show how to strengthen Theorem 15.2.1. The hypothesis which brings about the requirement (15.2.23) is (15.2.3). We are going to show how one can weaken this hypothesis. Note that we used (15.2.3) only at one point in the proof of Theorem 15.2.1. We used it to show that $w(s)$ given by (15.2.16) was in L^2. This implied (15.2.17). Let us write $Z_1(t)$ in a slightly different way:

$$Z_1(t)u = \int_0^\infty g_t(k) e^{-ikH} h(k)\, dk, \tag{15.4.1}$$

where

$$h(t) = [W_+ - W(t)]u. \tag{15.4.2}$$

Set

$$G(s) = (2\pi)^{-1/2} \int_0^\infty e^{isk} h(k)\, dk. \tag{15.4.3}$$

Then we have

Lemma 15.4.1. *If* $\|G'(s)\| \in L^1$ *and* $G(\pm\infty) = 0$, *then* (15.2.17) *holds.*

PROOF. We have

$$\begin{aligned}(Z_1(t)u, v) &= \int_0^\infty g_t(k)(h(k), e^{ikH}v)\, dk \\ &= \int \left(\int_0^\infty g_t(k) e^{-ik\lambda} h(k)\, dk, dE(\lambda)v \right) \\ &= \int \left(\int_I e^{-it\varphi(s)} G(s-\lambda)\, ds, dE(\lambda)v \right) \\ &= \int_I e^{-it\varphi(s)} \int (G(s-\lambda), dE(\lambda)v)\, ds \\ &= \int_I e^{-it\varphi(s)} \left[\int (G'(s-\lambda), E(\lambda)v)\, d\lambda \right] ds, \end{aligned} \tag{15.4.4}$$

where we used the spectral theorem, integration by parts, and Parseval's identity. Note that there are no boundary terms because $G(s)$ vanishes at $\pm\infty$. From this we see that

$$Z_1(t)u = \int_I e^{-it\varphi(s)} \int E(\lambda) G'(s-\lambda)\, d\lambda\, ds. \tag{15.4.5}$$

Since $G'(s) \in L^1$, the inner integral is bounded. Hence it is integrable over I. Thus, (15.2.17) follows from (15.2.2). □

Actually, we can do a bit better. In fact we have

Lemma 15.4.2. *If* $\|G(s)\| \in L^1$, *then*

$$\|Z_1(t)u\| \le (2\pi)^{-1/2} \int \|G(s)\|\, ds \tag{15.4.6}$$

and (15.2.17) *holds.*

PROOF. We have by (15.4.3),

$$h(k) = (2\pi)^{-1/2} \int e^{-isk} G(s)\, ds.$$

Hence by (15.4.1),

$$Z_1(t)u = (2\pi)^{-1/2}\int_0^\infty g_t(k)e^{-ikH}\int e^{-isk}G(s)\,ds\,dk$$
$$= (2\pi)^{-1/2}\int\left[\int_0^\infty g_t(k)e^{-ik(s+H)}\,dk\right]G(s)\,ds.$$

But

$$\int_0^\infty g_t(k)e^{-ik(s+H)}\,dk = \int\left[\int_0^\infty g_t(k)e^{-ik(s+\lambda)}\,dk\right]dE(\lambda).$$

By (15.2.12), the inner integral equals

$$e^{-it\varphi(s+\lambda)}\chi_I(s+\lambda),$$

which has absolute value ≤ 1. Hence

$$\left\|\int_0^\infty g_t(k)e^{-ik(s+H)}\,dk\,v\right\| \le \|v\| \tag{15.4.7}$$

[Theorem 1.10.1(g)]. This implies (15.4.6). Now suppose $G(s)$ is only known to be in L^1. Then there is a function $G_0(s) \in C_0^\infty$ such that

$$\int \|G(s) - G_0(s)\|\,ds < \varepsilon.$$

Set

$$Z_0(t)u = \int_0^\infty g_t(k)e^{-ikH}h_0(k)\,dk,$$

where

$$h_0(k) = (2\pi)^{-1/2}\int_{-\infty}^\infty e^{-isk}G_0(s)\,ds.$$

Then by Lemma 15.4.1,

$$Z_0(t)u \to 0 \quad \text{as} \quad t \to \infty.$$

Moreover,

$$\|Z_1(t)u - Z_0(t)u\| \le (2\pi)^{-1/2}\int \|G(s) - G_0\|\,ds$$

by (15.4.6). Thus, $\|Z_1(t)u\| < 2\varepsilon$ for t sufficiently large. Hence (15.2.17) holds. □

A sufficient condition for $G(s)$ to be in L^1 is given by

Lemma 15.4.3. *Assume that $h(t)$ is continuous, $h(0) = h(\infty) = 0$, $\|h'(t)\| \in L^2$, and $\|h'(t)\|[1 + \log(1 + t)] \in L^1$. Then $\|G(s)\| \in L^1$.*

PROOF. Integrating (15.4.3) by parts, we have

$$(2\pi)^{1/2}sG(s) = i\int_0^\infty e^{isk}h'(k)\,dk \tag{15.4.8}$$

[there are no boundary terms since $h(0) = h(\infty) = 0$]. Also,

$$\int_0^\infty h'(k)\,dk = h(\infty) - h(0) = 0.$$

Hence

$$(2\pi)^{1/2}|s|\,\|G(s)\| \le 2\int_0^\infty |\sin sk|\,\|h'(k)\|\,dk,$$

and consequently,

$$\begin{aligned}(2\pi)^{1/2}\int_{-1}^{1}\|G(s)\|\,ds &\le 2\int_0^\infty\left[\int_{-1}^{1}\frac{|\sin sk|}{|s|}\,ds\right]\|h'(k)\|\,dk\\ &\le 4\int_0^\infty\left[1+\log(1+k)\right]\|h'(k)\|\,dk.\end{aligned} \tag{15.4.9}$$

On the other hand,

$$\begin{aligned}\left(\int_1^\infty\|G(s)\|\,ds\right)^2 &= \left(\int_1^\infty s^{-1}s\|G(s)\|\,ds\right)^2\\ &\le \int_1^\infty s^{-2}\,ds\int_0^\infty\|h'(k)\|^2\,dk\end{aligned}$$

by (15.4.8) in view of Schwarz's inequality and Parseval's identity. A similar inequality holds for the interval $(-\infty, -1)$. Thus $\|G(s)\| \in L^1$. □

Now we are ready for

Theorem 15.4.1. *Suppose* $\varphi(s)$ *satisfies* (15.2.1) *and* (15.2.2). *Let* u *be an element of* $\mathcal{H}_{ac}(H_0) \cap M_+$ *such that*

$$d(E_0(s)u, u)/ds \quad \text{is bounded,} \tag{15.4.10}$$

$$\int_1^\infty \|W'(t)u\|^2\,dt < \infty, \tag{15.4.11}$$

and

$$\int_1^\infty \|W'(t)u\|(1+\log t)\,dt < \infty. \tag{15.4.12}$$

Then $u \in M_+(\varphi(H), \varphi(H_0))$ *and* (15.2.4) *holds.*

PROOF. All of the hypotheses of Theorem 15.2.1 are satisfied with the exception of (15.2.3). Thus it suffices to prove (15.2.7). Let $\sigma(k)$ be a function in C^∞ which equals 1 for $k \ge 2$ and vanishes for $k \le 1$. Set

$$Z_3(t)u = \int_0^\infty g_t(k)\sigma(k)f(k)u\,dk$$

and

$$Z_4(t)u = \int_0^\infty g_t(k)[1-\sigma(k)]f(k)u\,dk.$$

Thus $Z_1 = Z_3 + Z_4$. Now,

$$Z_4(t)u = \int_I e^{-it\varphi(s)} M(s)\, ds,$$

where

$$M(s) = (2\pi)^{-1} \int_0^\infty e^{isk} [1 - \sigma(k)] f(k) u\, dk.$$

Since $[1 - \sigma(k)]$ vanishes for $k \geq 2$, we see that $[1 - \sigma(k)] f(k) u \in L^2$. Thus, $M(s) \in L^2$, and consequently, $M(s)\chi_I \in L^1$. By hypothesis (15.2.2), we see that $Z_4(t)u \to 0$. Thus, it remains to prove

$$Z_3(t)u \to 0 \qquad \text{as} \quad t \to \infty. \tag{15.4.13}$$

But

$$Z_3(t)u = \int_0^\infty g_t(k) e^{-ikH} h(k)\, dk,$$

where

$$h(k) = \sigma(k)[W_+ - W(k)]u.$$

In particular, we have

$$h'(k) = -\sigma(k) W'(k) u + \sigma'(k)[W_+ - W(k)]u.$$

Since $\sigma'(k)$ vanishes for $k \geq 2$, (15.4.11) and (15.4.12) imply that $\|h'(k)\| \in L^2$ and $\|h'(k)\|[1 + \log(1 + k)]$. Since $u \in M_+$, we have $h(\infty) = 0$, and since $\sigma(k)$ vanishes for $k \leq 1$, we have $h(0) = 0$. Thus by Lemma 15.4.3 we see that $\|G(s)\| \in L^1$, where $G(s)$ is given by (15.4.3). Now we can apply Lemma 15.4.2 to conclude that (15.4.13) holds. □

Let us apply Theorem 15.4.1 to our Hamiltonian. Both (15.4.11) and (15.4.12) are implied by

$$\|W'(t)u\| = O(t^{-\delta}) \qquad \text{as} \quad t \to \infty \tag{15.4.14}$$

for some $\delta > 1$. When $H = H_0 + V$, this is equivalent to

$$\|Ve^{-itH_0}u\| = O(t^{-\delta}). \tag{15.4.15}$$

For the functions considered in Section 15.2 this is implied by

$$\int |V(x)(x - s)|^2 \exp\left\{ -\frac{(x - s)^2}{2(1 + t^2)} \right\} dx = O(t^\sigma) \tag{15.4.16}$$

for some $\sigma < 1$. This in turn is implied by

$$\int |V(x)|^2 |x - s|^\delta\, dx < \infty \tag{15.4.17}$$

for some $\delta > 1$. This holds if V is locally in L^2 and satisfies

$$V(x) = O(|x|^{-\tau}) \qquad \text{as} \quad |x| \to \infty \tag{15.4.18}$$

for some $\tau > 1$. Note that (15.4.15)–(15.4.18) are weaker than (15.2.22)

–(15.2.25), respectively. They are not much stronger than the hypotheses used in Chapter 6 to prove the existence of the wave operators. Summarizing, we have

Theorem 15.4.2. *Suppose* $\varphi(s)$ *satisfies* (15.2.1) *and* (15.2.2), *and* V *satisfies one of the statements* (15.4.16)–(15.4.18). *Then* (15.2.28) *and* (15.2.29) *hold.*

15.5. Another Form

In this section we shall show how even Theorem 15.4.1 can be strengthened if we are willing to assume that the relationship between H and H_0 is of the form considered in Chapters 8–14. We shall assume that there are closed operators A, B from $\mathcal{H}$ to $\mathcal{K}$ such that $D(H_0) \subset D(A)$, $D(H) \subset D(B)$, and

$$(u, Hv) = (H_0u, v) + (Au, Bv), \qquad u \in D(H_0), \quad v \in D(H). \tag{15.5.1}$$

We have

Theorem 15.5.1. *Assume that* (15.5.1) *holds and that* $\varphi(s)$ *satisfies* (15.2.1). *Let* $f \in \mathcal{H}_{ac}(H_0) \cap M_+$ *be such that*

$$\int_{-\infty}^{\infty} (\|Ae^{-itH_0}f\|^2 + \|BW_+e^{-itH_0}f\|^2)\, dt < \infty. \tag{15.5.2}$$

Then $f \in M_+(\varphi(H), \varphi(H_0))$ *and*

$$W_+(\varphi(H), \varphi(H_0))f = W_+ f. \tag{15.5.3}$$

PROOF. By (8.1.8),

$$a\int([R_0(z) - R(z)]f, R(z)g)\, ds = 2\pi a\int_0^{\infty} e^{-2at}([W(t) - 1]f, g)\, dt, \tag{15.5.4}$$

where $z = s + ia$, $a > 0$. On the other hand, (15.5.1) implies

$$(R(z)f - R_0(z)f, g) = (AR_0(z)f, BR(\bar{z})g)_{\mathcal{K}}.$$

Thus the left-hand side of (15.5.4) equals

$$\tfrac{1}{2}i\int(AR_0(z)f, B[R(z)-R(\bar{z})]g)\, ds.$$

But by (9.3.9) and (9.3.10),

$$\int \|AR_0(z)f - f_+(s)\|^2\, ds \to 0 \tag{15.5.5}$$

and

$$\int \|B[R(z) - R(\bar{z})]g - g(s)\|^2\, ds \to 0, \tag{15.5.6}$$

where

$$f_+(s) = -i\int_0^\infty e^{ist}Ae^{-itH_0}f\,dt \tag{15.5.7}$$

and

$$g(s) = -i\int_{-\infty}^\infty e^{ist}Be^{-itH}g\,dt. \tag{15.5.8}$$

Combining all of these formulas and taking the limit as $a \to 0$, we get

$$\int (f_+(s), g(s))\,ds = -2\pi i([W_+ - 1]f, g). \tag{15.5.9}$$

Next we note that $f_+(s)$ is $-i(2\pi)^{1/2}$ times the Fourier transform of $\chi_{(0,\infty)}Ae^{-itH_0}f$ and $g(s)$ is the same constant times the Fourier transform of $Be^{-itH}g$. If we apply Parseval's identity to (15.5.9), we find

$$\int_0^\infty (Ae^{-itH_0}f, Be^{-itH}g)\,dt = -i([W_+ - 1]f, g). \tag{15.5.10}$$

Now,

$$\begin{aligned}(A\delta_a(H_0 - s)f, g) &= \int \delta_a(\lambda - s)\,d(AE_0(\lambda)f, g)\\ &\to d(AE_0(s)f, g)/ds \quad \text{a.e.}\end{aligned} \tag{15.5.11}$$

But

$$\pi(A\delta_a(H_0 - s)f, g) = \tfrac{1}{2}i(A[R_0(z) - R_0(\bar{z})]f, g).$$

Moreover,

$$\int \|A[R_0(z) - R_0(\bar{z})]f - f(s)\|^2\,ds \to 0, \qquad a \to 0, \tag{15.5.12}$$

where

$$f(s) = -i\int_{-\infty}^\infty e^{ist}Ae^{-itH_0}f\,dt. \tag{15.5.13}$$

Hence

$$\pi d(AE_0(s)f, g)/ds = \tfrac{1}{2}i(f(s), g) \quad \text{a.e.} \tag{15.5.14}$$

Consequently, if $\alpha(\lambda)$ is any continuous function,

$$\begin{aligned}(A\alpha(H_0)f, g) &= \int \alpha(\lambda)\,d(AE_0(\lambda)f, g)\\ &= i(2\pi)^{-1}\int \alpha(\lambda)(f(\lambda), g)\,d\lambda.\end{aligned}$$

Thus,

$$A\alpha(H_0)f = i(2\pi)^{-1}\int \alpha(\lambda)f(\lambda)\,d\lambda. \tag{15.5.15}$$

In particular, we have

$$\begin{aligned}&\int_0^\infty \|Ae^{-i\sigma H_0 - it\varphi(H_0)}f\|^2\,d\sigma\\ &= (2\pi)^{-2}\int_0^\infty \left\|\int e^{-i\sigma\lambda - it\varphi(\lambda)}f(\lambda)\,d\lambda\right\|^2 d\sigma \to 0 \quad \text{as} \quad t \to \infty\end{aligned} \tag{15.5.16}$$

by (15.2.1). Now,

$$\left\|\left[e^{it\varphi(H)}e^{-it\varphi(H_0)} - W_+\right]f\right\|^2 = 2\|f\|^2 - 2\,\mathrm{Re}\left(e^{it\varphi(H)}e^{-it\varphi(H_0)}f,\, W_+ f\right). \tag{15.5.17}$$

By (15.5.10),

$$\left|\left(\left[W_+ - 1\right]e^{-it\varphi(H_0)}f,\, e^{-it\varphi(H)}W_+ f\right)\right|^2 \leq \int_0^\infty \|Ae^{-i\sigma H_0 - it\varphi(H_0)}f\|^2\, d\sigma \times \int_0^\infty \|Be^{-i\sigma H - it\varphi(H)}W_+ f\|^2\, d\sigma. \tag{15.5.18}$$

The first integral converges to 0 as $t \to \infty$ by (15.5.16). As far as the second integral is concerned, we note that

$$\int \|BW_+[R_0(z) - R_0(\bar{z})]f - h(s)\|^2\, ds \to 0, \quad a \to 0, \tag{15.5.19}$$

where

$$h(s) = -i\int_{-\infty}^{\infty} e^{ist}BW_+e^{-itH_0}f\, dt. \tag{15.5.20}$$

Reasoning as before, we see that

$$BW_+\alpha(H_0)f = i(2\pi)^{-1}\int \alpha(\lambda)h(\lambda)\, d\lambda. \tag{15.5.21}$$

Thus, by the intertwining relations,

$$\int_0^\infty \|Be^{-i\sigma H - it\varphi(H)}W_+ f\|^2\, d\sigma = (2\pi)^{-2}\int_0^\infty \left\|\int e^{-i\sigma\lambda - it\varphi(\lambda)}h(\lambda)\, d\lambda\right\|^2 d\sigma$$
$$\leq (2\pi)^{-1}\int \|h(\lambda)\|^2\, d\lambda = \int \|BW_+e^{-i\varphi H_0}f\|^2\, d\sigma. \tag{15.5.22}$$

Therefore the second integral is uniformly bounded in t. This shows that the left-hand side of (15.5.18) converges to 0 as $t \to \infty$. Thus,

$$\left(e^{it\varphi(H)}e^{-it\varphi(H_0)}f,\, W_+ f\right) = \left(e^{-it\varphi(H_0)}f,\, W_+e^{-it\varphi(H_0)}f\right) \to \|f\|^2$$

The desired result now follows from (15.5.17). □

To apply Theorem 15.5.1, let us assume that the hypotheses of Theorem 9.11.1 are satisfied. Then we know that

$$a\|AR_0(z)\|^2 + a\|BR(z)\|^2 \leq C_I, \qquad I \subset\subset \Gamma.$$

This implies

$$a\|AR_0(z)E_0(I)\|^2 + a\|BR(z)E(I)\|^2 \leq C_I'$$

(Lemma 9.3.3), which in turn gives

$$\int \left(\|Ae^{-itH_0}E_0(I)f\|^2 + \|Be^{-itH}E(I)g\|^2\right) dt < \infty$$

for any f, g, and $I \subset\subset \Gamma$ (Lemma 9.3.2). In particular, if we take $g = W_+ f$, we get

$$\int (\|Ae^{-itH_0}E_0(I)f\|^2 + \|BW_+e^{-itH_0}E_0(I)f\|^2)\, dt < \infty$$

for any f. Theorem 15.5.1 now tells us that $E_0(I)f \in M_+(\varphi(H), \varphi(H_0))$ for each $f \in \mathcal{H}_{ac}(H_0) \cap M_+$ and each $I \subset\subset \Gamma$. From this it follows that $E_0(\Gamma_1)f \in M_+(\varphi(H), \varphi(H_0))$ for each component Γ_1 of Γ. Thus, we have $E_0(\Gamma)f \in M_+(\varphi(H), \varphi(H_0))$. Since $\Lambda - \Gamma$ has measure 0 and $f \in \mathcal{H}_{ac}(H_0)$, we see that $E_0(\Lambda)f \in M_+(\varphi(H), \varphi(H_0))$ and (15.5.3) holds for $E_0(\Lambda)f$. Thus we have

Theorem 15.5.2. *Under the hypotheses of Theorem* 9.11.2, *every* $f \in \mathcal{H}_{ac}(H_0)$ *is in* $M_+(\varphi(H), \varphi(H_0))$ *and* (15.5.3) *holds.*

Returning to our Hamiltonian, we have by Theorem 9.14.1

Theorem 15.5.3. *If* $V \in L^1$ *and* $\varphi(s)$ *satisfies* (15.2.1), *then* (15.2.28) *and* (15.2.29) *hold.*

Exercises

1. Prove (15.2.7) and (15.2.18).

2. Prove (15.2.10) and (15.2.12).

3. Prove (15.2.15).

4. Show that (15.2.24) implies (15.2.23).

5. Show that (15.2.25) implies (15.2.24).

6. Prove (15.4.9).

7. Show that (15.4.16) implies (15.4.15).

8. Define

$$g_+(s) = -i\int_0^\infty e^{ist}Be^{-itH}g\, dt$$

and determine what the expression

$$\int (f(s), g_+(s))\, ds$$

represents.

9. Show that (15.2.1) implies that the right-hand side of (15.5.16) converges to 0.

16
Trace Class Operators

16.1. The Abstract Theorem

A very useful method for treating scattering theory, at least in one dimension, involves the use of so called *trace class* operators (sometimes called *nuclear* operators). They are defined as follows. An operator T on a Hilbert space $\mathcal{H}$ is said to be of trace class if there are orthonormal sequences $\{\varphi_k\}$, $\{\psi_k\}$ in $\mathcal{H}$ and a sequence $\{\alpha_k\}$ of scalars such that

$$Tu = \sum_1^\infty \alpha_k(u, \varphi_k)\psi_k, \qquad u \in \mathcal{H}, \tag{16.1.1}$$

and

$$\sum_1^\infty |\alpha_k| < \infty. \tag{16.1.2}$$

It is easily shown that a trace class operator is compact. The importance of such operators can be seen by

Theorem 16.1.1. *Let H_0 be a self-adjoint operator on $\mathcal{H}$ and T a Hermitian trace class operator on $\mathcal{H}$. Then the operator $H = H_0 + T$ is self-adjoint and the wave operators*

$$W_\pm u = \lim_{t\to\pm\infty} e^{itH}e^{-itH_0}u \tag{16.1.3}$$

exist for each $u \in \mathcal{H}_{ac}(H_0)$. Thus $\mathcal{H}_{ac}(H_0) \subset M_\pm$.

PROOF. H is self-adjoint by Corollary 2.7.1. Let f, g be elements in $\mathcal{H}_{ac}(H_0) \cap D(H_0)$ which satisfy

$$d(E_0(\lambda)u, u)/d\lambda \le m^2 \tag{16.1.4}$$

and let L be an operator which commutes with H. Setting $f_t = e^{-itH_0}f$, we have

$$d(Lf_t, g_t)/dt = i(Lf_t, H_0 g_t) - i(LH_0 f_t, g_t)$$
$$= i(LTf_t, g_t) - i(Lf_t, Tg_t) \qquad (16.1.5)$$

since H is self-adjoint and commutes with L. Thus we have

$$\begin{aligned}(W(t)f_b, W(s)f_b) &- (W(t)f, W(s)f)\\ &= \left(e^{i(t-s)H}f_{t+b}, f_{s+b}\right) - \left(e^{i(t-s)H}f_t, f_s\right)\\ &= i\int_0^b \left\{\left(e^{i(t-s)H}f_{t+\sigma}, Tf_{s+\sigma}\right) - \left(e^{i(t-s)H}Tf_{t+\sigma}, f_{s+\sigma}\right)\right\} d\sigma, \qquad (16.1.6)\end{aligned}$$

where we took $L = \exp\{i(t-s)H\}$ in (16.1.5). Now for any $h \in \mathcal{H}$,

$$\begin{aligned}([W(t) - W(s)]f_b, h) &= \int_s^t \frac{d}{d\sigma}(W(\sigma)f_b, h)\, d\sigma\\ &= -i\int_s^t \left(e^{i\sigma H}Tf_{b+\sigma}, h\right) d\sigma\\ &= -i\int_s^t \Sigma\alpha_k(f_{b+\sigma}, \psi_k)\left(\varphi_k, e^{-i\sigma H}h\right) d\sigma.\end{aligned}$$

The absolute value of this is bounded by

$$\begin{aligned}&\|h\|\int_s^t \Sigma|\alpha_k|\, |(f_{b+\sigma}, \varphi_k)|\, d\sigma\\ &\leq \|h\|\int_s^t \left(\Sigma|\alpha_k|\, |(f_{b+\sigma}, \varphi_k)|^2\right)^{1/2}\left(\Sigma|\alpha_k|\right)^{1/2} d\sigma\\ &\leq \|h\|\left[(t-s)\Sigma|\alpha_k|\int_{s+b}^{t+b}\Sigma|\alpha_k|\, |(f_\tau, \varphi_k)|^2\, d\tau\right]^{1/2}.\end{aligned}$$

Since

$$\int_0^\infty \Sigma|\alpha_k|\, |(f_\tau, \varphi_k)|^2\, d\tau \leq 2\pi m^2 \Sigma|\alpha_k| \qquad (16.1.7)$$

by Lemma 15.2.1, we see that

$$\|[W(t) - W(s)]f_b\| \to 0 \quad \text{as} \quad b \to \infty. \qquad (16.1.8)$$

In particular, this gives

$$([W(t) - W(s)]f_b, W(s)f_b) \to 0 \quad \text{as} \quad b \to \infty. \qquad (16.1.9)$$

Combining this with (16.1.6), we obtain

$$\begin{aligned}&(W(t)f, [W(t) - W(s)]f)\\ &= i\int_0^\infty \left\{\left(e^{i(t-s)H}f_{t+\sigma}, Tf_{s+\sigma}\right) - \left(e^{i(t-s)H}Tf_{t+\sigma}, f_{s+\sigma}\right)\right\} d\sigma. \qquad (16.1.10)\end{aligned}$$

The first term on the right-hand side of (16.1.10) is bounded by

$$\int_0^\infty \Sigma|\alpha_k(e^{i(t-s)H}f_{t+\sigma},\psi_k)(\varphi_k,f_{s+\sigma})|\,d\sigma$$

$$\le \int_0^\infty (\Sigma|\alpha_k|\,|(e^{i(t-s)H}f_{t+\sigma},\varphi_k)|^2)^{1/2}(\Sigma|\alpha_k|\,|(\varphi_k,f_{s+\sigma})|^2)^{1/2}\,d\sigma$$

$$\le \left(\int_0^\infty \Sigma|\alpha_k|\,|(e^{i(t-s)H}f_{t+\sigma},\psi_k)|^2\,d\sigma\right)^{1/2}\left(\int_s^\infty \Sigma|\alpha_k|\,|(\varphi_k,f_\tau)|^2\,d\tau\right)^{1/2}.$$

This converges to 0 as $s\to\infty$ in view of (16.1.7). The same reasoning shows that the second term on the right of (16.1.10) tends to 0 as $t\to\infty$. Thus,

$$(W(t)f,[W(t)-W(s)]f)\to 0 \quad \text{as} \quad s,t\to\infty. \tag{16.1.11}$$

Similar reasoning gives

$$(W(s)f,[W(t)-W(s)]f)\to 0 \quad \text{as} \quad s,t\to\infty,$$

which gives

$$\|[W(t)-W(s)]f\|\to 0 \quad \text{as} \quad s,t\to\infty.$$

This shows that $f\in M_+$ for f in the class considered. Now suppose f is any element in $\mathcal{H}_{ac}(H_0)$. For each integer $n>0$, let Γ_n be the set of those real λ such that either $|\lambda|>n$ or

$$d(E_0(\lambda)f,f)/d\lambda > n, \tag{16.1.12}$$

and set

$$f_n = [1 - E_0(\Gamma_n)]f.$$

Then

$$\|E_0(\lambda)f_n\|^2 = \|E_0(\lambda)[1-E_0(\Gamma_n)]f\|^2$$

$$= \int_{-\infty}^{\lambda}[1-\chi_{\Gamma_n}(\mu)]\,d\|E_0(\mu)f\|^2,$$

where χ_{Γ_n} is the characteristic function of Γ_n. Thus,

$$\frac{d}{d\lambda}\|E_0(\lambda)f_n\|^2 = [1-\chi_{\Gamma_n}(\lambda)]\frac{d}{d\lambda}\|E_0(\lambda)f\|^2 \le n$$

since $\chi_{\Gamma_n}(\lambda)=1$ when (16.1.12) holds. Moreover,

$$\int \lambda^2\,d\|E_0(\lambda)f_n\|^2 = \int_{-n}^{n}\lambda^2[1-\chi_{\Gamma_n}(\lambda)]\,d\|E_0(\lambda)f\|^2 \le n\|f\|^2$$

by Theorem 1.10.1. Hence $f_n\in D(H_0)$ and satisfies (16.1.4) for $m^2=n$. Thus, $f_n\in M_+$ for each n by what we have just proved. Since

$$\|f-f_n\|^2 = \int[1-\chi_{\Gamma_n}(\lambda)]\,d\|E_0(\lambda)f\|^2$$

and

$$\chi_{\Gamma_n}(\lambda) \to 1 \quad \text{as} \quad n \to \infty$$

for each real λ, we see that $f_n \to f$. Thus $f \in M_+$. Similar reasoning works for M_-. □

16.2. Some Consequences

Theorem 16.1.1 is very strong in several respects. For instance, an immediate consequence is

Theorem 16.2.1. *Under the hypotheses of Theorem* 16.1.1, $\mathcal{H}_{ac}(H) \subset R_\pm$.

PROOF. We merely interchange H and H_0 in Theorem 16.1.1. This shows that $\mathcal{H}_{ac}(H) \subset M_\pm(H_0, H) = R_\pm(H, H_0)$ by Lemma 9.1.2. □

In particular, it follows that the wave operators (16.1.3) are complete if $\mathcal{H}_{ac}(H_0) = \mathcal{H}_{ac}(H)$ (cf. the definition given at the beginning of Chapter 9). Another strength of Theorem 16.1.1 is that it makes no assumptions on H_0. However, it has the weakness that it is rare in applications (especially those we have in mind) that the perturbing operator T is of trace class. This can be remedied by the following considerations.

Theorem 16.2.2. *Let H, H_0 be self-adjoint operators on a Hilbert space $\mathcal{H}$ having the interval $(-\infty, \lambda]$ in both of their resolvent sets. Suppose $T = R(\lambda) - R_0(\lambda)$ is of trace class. Then $\mathcal{H}_{ac}(H_0) \subset M_\pm$ and $\mathcal{H}_{ac}(H) \subset R_\pm$.*

In proving Theorem 16.2.2 we shall make use of

Theorem 16.2.3. *The invariance principle holds under the hypotheses of Theorem* 16.1.1 *if $\varphi(\lambda)$ satisfies* (15.2.1).

Before proving Theorem 16.2.3 we show how it implies Theorem 16.2.2.

PROOF OF THEOREM 16.2.2. By Theorem 16.1.1 the wave operators

$$\tilde{W}f = \lim_{t \to \pm\infty} e^{itR(\lambda)} e^{-itR_0(\lambda)} f \tag{16.2.1}$$

exist for $f \in \mathcal{H}_{ac}(R_0(\lambda))$. Since $\lambda \in \rho(H_0) \cap \rho(H)$, there is a positive constant c such that $(-\infty, \lambda + c) \subset \rho(H_0) \cap \rho(H)$. Set

$$\varphi(\tau) = \begin{matrix} \lambda - \tau^{-1}, & \tau \leq -c^{-1} \\ \lambda - c^{-1}, & \tau > -c^{-1}. \end{matrix}$$

Then $\varphi(\tau)$ satisfies (15.2.1) (Theorem 14.5.1). Thus the wave operators

$$\hat{W}f = \lim_{t \to \pm\infty} e^{it\varphi(R(\lambda))} e^{-it\varphi(R_0(\lambda))} f \tag{16.2.2}$$

exist for $f \in \mathcal{H}_{ac}(R_0(\lambda))$. Now, $\varphi([\lambda - \mu]^{-1}) = \mu$ for $\mu \geq \lambda + c$. Since the line $[\lambda + c, \infty)$ contains the spectra of both H and H_0, we have $\varphi(R(\lambda)) = H$ and $\varphi(R_0(\lambda)) = H_0$. Thus the wave operators $W_\pm(H, H_0)$ exist on $\mathcal{H}_{ac}(R_0(\lambda))$. But if $f \in \mathcal{H}_{ac}(H_0)$, then $f \in \mathcal{H}_{ac}(R_0(\lambda))$. To see this, let $\hat{E}(\mu)$ be the spectral projection for $R_0(\lambda)$. Then

$$(\hat{E}_I f, f) = (\chi_I(R_0(\lambda))f, f) = \int \chi_I[(\lambda - \mu)^{-1}]\, d(E_0(\mu)f, f)$$
$$= \int \chi_I(s)\, d(E_0(\lambda - s^{-1})f, f). \tag{16.2.3}$$

From this it follows that

$$(\hat{E}(s)f, f) = (E_0(\lambda - s^{-1})f, f), \tag{16.2.4}$$

and this is an absolutely continuous function of s. Hence $\mathcal{H}_{ac}(H_0) \subset M_\pm$. The second statement follows by interchanging H and H_0. □

Next we give the

PROOF OF THEOREM 16.2.3. We apply Theorem 15.5.1. Define

$$Au = \Sigma|\alpha_k|^{1/2}(u, \varphi_k)\varphi_k,$$
$$Bv = \Sigma\alpha_k|\alpha_k|^{-1/2}(v, \psi_k)\varphi_k.$$

Then

$$(Au, Bv) = \Sigma\alpha_k(u, \varphi_k)(\psi_k, v) = (Tu, v)$$
$$= (u, Hv) - (H_0u, v).$$

Moreover,

$$\|Au\|^2 = \Sigma|\alpha_k|\,|(u, \varphi_k)|^2,$$
$$\|BW_+u\|^2 = \Sigma|\alpha_k|\,|(u, W_+^*\varphi_k)|^2.$$

Thus if f satisfies (16.1.4), we see that (15.5.2) holds in view of Lemma 15.2.1. This completes the proof. □

ANOTHER PROOF OF THEOREM 16.2.3. Another proof can be given without appealing to the theorems of Chapter 15. We can reason as follows.

$$\|[W_+ - 1]f\|^2 = 2\|f\|^2 - 2\,\mathrm{Re}(f, W_+ f)$$
$$= 2\,\mathrm{Re}\,([W_+ - 1]f, W_+ f)$$
$$= 2\,\mathrm{Re}\int_0^\infty (W'(\sigma)f, W_+ f)\, d\sigma$$
$$= 2\,\mathrm{Re}\, i\int_0^\infty (Tf_\sigma, W_+ f_\sigma)\, d\sigma$$
$$= -2\,\mathrm{Im}\int_0^\infty \Sigma\alpha_k(f_\sigma, \varphi_k)(\psi_k, W_+ f_\sigma)\, d\sigma, \tag{16.2.5}$$

where we used the intertwining relations (see Theorem 6.3.3). Thus, if f

satisfies (16.1.4), we have

$$\|[W_+ - 1]f\|^2 \le 2m\|f\|(2\pi\Sigma|\alpha_k|)^{1/2}\left(\int_0^\infty \Sigma|\alpha_k|\,|(f_\sigma, \varphi_k)|^2\,d\sigma\right)^{1/2}. \tag{16.2.6}$$

This gives

$$\|[W_+ - 1]e^{-it\varphi(H_0)}f\|^2 \le C\left(\Sigma|\alpha_k|\int_0^\infty\left|\int e^{-it\varphi(\lambda)-i\sigma\lambda}m_0(\lambda, f, \varphi_k)\,d\lambda\right|^2 d\sigma\right)^{1/2}, \tag{16.2.7}$$

where

$$m_0(\lambda, f, g) = d(E_0(\lambda)f, g)/d\lambda. \tag{16.2.8}$$

By (16.2.1) each integral on the right-hand side of (16.2.7) tends to 0 as $t \to \infty$. In view of (16.1.2) this implies that

$$[W_+ - 1]e^{-it\varphi(H_0)}f \to 0 \quad \text{as} \quad t \to \infty. \tag{16.2.9}$$

This implies the invariance principle. □

16.3. Hilbert–Schmidt Operators

If one wants to use Theorem 16.1.1 or Theorem 16.2.2 in an application, one must be able to verify that an operator is of trace class. This is not difficult in practice if one makes use of several facts concerning compact operators. We describe them in this section. Recall that an orthonormal sequence $\{\varphi_k\}$ is called complete if every element f satisfies

$$f = \sum_{k=1}^{\infty}(f, \varphi_k)\,\varphi_k. \tag{16.3.1}$$

First we note

Lemma 16.3.1. *L^2 has a complete orthonormal sequence.*

PROOF. Set $S = e^{-|x|^2}R_0(-1)$, where $R_0(z)$ is the resolvent operator for H_0. Since resolvent operators are one-to-one, the same is true of S. Moreover, S is compact by Lemma 9.14.3. Thus S^*S is self-adjoint, compact, and one-to-one. The result now follows from Corollary 13.3.1. □

Next let $K(x, y)$ be a function satisfying

$$\iint |K(x, y)|^2\,dx\,dy < \infty \tag{16.3.2}$$

and define the operator K by

$$Ku(x) = \int K(x,y)u(y)\,dy. \tag{16.3.3}$$

Such operators are called *Hilbert-Schmidt* operators. We have

Lemma 16.3.2. *K is a compact operator on L^2 and*

$$\sum_{k=1}^{\infty} \|K\varphi_k\|^2 \le \iint |K(x,y)|^2\,dx\,dy \tag{16.3.4}$$

for any orthonormal sequence $\{\varphi_k\}$.

PROOF. Note that

$$|Ku(x)|^2 \le \int |K(x,y)|^2\,dy\,\|u\|^2. \tag{16.3.5}$$

This shows that $D(K) = L^2$ and K is bounded. Moreover, if u_n converges weakly to 0, then

$$Ku_n(x) = \int K(x,y)u_n(y)\,dy \to 0 \tag{16.3.6}$$

for a.e. x. The functions (16.3.6) are majorized by an L^2-function in view of (16.3.5). Hence $Ku_n \to 0$ in L^2 by the Lebesgue dominated convergence theorem. By Theorem 9.7.2, K is compact. To prove (16.3.4), let $\{\psi_k\}$ be a complete orthonormal sequence in L^2 (Lemma 16.3.2). Then

$$\sum_k \|K\varphi_k\|^2 = \sum_{j,k} |(K\varphi_k, \psi_j)|^2$$

$$= \sum_{j,k} \left| \iint K(x,y)\varphi_k(y)\psi_j(x)^*\,dx\,dy \right|^2. \tag{16.3.7}$$

Note that the sequence $\{\varphi_k(y)\psi_j(x)^*\}$ is orthonormal in $L^2(E^2)$. Thus the right hand side of (16.3.7) is bounded by (16.3.2). □

Lemma 16.3.3. *Let $T = K_2^* B K_1$ be Hermitian, where B is bounded and the K_j are Hilbert-Schmidt operators. Then there are an orthonormal sequence $\{\varphi_k\}$ and numbers $\{\alpha_k\}$ such that*

$$Tu = \Sigma \alpha_k (u, \varphi_k)\varphi_k \tag{16.3.8}$$

and

$$\Sigma |\alpha_k| < \infty. \tag{16.3.9}$$

PROOF. Since K_1 is compact, the same is true of T. Then T can be written in the form (16.3.8) by Lemma 13.3.2. But

$$|\alpha_k| = |(T\varphi_k, \varphi_k)| = |(BK_1\varphi_k, K_2\varphi_k)|$$
$$\le \|B\|\,\|K_1\varphi_k\|\,\|K_2\varphi_k\|.$$

Thus

$$\Sigma|\alpha_k| \le \|B\|(\Sigma\|K_1\varphi_k\|^2)^{1/2}(\Sigma\|K_2\varphi_k\|^2)^{1/2}$$

$$\le \|B\|\left(\iint|K_1(x,y)|^2\,dx\,dy\iint|K_2(x,y)|^2\,dx\,dy\right)^{1/2}.$$

This gives (16.3.9). □

16.4. Verification for the Hamiltonian

In this section we shall show how Theorem 16.2.2 can be used to prove Theorem 9.14.1. We define A, B, H as in Section 9.14. Note that $D(|H|^{1/2}) = D(H_0^{1/2})$ and that A, B are $H_0^{1/2}$- and $|H|^{1/2}$—bounded. Since both H and H_0 are bounded from below, there is a negative number $\lambda \in \rho(H) \cap \rho(H_0)$. Then

$$R(\lambda) - R_0(\lambda) = [BR(\lambda)]^*AR_0(\lambda) = [AR(\lambda)]^*BR_0(\lambda). \quad (16.4.1)$$

Thus,

$$AR(\lambda) = AR_0(\lambda) + A[BR(\lambda)]^*AR_0(\lambda)$$
$$= G(\lambda)AR_0(\lambda).$$

Substituting this into the second part of (16.4.1), we obtain

$$R(\lambda) - R_0(\lambda) = [AR_0(\lambda)]^*G(\lambda)^*BR_0(\lambda). \quad (16.4.2)$$

The left-hand side of (16.4.2) is Hermitian, and $G(\lambda)$ is a bounded operator. I claim that the operators $AR_0(\lambda)$, $BR_0(\lambda)$ are Hilbert–Schmidt. To see this note that

$$AR_0(\lambda)u(x) = \frac{1}{2i\sqrt{\lambda}}\int e^{i\lambda^{1/2}|x-y|}A(x)u(y)\,dy.$$

This is of the form (16.3.3) with

$$K(x,y) = \frac{1}{2i\sqrt{\lambda}}e^{i\lambda^{1/2}|x-y|}A(x).$$

Hence

$$\iint|K(x,y)|^2\,dx\,dy = \frac{1}{4|\lambda|}\iint e^{-2|\lambda|^{1/2}|x-y|}|A(x)|^2\,dx\,dy.$$

Since $V \in L^1$, (16.3.2) is clearly satisfied. The same reasoning applies to $BR_0(\lambda)$. Hence these operators are Hilbert-Schmidt. We can now apply Lemma 16.3.3 to conclude that the right hand side of (16.4.2) is of trace class. Thus the conclusions of Theorem 16.2.2 hold. Since $\mathcal{H}_c(H_0) = \mathcal{H}_{ac}(H_0) = L^2$ (Theorem 9.12.3), we have $\mathcal{H}_c(H_0) = M_\pm$. We also have $\mathcal{H}_{ac}(H) = R_\pm$. One inclusion follows from Theorem 16.2.2. The other follows from Lemma 9.13.2. Thus the wave operators are complete.

Exercises

1. Show that a trace class operator is compact.

2. Use the above exercise to give an alternative proof of (16.1.8).

3. Show that the sum of two trace class operators is of trace class.

4. Fill in the details in the proof of (16.1.11).

5. Show that $\varphi(R(\lambda)) = H$ in Section 16.2.

6. Show that (16.2.3) implies (16.2.4).

7. Show that (16.2.4) is absolutely continuous.

8. Prove that the right-hand side of (16.2.7) tends to 0 as $t \to \infty$.

9. Verify the last two statements in the proof of Lemma 16.3.2.

Appendix A
The Fourier Transform

In this Appendix we shall prove properties (a)–(f) of Section 1.3. For this purpose it is convenient to introduce the class $\mathcal{S}$ of functions $u \in C^\infty$ for which $x^n u^{(m)}(x)$ is bounded for each $m, n \geq 0$. Clearly, $C_0^\infty \subset \mathcal{S}$ and hence $\mathcal{S}$ is dense in L^2. If

$$\hat{u}(k) = (2\pi)^{-1/2} \int e^{-ikx} u(x)\, dx \tag{A.1}$$

and $u \in \mathcal{S}$, we can differentiate as many times as we like under the integral sign to obtain

$$\hat{u}^{(m)}(k) = (-i)^m [x^m u]\hat{}.$$

This gives (c). Moreover, if we integrate by parts, we obtain

$$\int_{-R}^{R} e^{-ikx} u'(x)\, dx = ik \int_{-R}^{R} e^{-ikx} u(x)\, dx$$

$$+ e^{-ikR} u(R) - e^{ikR} u(-R) \to (2\pi)^{1/2} ik\hat{u}$$

as $R \to \infty$. This gives (e). In proving the rest it will be convenient to use the Riemann–Lebesgue lemma:

Lemma A.1. *If $v \in L^1$, then*

$$\int v(x) e^{iRx}\, dx \to 0 \qquad \textit{as} \quad R \to \infty. \tag{A.2}$$

Before proving the lemma, let us use it in proving

$$u = [\hat{u}]\check{}. \tag{A.3}$$

Set

$$G_R(x) = (2\pi)^{-1/2} \int_{-R}^{R} e^{ikx} \hat{u}(k)\, dk.$$

Then by (A.1),

$$G_R(x) = \tfrac{1}{2}\pi^{-1}\int_{-\infty}^{\infty} u(y)\int_{-R}^{R} e^{ik(x-y)}\,dy\,dk. \tag{A.4}$$

Thus

$$G_R(x) - u(x) = \pi^{-1}\int_{-\infty}^{\infty} \frac{u(y)-u(x)}{y-x}\sin R(y-x)\,dy. \tag{A.5}$$

Now if $u \in \mathcal{S}$, then $[u(y) - u(x)]/(y - x) \in L^1$ for each fixed x. Thus $G_R(x) \to u(x)$ by Lemma A.1. This proves (a). To prove (b) we note that

$$\begin{aligned}(u, v) &= (2\pi)^{-1/2}\int\left[\int e^{ixk}\hat{u}(k)\,dk\right]v(x)^*\,dx \\ &= \int \hat{u}(k)\left[\int e^{-ixk}v(x)\,dx\right]^* dk = (\hat{u}, \hat{v}).\end{aligned} \tag{A.6}$$

Note that (d) and (f) follow from (c) and (e) if we make use of the inverse Fourier transform.

PROOF OF LEMMA A.1. We first note that the step functions are dense in L^1. This is proved in the same way as Theorem 14.6.1. Hence it suffices to prove Lemma A.1 for step functions. For if u is any function in L^1, let $\varepsilon > 0$ be given. Then there is a step function v such that $\|u - v\|_1 < \varepsilon$. Thus

$$\left|\int u(x)e^{iRx}\,dx\right| \le \|u - v\|_1 + \left|\int v(x)e^{iRx}\,dx\right|.$$

If the lemma is true for step functions, the last term can be made $< \varepsilon$ for R sufficiently large. Thus

$$\left|\int u(x)e^{iRx}\,dx\right| < 2\varepsilon$$

for R sufficiently large. Thus the lemma holds for u as well. To prove it for step functions, let v be the characteristic function of a bounded interval (a, b). Then

$$\int v(x)e^{iRx}\,dx = \int_a^b e^{iRx}\,dx = (iR)^{-1}\left[e^{iRb} - e^{iRa}\right]$$

and this goes to 0 as $R \to \infty$. But every step function is a linear combination of characteristic functions, and hence the lemma holds for them as well. □

Now let us examine precisely what functions will satisfy the various formulas. The Fourier transform $\hat{u}$ is defined if u is in L^1. We can use Parseval's identity to define it for $u \in L^2$. For there is a sequence $\{u_n\} \subset \mathcal{S}$ converging to u in L^2, and Parseval's identity shows that $\{\hat{u}_n\}$ converges to a function w in L^2. Moreover, the limit function is independent of the particular sequence $\{u_n\}$. We can define $\hat{u}$ to be w. Then Parseval's identity will hold for functions in L^2. Since $\mathcal{S}$ is dense in L^2, the inverse transform formula will also hold for all functions in L^2. The other formulas will hold as long as the function of which the Fourier transform or inverse Fourier transform is taken is in L^2.

Now we turn to the proof of Theorem 10.5.2. It is convenient to use the M. Riesz–Thorin interpolation theorem. A simplified version of this theorem can be stated as follows:

Theorem A.1. *Let T be a linear operator defined on the step functions such that*

$$\|Tf\|_{q_1} \le M_1\|f\|_{p_1}, \qquad \|Tf\|_{q_2} \le M_2\|f\|_{p_2} \tag{A.7}$$

for some p_j, q_j in the interval $[1, \infty)$. Let θ be any number satisfying $0 \le \theta \le 1$, and let

$$\frac{1}{p} = \frac{1-\theta}{p_1} + \frac{\theta}{p_2}, \qquad \frac{1}{q} = \frac{1-\theta}{q_1} + \frac{\theta}{q_2}. \tag{A.8}$$

Then

$$\|Tf\|_p \le M_1^{1-\theta}M_2^{\theta}\|f\|_q. \tag{A.9}$$

Before we indicate the proof of this theorem, let us show how it can be used to obtain (10.5.5). From (1.3.1) we have the estimate

$$\|\hat{f}\|_\infty \le (2\pi)^{-1/2}\|f\|_1.$$

Moreover, by (1.3.6), we have

$$\|\hat{f}\|_2 = \|f\|_2$$

Thus, we may take $p_1 = 1$, $p_2 = 2$, $q_1 = \infty$, $q_2 = 2$, $M_1 = (2\pi)^{-1/2}$, $M_2 = 1$, $\theta = 2/p'$ in Theorem A.1. Then q turns out to be p'. The desired result now follows from (A.9).

Theorem A.1 can be made to depend on the three-line theorem, which can be stated as follows:

Theorem A.2. *Let $f(z)$ be continuous and bounded in the strip B: $0 \le x \le 1$ and analytic in the interior. Suppose $|f(iy)| \le M_1$, $|f(1 + iy)| \le M_2$, y real. Then*

$$|f(x + iy)| \le M_1^{1-x}M_2^{x}, \qquad x + iy \in B.$$

This theorem can easily be proved by applying the maximum principle to the function $F(z) = f(z)e^{\delta z^2}/M_1^{1-z}M_2^{z}$ in a rectangle and letting $\delta \to 0$.

Once we have Theorem A.2 we can give the

PROOF OF THEOREM A.1. Let u, v, be any step functions. Set $F(z) = |u|^{(p/p_z)-1}u$, $G(z) = |v|^{(q'/q'_z)-1}v$, where

$$\frac{1}{p_z} = \frac{1-z}{p_1} + \frac{z}{p_2}, \qquad \frac{1}{q_z} = \frac{1-z}{q_1} + \frac{z}{q_2}.$$

It is easily verified that $F(z)$, $G(z)$ depend analytically on z. Thus,

$$f(z) \equiv (TF(z), G(\bar{z})) \tag{A.10}$$

is analytic in the strip B. Moreover, it is bounded there. Now,

$$\begin{aligned} |f(iy)| &\le \|TF(iy)\|_{q_1}\|G(-iy)\|_{q'_1} \\ &\le M_1\|F(iy)\|_{p_1}\|v\|_{q'_1} = M_1\|u\|_p^{p/p_1}\|v\|_{q'}^{q'/q'_1} \end{aligned} \tag{A.11}$$

and

$$\begin{aligned} |f(1+iy)| &\le \|TF(1+iy)\|_{q_2}\|G(1-iy)\|_{q'_2} \\ &\le M_2\|u\|_p^{p/p_2}\|v\|_{q'}^{q'/q'_2}. \end{aligned} \tag{A.12}$$

Theorem A.2 now gives

$$|f(\theta)| \le M_1^{1-\theta}M_2^{\theta}\|u\|_p\|v\|_{q'} \tag{A.13}$$

since $p = p_\theta$, $q = q_\theta$. But $f(\theta) = (Tu, v)$. Since this is true for all step functions u, v, the result follows. □

Exercises

1. Prove (A.4) and (A.5).

2. Justify the change in order of integration in (A.6).

3. Prove that step functions are dense in L^1.

4. Show that the limit w in L^2 of the sequence $\{\hat{u}_n\}$ is unique if $u_n \to u$ in L^2.

5. Prove Theorem A.2.

6. Show that $f(z)$ given by (A.10) is bounded and analytic in B.

7. Prove (A.11), (A.12), and (A.13).

Appendix B
Hilbert Space

We shall prove several theorems in Hilbert space which were used in the text. First we note that the parallelogram law

$$\|u - v\|^2 + \|u + v\|^2 = 2\|u\|^2 + 2\|v\|^2 \tag{B.1}$$

holds in every Hilbert space. This follows from expanding the left-hand side of (B.1) in terms of the scalar product and canceling some terms. We also have

Lemma B.1. *Let N be a closed subspace of a Hilbert space $\mathcal{H}$, and let u be an element of $\mathcal{H}$ which is not in N. Set*

$$d = \inf_{v \in N} \|u - v\|. \tag{B.2}$$

Then there is an element $v \in N$ such that $\|u - v\| = d$.

PROOF. There is a sequence $\{v_n\} \subset N$ such that $\|u - v_n\| \to d$. By (B.1),

$$\|(u - v_n) - (u - v_m)\|^2 + \|(u - v_n) + (u - v_m)\|^2$$
$$= 2\|u - v_n\|^2 + 2\|u - v_m\|^2.$$

Thus,

$$4\|u - \tfrac{1}{2}(v_n + v_m)\|^2 + \|v_n - v_m\|^2 = 2\|u - v_n\|^2 + 2\|u - v_m\|^2 \to 4d^2. \tag{B.3}$$

But $\frac{1}{2}(u_n + u_m) \in N$ (this is where the fact that N is a subspace is used). Thus the left-hand side of (B.3) is $\geq 4d^2 + \|v_n - v_m\|^2$. This shows that

$$\|v_n - v_m\| \to 0.$$

Thus $\{v_n\}$ is a Cauchy sequence in $\mathcal{H}$ and consequently converges to an element $v \in H$. Since N is closed, $v \in N$. Also,

$$\|u - v\| = \lim\|u - v_n\| = d. \qquad \square$$

Corollary B.1. *Under the same hypotheses, there are $v \in N$ and $w \perp N$ such that $u = v + w$.*

PROOF. Let v be the element of N given by Lemma B.1, and set $w = u - v$. It remains to show that $w \perp N$. Let y be any element of N and set $\alpha = (w, y)/\|y\|^2$. Then

$$d^2 \leq \|w - \alpha y\|^2 = \|w\|^2 - 2 \operatorname{Re} \bar{\alpha}(w, y) + |\alpha|^2\|y\|^2$$

$$= d^2 - \frac{|(w, y)|^2}{\|y\|^2}.$$

The only way this can happen is if $(w, y) = 0$. Since y was any element of N, we see that $w \perp N$. $\square$

Now we can give the

PROOF OF THEOREM 1.9.1. If $Fu = 0$ for every u, we merely take $f = 0$ and everything is proved. Otherwise the set $N = \{u \in \mathcal{H} | Fu = 0\}$ is a closed subspace not the whole of $\mathcal{H}$. By Corollary B.1 there is a $w \neq 0$ such that $w \perp N$. Now for each $u \in \mathcal{H}$, the element $(Fw)u - (Fu)w \in N$ since

$$F[(Fw)u - (Fu)w] = (Fw)(Fu) - (Fu)(Fw) = 0.$$

Hence

$$([(Fw)u - (Fu)w], w) = 0,$$

or

$$Fu = (u, (Fw)^*\|w\|^{-2}w).$$

If we take

$$f = (Fw)^*\|w\|^{-2}w,$$

we see that (1.9.2) holds. The uniqueness of f is trivial, for if f_1 is any other element having the same properties, then

$$(u, f - f_1) = 0, \qquad u \in \mathcal{H}.$$

If we take $u = f - f_1$, we get $f = f_1$. To show that $\|f\| = \|F\|$, note that

$$|Fu| = |(u, f)| \leq \|u\| \, \|f\|.$$

Hence

$$\|F\| = \sup \frac{|Fu|}{\|u\|} \leq \|f\|$$

and

$$\|F\| \geq \frac{|Ff|}{\|f\|} = \|f\|. \qquad \square$$

Next we turn to the

PROOF OF THEOREM 2.6.2. For each n let

$$L_n = 1 + M + \cdots + M^n.$$

Then for $m < n$

$$\|L_n u - L_m u\| \leq \sum_{m+1}^{n} \|M^j u\| \leq \|u\| \sum_{m+1}^{n} \theta^j \leq \theta^{m+1}(1-\theta)^{-1}\|u\|,$$

where $\theta = \|M\| < 1$. Thus for each u, $L_n u$ converges to an element $w \in \mathcal{H}$. We define the operator L by $Lu = w$. Clearly, L is a linear operator defined everywhere on $\mathcal{H}$. It is also bounded. To see this we note that

$$\|L_n u\| \leq (1 + \theta + \cdots + \theta^n)\|u\| \leq (1-\theta)^{-1}\|u\|.$$

Hence

$$\|Lu\| \leq (1-\theta)^{-1}\|u\|.$$

Now,

$$L_n(1-M)u = (1-M)L_n = u - M^{n+1}u.$$

Taking the limits as $n \to \infty$, we get

$$L(1-M) = (1-M) = 1.$$

This gives the theorem. $\square$

We note that Corollary B.1 is equivalent to Theorem 3.5.1. Next we turn to the

PROOF OF THEOREM 3.6.3. Let $\{v_k\} \subset \mathcal{H}$ satisfy $\|v_k\| \leq C_0$. For each fixed j the complex numbers (v_n, v_j) are bounded. By the usual diagonalization procedure we can find a subsequence $\{\tilde{v}_n\}$ such that $(\tilde{v}_n, v_j)$ converges for each fixed j. Thus, $(\tilde{v}_n, f)$ converges if $f \in S$, the set of linear combinations of the v_j. It also converges if $f \in \bar{S}$. For let $\varepsilon > 0$ be given. Then there is a $g \in S$ such that $\|f - g\| < \varepsilon$. Then

$$\begin{aligned} |(\tilde{v}_n - \tilde{v}_m, f)| &\leq |(\tilde{v}_n - \tilde{v}_m, f - g)| + |(\tilde{v}_n - \tilde{v}_m, g)| \\ &\leq 2C_0\varepsilon + |(\tilde{v}_n - \hat{v}_m, g)| < (2C_0 + 1)\varepsilon \end{aligned}$$

for m, n sufficiently large. Now, for each $w \in \mathcal{H}$ there is an $f \in \bar{S}$ such that $w - f \perp \bar{S}$ (Corollary B.1). Thus,

$$(\tilde{v}_n, w) = (\tilde{v}_n, w - f) + (\tilde{v}_n, f) = (\tilde{v}_n, f).$$

Thus, $(\hat{v}_n, w)$ converges for each $w \in \mathcal{H}$. Set

$$Fw = \lim(w, \tilde{v}_n).$$

It is easily checked that F is a bounded linear functional on $\mathcal{H}$ (in fact we have $\|F\| \le C_0$). Thus by the Riesz representation theorem (Theorem 1.9.1), there is an element $v \in \mathcal{H}$ such that $Fw = (w, v)$. Hence

$$(\tilde{v}_n - v, w) \to 0, \qquad w \in \mathcal{H},$$

and the theorem is proved. □

We shall also prove the Banach–Steinhaus theorem:

Theorem B.1. *Let $\{F_n\}$ be a sequence of bounded linear functionals on $\mathcal{H}$ such that for each $u \in \mathcal{H}$*

$$\sup_n |F_n u| < \infty. \tag{B.4}$$

Then there is a constant C such that

$$|F_n u| \le C\|u\|, \qquad u \in \mathcal{H}. \tag{B.5}$$

PROOF. It suffices to prove that there are $v_0 \in \mathcal{H}$ and constants $K, \delta > 0$ such that

$$|F_n v| \le K \qquad \text{for} \quad \|v - v_0\| < \delta. \tag{B.6}$$

For, if (B.6) holds and u is any element of $\mathcal{H}$, set $v = v_0 + (\delta u/2\|u\|)$. Then $\|v - v_0\| < \delta$, and consequently, $|F_n v| \le K$. But

$$F_n v = F_n v_0 + (\delta F_n u/2\|u\|),$$

and consequently,

$$|F_n u| \le 2\|u\|\delta^{-1}(|F_n v| + |F_n v_0|) \le 4K\delta^{-1}\|u\|.$$

Thus we can take $C = 4K\delta^{-1}$ in (B.5).

Suppose (B.6) did not hold. Then there would be a $u_1 \in \mathcal{H}$ and an integer n_1 such that

$$\|u_1\| = 1 \qquad \text{and} \qquad |F_{n_1} u_1| > 1.$$

Since F_{n_1} is continuous, there is a $\delta_1 > 0$ such that

$$|F_{n_1} u| > 1 \qquad \text{for} \quad \|u - u_1\| < \delta_1.$$

We can take $\delta_1 < 1$. There must also be a $u_2 \in \mathcal{H}$ and an integer $n_2 > n_1$ such that

$$\|u_2 - u_1\| < \delta_1 \qquad \text{and} \qquad |F_{n_2} u_2| > 2,$$

for otherwise (B.6) will hold with $v_0 = u_1$, $\delta = \delta_2$, and $K = 2$. By continuity, there is a $\delta_2 > 0$ such that

$$|F_{n_2} u| > 2 \qquad \text{for} \quad \|u - u_2\| > \delta_2.$$

Take $\delta_2 < \min(\frac{1}{2}, \delta_1 - \|u_1 - u_2\|)$. This is done to guarantee that $\|u - u_2\| < \delta_2$ implies $\|u - u_1\| < \delta_1$. Continuing, we see that there are a $u_3 \in \mathcal{H}$ and an $n_3 > n_2$ such that

$$\|u_3 - u_2\| < \delta_2 \qquad \text{and} \qquad |F_{n_3}u_3| > 3.$$

Thus there is a $\delta_3 > 0$ such that

$$|F_{n_3}u| > 3 \qquad \text{for} \quad \|u - u_3\| < \delta_3.$$

We take $\delta_3 < \min(1/3, \delta_2 - \|u_3 - u_2\|)$. Inductively, there are a $u_j \in \mathcal{H}$, an $n_j > n_{j-1}$ and a $\delta_j > 0$ such that

$$\delta_j < \min(1/j, \delta_{j-1} - \|u_j - u_{j-1}\|)$$

and

$$|F_{n_j}u| > j \qquad \text{for} \quad \|u - u_j\| < \delta_j. \tag{B.7}$$

Now, if $j > k$, we have

$$\|u_j - u_k\| \le \sum_{i=k}^{j-1} \|u_{i+1} - u_i\| \le \sum_{i=k}^{j-1} \delta_i - \delta_{i+1} = \delta_k - \delta_j < k^{-1}. \tag{B.8}$$

Hence $\{u_j\}$ is a Cauchy sequence. Let u_0 be its limit. Holding k fixed in (B.8) and letting $j \to \infty$, we get

$$\|u_0 - u_k\| \le \delta_k.$$

By (B.7), this implies

$$|F_{n_k}u_0| \ge k.$$

Hence

$$\sup_n |F_n u_0| = \infty,$$

contradicting (B.4). Hence (B.6) holds, and the proof is complete. □

Next we give the

PROOF OF THEOREM 2.7.2. Suppose A maps $\mathcal{H}_1$ into $\mathcal{H}_2$. Let D be the set of those $w \in \mathcal{H}_2$ for which there is a $w^* \in \mathcal{H}_1$ satisfying

$$(w, Au) = (w^*, u), \qquad u \in \mathcal{H}_1. \tag{B.9}$$

Note that w^* is unique, for if

$$(w, Au) = (y, u), \qquad u \in \mathcal{H}_1,$$

then

$$(w^* - y, u) = 0, \qquad u \in \mathcal{H}_1,$$

and consequently, $y = w^*$. Clearly, D is a subspace of $\mathcal{H}_2$. We shall prove

a. there is a constant C such that

$$\|w^*\| \le C\|w\|, \qquad w \in D, \tag{B.10}$$

b. D is dense in $\mathcal{H}_2$.

These two statements imply the theorem. For if A were not bounded, there would be a sequence $\{u_n\} \subset \mathcal{H}_1$ such that

$$\|u_n\| = 1, \qquad \|Au_n\| \to \infty \qquad \text{as} \quad n \to \infty. \tag{B.11}$$

Then by (B.9) and (B.10),

$$|(w, Au_n)| = |(w^*, u_n)| \le \|w^*\| \, \|u_n\| \le C\|w\|.$$

Thus,

$$|(w, Au_n)| \le C\|w\|, \qquad w \in D, \quad \text{all } n.$$

Since D is dense in $\mathcal{H}_2$, this holds for all $y \in \mathcal{H}_2$. To see this, let $\{w_k\}$ be a sequence of elements in D converging to y in $\mathcal{H}_2$. Then

$$|(w_k, Au_n)| \le C\|w_k\|$$

Taking the limit, we get

$$|(y, Au_n)| \le C\|y\|, \qquad y \in \mathcal{H}_2, \quad \text{all } n.$$

But this implies

$$\|Au_n\|^2 \le C\|Au_n\|,$$

or

$$\|Au_n\| \le C,$$

contradicting (B.11). Thus it remains only to prove (a) and (b). To prove (a), suppose there were a sequence $\{w_n\} \subset D$ such that

$$\|w_n\| = 1, \qquad \|w_n^*\| \to \infty \qquad \text{as} \quad n \to \infty. \tag{B.12}$$

Set

$$F_n u = (Au, w_n) = (u, w_n^*).$$

Since

$$|F_n u| \le \|w_n^*\| \, \|u\|,$$

we see that $F_n u$ is a bounded linear functional on $\mathcal{H}_1$ for each n. On the other hand,

$$|F_n u| \le \|Au\| \, \|w_n\| = \|Au\|.$$

Hence

$$\sup_n |F_n u| \le \|Au\| < \infty.$$

Now we can apply Theorem B.1 to conclude that

$$|F_n u| \le C\|u\|, \qquad u \in \mathcal{H}_1.$$

Hence

$$|(u, w_n^*)| \le C\|u\|, \qquad u \in \mathcal{H}_1.$$

Taking $u = w_n^*$, we get

$$\|w_n^*\| \le C,$$

contradicting (B.12). This proves (B.10).

Let us turn to (b). Let $\mathcal{H}_3$ be the set of ordered pairs $\{u, v\}$, where $u \in \mathcal{H}_1$ and $v \in \mathcal{H}_2$. If we define

$$\alpha\{u, v\} = \{\alpha u, \alpha v\},$$

$$\{u_1, v_1\} + \{u_2, v_2\} = \{u_1 + u_2, v_1 + v_2\},$$

$$(\{u_1, v_1\}, \{u_2, v_2\}) = (u_1, u_2) + (v_1, v_2),$$

then $\mathcal{H}_3$ becomes a Hilbert space (it is called the *cartesian product* of $\mathcal{H}_1$ and $\mathcal{H}_2$). Next, let G_A be the set of those elements of $\mathcal{H}_3$ which are of the form $\{u, Au\}$ (it is called the *graph* of A). It is a closed subspace of $\mathcal{H}_3$ iff A is a closed operator. Moreover, $G_A^\perp$ consists of those elements of $\mathcal{H}_3$ which are of the form $\{-w^*, w\}$, $w \in D$. Clearly, such elements are in $G_A^\perp$ by (B.9). Conversely, if $\{h, v\} \perp G_A$, then

$$(h, u) + (v, Au) = 0, \qquad u \in \mathcal{H}_1.$$

Hence $v \in D$ are $h = -v^*$. Now suppose $y \perp \bar{D}$ in $\mathcal{H}_2$. Then $y \perp D$ and

$$(\{0, y\}, \{-w^*, w\}) = -(0, w^*) + (y, w) = 0.$$

Thus $\{0, y\} \perp G_A^\perp$. Since G_A is a closed subspace, we see that $\{0, y\} \in G_A$. But if $\{u, Au\} = \{0, y\}$, we must have $u = 0$ and $y = 0$. Thus the only vector in $\mathcal{H}_2$ orthogonal to $\bar{D}$ is 0. This implies that $\bar{D} = \mathcal{H}_2$ (Corollary B.1). This completes the proof. □

Exercises

1. Prove (B.1).

2. Show that the set N in the proof of Theorem 1.9.1 is a closed subspace.

3. If N is a closed subspace of a Hilbert space, show that $(N^\perp)^\perp = N$.

Appendix C
Hölder's Inequality and Banach Space

We shall prove inequality (7.3.27) (Hölder's inequality). Our proof is based on the inequality

$$\sigma^{\theta}\tau^{1-\theta} \leq \theta\sigma + (1-\theta)\tau, \qquad 0 < \theta < 1, \quad \sigma \geq 0, \quad \tau \geq 0 \tag{C.1}$$

where all quantities are nonnegative. To prove (C.1) note that

$$f(t) = t^{\theta} - \theta t$$

is increasing in the interval [0, 1] since $f'(t) = \theta t^{\theta-1} - \theta > 0$. Thus, it has its maximum in that interval at $t = 1$. Thus

$$t^{\theta} - \theta t \leq 1 - \theta, \qquad 0 \leq t \leq 1. \tag{C.2}$$

The same reasoning gives

$$t^{1-\theta} - (1-\theta)t \leq \theta, \qquad 0 \leq t \leq 1. \tag{C.3}$$

If $\sigma \leq \tau$, set $t = \sigma/\tau$ in (C.2). If $\sigma > \tau$, put $t = \tau/\sigma$ in (C.3). This gives (C.1). Once we have this, we take $\theta = 1/p$, $\sigma = |f(x)|^p/\|f\|_p^p$, and $\tau = |g(x)|^{p'}/\|g\|_{p'}^{p'}$, where

$$\|f\|_p^p = \int |f(x)|^p \, dx. \tag{C.4}$$

Since $1 - \theta = 1/p'$, (C.1) gives

$$|f(x)g(x)| \leq \|f\|_p \|g\|_{p'} \big(\theta \|f\|_p^{-p} |f(x)|^p + (1-\theta)\|g\|_{p'}^{-p'} |g(x)|^{p'}\big).$$

If we now integrate with respect to x, we get the desired inequality. Note that we have tacitly assumed that $\|f\|_p \neq 0$ and $\|g\|_{p'} \neq 0$. If either of these vanished, then the corresponding function would vanish a.e. and so would the product fg.

Next we turn our attention to Theorem 9.7.1. That theorem was proved in Hilbert space. However, in Section 12.5 we needed the same theorem in an arbitrary Banach space. The proof is a bit harder; we shall sketch it here. An important ingredient is the following lemma due to F. Riesz.

Lemma C.1. *Let M be a closed subspace of a Banach space X. If $M \neq X$, then for each number θ satisfying $0 < \theta < 1$ there is an element $x_\theta \in X$ such that*

$$\|x_\theta\| = 1, \qquad \|x_\theta - x\| > \theta, \qquad x \in M.$$

PROOF. There is an $x_1 \in X$ which is not in M. Since M is closed, there is a $d > 0$ such that $\|x_1 - x\| \geq d$ for $x \in M$, but for any $\varepsilon > 0$ there is an $x_0 \in M$ such that $\|x_1 - x_0\| < d + \varepsilon$. Take $\varepsilon = d(1-\theta)/\theta$. Then $d + \varepsilon = d/\theta$. Set $x_\theta = (x_1 - x_0)/\|x_1 - x_0\|$. Then $\|x_\theta\| = 1$ and for any $x \in M$

$$\begin{aligned}\|x - x_\theta\| &= \|(\|x_1 - x_0\|x + x_0) - x_1\|/\|x_1 - x_0\| \\ &\geq d/\|x_1 - x_0\| \geq \theta\end{aligned}$$

since $\|x_1 - x_0\|x + x_0 \in M$. □

The main result we need is

Lemma C.2. *If K is a compact operator on a Banach space X, $A = 1 - K$, and $N(A) = \{0\}$, then $R(A) = X$.*

PROOF. First we note that there is a constant C such that

$$\|x\| \leq C\|Ax\|, \qquad x \in X. \tag{C.5}$$

For if this were not so, there would be a sequence $\{x_n\}$ such that

$$\|x_n\| = 1, \qquad Ax_n \to 0.$$

In particular, there is a subsequence (also denoted by $\{x_n\}$) such that $Kx_n \to y$. Hence $x_n = Ax_n + Kx_n \to y$. This shows that $Ay = 0$. But A is one-to-one. Hence $y = 0$. But $\|y\| = \lim\|x_n\| = 1$. Thus (C.5) holds. From this we see that $R(A)$ is closed. For if $Ax_n \to y$, then $\{x_n\}$ is a Cauchy sequence by (C.5). Hence $x_n \to x$, and consequently, $Ax = y$. Note that $R(A^n) = 1 - K_n$, where K_n is a compact operator. Hence $R(A^n)$ is closed for each n. In general, we have $R(A^{n+1}) \subset R(A^n)$. I claim that $R(A) = X$. To see this, suppose it were not. Then there cannot be an $n > 0$ such that $R(A^{n+1}) = R(A^n)$. Otherwise, for each $x \in X$ there would be a $y \in X$ such that $A^n x = A^{n+1}y$. Since A is one-to-one, we would have $Ay = x$, showing that $R(A) = X$. Hence $R(A^{n+1}) \neq R(A^n)$ for each n. By Lemma C.1, for each n there is an $x_n \in R(A^n)$ such that $\|x_n\| = 1$ and $\|x_n - x\|$

$\geq \frac{1}{2}$ for $x \in R(A^{n+1})$. If $n > m$,

$$Kx_m - Kx_n = x_m - (x_n + Ax_m - Ax_n) = x_m - x,$$

where $x \in R(A^{m+1})$. Hence $\|Kx_m - Kx_n\| \geq \frac{1}{2}$. This shows that $\{Kx_m\}$ cannot have a convergent subsequence, contrary to the assumption that K is compact. Hence $R(A) = X$. Since A is one-to-one, it has an inverse defined on the whole of X. This inverse is bounded by (C.5). □

Bibliography

Agmon, Shmuel, Spectral properties of Schrödinger operators and scattering theory, *Ann. Sculoa Norm. Sup. Pisa* 2, pp. 151–218, 1975.

Akhiezer, N. I., and Glazman, I. M., Theory of Linear Operators in Hilbert Space, New York: Frederick Ungar Publishing Company, 1961.

Alsholm, P. K., Wave operators for long-range scattering, *J. Math. Anal. Appl.* 59, pp. 550–572, 1977.

Alsholm, P. K., and Kato, Tosio, Scattering with long range potentials, Proceedings of Symposium in Pure Math., Am. Math. Soc. 23, pp. 393–399, 1973.

Alsholm, P. K. and Schmidt, G., Spectral and scattering theory for Schrödinger operators, *Arch. Rat. Mech. Anal.* 40, pp. 281–311, 1971.

Amrein, W. O., Georgescu, V., and Jauch, J. M., Stationary state scattering theory, *Helv. Phys. Acta.* 44, pp. 407–434, 1971.

Baetman, M. L., and Chadan, K., Scattering theory with highly singular oscillating potentials, *Ann. Inst. Henri Poincaré* 24, pp. 1–16, 1976.

Birman, M. S., Existence conditions for wave operators, *Izv. Akad. Nauk SSSR* 27, pp. 883–906, 1963; The spectrum of singular boundary problems, *Mat. Sb.* 55, pp. 125–174, 1961.

Brownell, F. H., Finite dimensionality of the Schrödinger bottom, *Arch. Rat. Mech. Anal.* 8, pp. 59–67, 1961.

Combescure, M., and Ginibre, J., Spectral and scattering theory for the Schrödinger operator with strongly oscillating potentials, *Ann. Inst. Henri Poincaré* 24, pp. 17–29, 1976.

Deift, Percy, and Simon, Barry, On the decoupling of finite singularities from the question of asymptotic completeness, *J. Func. Anal.* 23, pp. 218–238, 1976.

Dollard, J. D., Asymptotic convergence and the Coulomb interaction, *J. Math. Phys.* 5, pp. 729–738, 1964.

Donaldson, J. A., Gibson, A. G., and Hersh, Reuben, On the invariance principle of scattering theory, *J. Func. Anal.* 14, pp. 131–145, 1973.

Faris, W. G., Self-Adjoint Operators, Lecture Notes in Mathematics 433, New York: Springer-Verlag, 1975.

Feller, William, An Introduction to Probability Theory and its Applications, New York: John Wiley & Sons, 1950.

Glazman, I. M., Direct Methods of the Qualitative Spectral Analysis of Singular Differential Operators, Jerusalem: Israel Program for Scientific Translations, 1965.

Haak, M. N., On convergence to the Møller wave operators, *Nuovo Cimento* 13, pp. 231–236, 1959.

Hille, Einar, and Phillips, R. S., Functional Analysis and Semi-Groups, Providence: American Mathematics Society, 1957.

Hormander, Lars, The existence of wave operators in scattering theory, *Math. Z.* 146, pp. 69–91, 1976.

Ikebe, Teruo, Eigenfunction expansions associated with Schrödinger operators and their applications to scattering theory, *Arch. Rat. Mech. Anal.* 5, pp. 1–34, 1960.

Jauch, J. M., Theory of the scattering operator, *Helv. Phys. Acta* 31, pp. 127–158, 1958.

Jörgens, Konrad, and Weidmann, Joachim, Spectral Properties of Hamiltonian Operators, Lecture Notes in Mathematics 313, New York: Springer-Verlag, 1973.

Kato, Tosio, Wave operators and similarity for non-selfadjoint operators, *Math. Ann.* 162, pp. 258–279, 1966; Perturbation Theory for Linear Operators, New York: Springer-Verlag, 1966.

Kato, Tosio, and Kuroda, S. T., The abstract theory of scattering, *Rocky Mountain J. Math.* 1, pp. 127–171, 1971.

Kuroda, S. T., On the existence and unitary property of the scattering operator, *Nuovo cimento* 12, pp. 431–454, 1959; Some remarks on scattering for Schrödinger operators, *J. Fac. Sci. Univ. Tokyo* 17, pp. 315–329, 1970; Scattering theory for differential operators I, II, *J. Math. Soc. Japan* 25, pp. 75–104, 222–234, 1973.

Lavine, Richard, Commutators and scattering II, *Indiana Math. J.* 21, pp. 643–656, 1972.

Matveev, V. B., Invariance principle for generalized wave operators, *Teoret. Mat. Fiz.* 8, pp. 49–54, 1971.

Matveev, V. B., and Skriganov, M. M., Wave operators for the Schrödinger equation with rapidly oscillating potentials, *Dokl. Adad. Nauk SSSR* 206, pp. 755–757, 1972.

Müller-Pfeiffer, Erich, Spektraleigen-schaften ein dimensionaler Differential-Operatoren hoherer Ordhung, *Studia Math.* 34, pp. 183–196, 1970; Eine Bemerkung über das Spektrum des Schrödinger-Operators, *Math. Nach.* 58, pp. 299–303, 1973.

Najman, Branko, Scattering for the Dirac operator, *Glasnik Mat.* 11, pp. 63–80, 1976.

Pearson, D. B., A generalization of the Birman trace theorem, *J. Func. Anal.* 28, pp. 182–186, 1978.

Povzner, A. Ja., The expansion of arbitrary functions in terms of eigenfunctions of the operator $-\Delta u + cu$, *Mat. Sb.* 32, pp. 109–156, 1953.

Prugovecki, Eduard, Quantum Mechanics in Hilbert Space, New York: Academic Press, 1971.

Reed, Michael, and Simon, Barry, Methods of Modern Mathematical Physics, New York: Academic Press, I, 1972; II, 1975.

Rejto, P. A., On a limiting case of a theorem of Kato and Kuroda, *J. Math. Anal. Appl.* 39, pp. 541–557, 1972.

Sakhnovich, L. A., The invariance principle for generalized wave operators, *Funk. Anal. Pril.* 5, pp. 61–68, 1971.

Schechter, Martin, Principles of Functional Analysis, New York: Academic Press, 1971; Spectra of Partial Differential Operators, Amsterdam: North-Holland –Elsevier, 1971; Scattering theory for elliptic operators of arbitrary order, *Comm. Math. Helv.* 49, pp. 84–113, 1974; A unified approach to scattering, *J. Math. Pures Appl.* 53, pp. 373–396, 1974; Scattering theory for second order elliptic operators, *Ann. Mat. Pure Appl.* 105, pp. 313–331, 1975; On the spectra of singular elliptic operators, *Mathematika* 23, pp. 107–115, 1976.

Simon, Barry, Quantum Mechanics for Hamiltonians Defined as Quadratic Forms, Princeton: Princeton University Press, 1971.

Skriganov, M. M., Spectrum of the Schrödinger operator with strongly oscillating potentials, *Trudy Steklov Math. Inst.* 125, pp. 187–195, 1973.

Weidmann, Joachim, Lineare Operatoren in Hilbertraumen, Stuttgart: B. G. Teubner, 1976.

Wilcox, C. H., Scattering states and wave operators in the abstract theory of scattering, *J. Func. Anal.* 12, pp. 257–274, 1973.

Yafaev, D. R., On the spectrum of the perturbed polyharmonic operator, *Topics Math. Phys.* 5, pp. 107–112, 1972.

Index

A CATALOG OF SELECTED

DOVER BOOKS

IN SCIENCE AND MATHEMATICS

METAL FATIGUE, N. E. Frost, K. J. Marsh, and L. P. Pook. Definitive, clearly written, and well-illustrated volume addresses all aspects of the subject, from the historical development of understanding metal fatigue to vital concepts of the cyclic stress that causes a crack to grow. Includes 7 appendixes. 544pp. 5⅜ x 8½. 40927-9

STATISTICAL MECHANICS: Principles and Applications, Terrell L. Hill. Standard text covers fundamentals of statistical mechanics, applications to fluctuation theory, imperfect gases, distribution functions, more. 448pp. 5⅜ x 8½. 65390-0

THE VARIATIONAL PRINCIPLES OF MECHANICS, Cornelius Lanczos. Graduate level coverage of calculus of variations, equations of motion, relativistic mechanics, more. First inexpensive paperbound edition of classic treatise. Index. Bibliography. 418pp. 5⅜ x 8½. 65067-7

THE VARIOUS AND INGENIOUS MACHINES OF AGOSTINO RAMELLI: A Classic Sixteenth-Century Illustrated Treatise on Technology, Agostino Ramelli. One of the most widely known and copied works on machinery in the 16th century. 194 detailed plates of water pumps, grain mills, cranes, more. 608pp. 9 x 12. 28180-9

ORDINARY DIFFERENTIAL EQUATIONS AND STABILITY THEORY: An Introduction, David A. Sánchez. Brief, modern treatment. Linear equation, stability theory for autonomous and nonautonomous systems, etc. 164pp. 5⅜ x 8¼. 63828-6

ROTARY WING AERODYNAMICS, W. Z. Stepniewski. Clear, concise text covers aerodynamic phenomena of the rotor and offers guidelines for helicopter performance evaluation. Originally prepared for NASA. 537 figures. 640pp. 6⅛ x 9¼. 64647-5

INTRODUCTION TO SPACE DYNAMICS, William Tyrrell Thomson. Comprehensive, classic introduction to space-flight engineering for advanced undergraduate and graduate students. Includes vector algebra, kinematics, transformation of coordinates. Bibliography. Index. 352pp. 5⅜ x 8½. 65113-4

HISTORY OF STRENGTH OF MATERIALS, Stephen P. Timoshenko. Excellent historical survey of the strength of materials with many references to the theories of elasticity and structure. 245 figures. 452pp. 5⅜ x 8½. 61187-6

ANALYTICAL FRACTURE MECHANICS, David J. Unger. Self-contained text supplements standard fracture mechanics texts by focusing on analytical methods for determining crack-tip stress and strain fields. 336pp. 6⅛ x 9¼. 41737-9

Mathematics

HANDBOOK OF MATHEMATICAL FUNCTIONS WITH FORMULAS, GRAPHS, AND MATHEMATICAL TABLES, edited by Milton Abramowitz and Irene A. Stegun. Vast compendium: 29 sets of tables, some to as high as 20 places. 1,046pp. 8 x 10½. 61272-4

APPLIED COMPLEX VARIABLES, John W. Dettman. Step-by-step coverage of fundamentals of analytic function theory–plus lucid exposition of five important applications: Potential Theory; Ordinary Differential Equations; Fourier Transforms; Laplace Transforms; Asymptotic Expansions. 66 figures. Exercises at chapter ends. 512pp. 5⅜ x 8½. 64670-X

INTRODUCTION TO LINEAR ALGEBRA AND DIFFERENTIAL EQUATIONS, John W. Dettman. Excellent text covers complex numbers, determinants, orthonormal bases, Laplace transforms, much more. Exercises with solutions. Undergraduate level. 416pp. 5⅜ x 8½. 65191-6

MATHEMATICAL METHODS IN PHYSICS AND ENGINEERING, John W. Dettman. Algebraically based approach to vectors, mapping, diffraction, other topics in applied math. Also generalized functions, analytic function theory, more. Exercises. 448pp. 5⅜ x 8¼. 65649-7

CALCULUS OF VARIATIONS WITH APPLICATIONS, George M. Ewing. Applications-oriented introduction to variational theory develops insight and promotes understanding of specialized books, research papers. Suitable for advanced undergraduate/graduate students as primary, supplementary text. 352pp. 5⅜ x 8½. 64856-7

COMPLEX VARIABLES, Francis J. Flanigan. Unusual approach, delaying complex algebra till harmonic functions have been analyzed from real variable viewpoint. Includes problems with answers. 364pp. 5⅜ x 8½. 61388-7

AN INTRODUCTION TO THE CALCULUS OF VARIATIONS, Charles Fox. Graduate-level text covers variations of an integral, isoperimetrical problems, least action, special relativity, approximations, more. References. 279pp. 5⅜ x 8½. 65499-0

CATASTROPHE THEORY FOR SCIENTISTS AND ENGINEERS, Robert Gilmore. Advanced-level treatment describes mathematics of theory grounded in the work of Poincaré, R. Thom, other mathematicians. Also important applications to problems in mathematics, physics, chemistry and engineering. 1981 edition. References. 28 tables. 397 black-and-white illustrations. xvii + 666pp. 6⅛ x 9¼. 67539-4

INTRODUCTION TO DIFFERENCE EQUATIONS, Samuel Goldberg. Exceptionally clear exposition of important discipline with applications to sociology, psychology, economics. Many illustrative examples; over 250 problems. 260pp. 5⅜ x 8½. 65084-7

NUMERICAL METHODS FOR SCIENTISTS AND ENGINEERS, Richard Hamming. Classic text stresses frequency approach in coverage of algorithms, polynomial approximation, Fourier approximation, exponential approximation, other topics. Revised and enlarged 2nd edition. 721pp. 5⅜ x 8½. 65241-6

INTRODUCTION TO NUMERICAL ANALYSIS (2nd Edition), F. B. Hildebrand. Classic, fundamental treatment covers computation, approximation, interpolation, numerical differentiation and integration, other topics. 150 new problems. 669pp. 5⅜ x 8½. 65363-3

POPULAR LECTURES ON MATHEMATICAL LOGIC, Hao Wang. Noted logician's lucid treatment of historical developments, set theory, model theory, recursion theory and constructivism, proof theory, more. 3 appendixes. Bibliography. 1981 edition. ix + 283pp. 5⅜ x 8½. 67632-3

CALCULUS OF VARIATIONS, Robert Weinstock. Basic introduction covering isoperimetric problems, theory of elasticity, quantum mechanics, electrostatics, etc. Exercises throughout. 326pp. 5⅜ x 8½. 63069-2

THE CONTINUUM: A Critical Examination of the Foundation of Analysis, Hermann Weyl. Classic of 20th-century foundational research deals with the conceptual problem posed by the continuum. 156pp. 5⅜ x 8½. 67982-9

CHALLENGING MATHEMATICAL PROBLEMS WITH ELEMENTARY SOLUTIONS, A. M. Yaglom and I. M. Yaglom. Over 170 challenging problems on probability theory, combinatorial analysis, points and lines, topology, convex polygons, many other topics. Solutions. Total of 445pp. 5⅜ x 8½. Two-vol. set.
Vol. I: 65536-9 Vol. II: 65537-7

A SURVEY OF NUMERICAL MATHEMATICS, David M. Young and Robert Todd Gregory. Broad self-contained coverage of computer-oriented numerical algorithms for solving various types of mathematical problems in linear algebra, ordinary and partial, differential equations, much more. Exercises. Total of 1,248pp. 5⅜ x 8½. Two volumes. Vol. I: 65691-8 Vol. II: 65692-6

INTRODUCTION TO PARTIAL DIFFERENTIAL EQUATIONS WITH APPLICATIONS, E. C. Zachmanoglou and Dale W. Thoe. Essentials of partial differential equations applied to common problems in engineering and the physical sciences. Problems and answers. 416pp. 5⅜ x 8½. 65251-3

THE THEORY OF GROUPS, Hans J. Zassenhaus. Well-written graduate-level text acquaints reader with group-theoretic methods and demonstrates their usefulness in mathematics. Axioms, the calculus of complexes, homomorphic mapping, p-group theory, more. Many proofs shorter and more transparent than older ones. 276pp. 5⅜ x 8½. 40922-8

DISTRIBUTION THEORY AND TRANSFORM ANALYSIS: An Introduction to Generalized Functions, with Applications, A. H. Zemanian. Provides basics of distribution theory, describes generalized Fourier and Laplace transformations. Numerous problems. 384pp. 5⅜ x 8½. 65479-6

Math–Decision Theory, Statistics, Probability

ELEMENTARY DECISION THEORY, Herman Chernoff and Lincoln E. Moses. Clear introduction to statistics and statistical theory covers data processing, probability and random variables, testing hypotheses, much more. Exercises. 364pp. 5⅜ x 8½. 65218-1

STATISTICS MANUAL, Edwin L. Crow et al. Comprehensive, practical collection of classical and modern methods prepared by U.S. Naval Ordnance Test Station. Stress on use. Basics of statistics assumed. 288pp. 5⅜ x 8½. 60599-X

SOME THEORY OF SAMPLING, William Edwards Deming. Analysis of the problems, theory and design of sampling techniques for social scientists, industrial managers and others who find statistics important at work. 61 tables. 90 figures. xvii +602pp. 5⅜ x 8½. 64684-X

STATISTICAL ADJUSTMENT OF DATA, W. Edwards Deming. Introduction to basic concepts of statistics, curve fitting, least squares solution, conditions without parameter, conditions containing parameters. 26 exercises worked out. 271pp. 5⅜ x 8½. 64685-8

LINEAR PROGRAMMING AND ECONOMIC ANALYSIS, Robert Dorfman, Paul A. Samuelson and Robert M. Solow. First comprehensive treatment of linear programming in standard economic analysis. Game theory, modern welfare economics, Leontief input-output, more. 525pp. 5⅜ x 8½. 65491-5

DICTIONARY/OUTLINE OF BASIC STATISTICS, John E. Freund and Frank J. Williams. A clear concise dictionary of over 1,000 statistical terms and an outline of statistical formulas covering probability, nonparametric tests, much more. 208pp. 5⅜ x 8½. 66796-0

PROBABILITY: An Introduction, Samuel Goldberg. Excellent basic text covers set theory, probability theory for finite sample spaces, binomial theorem, much more. 360 problems. Bibliographies. 322pp. 5⅜ x 8½. 65252-1

GAMES AND DECISIONS: Introduction and Critical Survey, R. Duncan Luce and Howard Raiffa. Superb nontechnical introduction to game theory, primarily applied to social sciences. Utility theory, zero-sum games, n-person games, decision-making, much more. Bibliography. 509pp. 5⅜ x 8½. 65943-7

FIFTY CHALLENGING PROBLEMS IN PROBABILITY WITH SOLUTIONS, Frederick Mosteller. Remarkable puzzlers, graded in difficulty, illustrate elementary and advanced aspects of probability. Detailed solutions. 88pp. 5⅜ x 8½. 65355-2

PROBABILITY THEORY: A Concise Course, Y. A. Rozanov. Highly readable, self-contained introduction covers combination of events, dependent events, Bernoulli trials, etc. 148pp. 5⅜ x 8¼. 63544-9

STATISTICAL METHOD FROM THE VIEWPOINT OF QUALITY CONTROL, Walter A. Shewhart. Important text explains regulation of variables, uses of statistical control to achieve quality control in industry, agriculture, other areas. 192pp. 5⅜ x 8½. 65232-7

THE COMPLEAT STRATEGYST: Being a Primer on the Theory of Games of Strategy, J. D. Williams. Highly entertaining classic describes, with many illustrated examples, how to select best strategies in conflict situations. Prefaces. Appendices. 268pp. 5⅜ x 8½. 25101-2

Math–Geometry and Topology

ELEMENTARY CONCEPTS OF TOPOLOGY, Paul Alexandroff. Elegant, intuitive approach to topology from set-theoretic topology to Betti groups; how concepts of topology are useful in math and physics. 25 figures. 57pp. 5⅜ x 8½. 60747-X

COMBINATORIAL TOPOLOGY, P. S. Alexandrov. Clearly written, well-organized, three-part text begins by dealing with certain classic problems without using the formal techniques of homology theory and advances to the central concept, the Betti groups. Numerous detailed examples. 654pp. 5⅜ x 8½. 40179-0

EXPERIMENTS IN TOPOLOGY, Stephen Barr. Classic, lively explanation of one of the byways of mathematics. Klein bottles, Moebius strips, projective planes, map coloring, problem of the Koenigsberg bridges, much more, described with clarity and wit. 43 figures. 210pp. 5⅜ x 8½. 25933-1

CONFORMAL MAPPING ON RIEMANN SURFACES, Harvey Cohn. Lucid, insightful book presents ideal coverage of subject. 334 exercises make book perfect for self-study. 55 figures. 352pp. 5⅝ x 8¼. 64025-6

THE GEOMETRY OF RENÉ DESCARTES, René Descartes. The great work founded analytical geometry. Original French text, Descartes's own diagrams, together with definitive Smith-Latham translation. 244pp. 5⅜ x 8½. 60068-8

THE THIRTEEN BOOKS OF EUCLID'S ELEMENTS, translated with introduction and commentary by Sir Thomas L. Heath. Definitive edition. Textual and linguistic notes, mathematical analysis. 2,500 years of critical commentary. Unabridged. 1,414pp. 5⅜ x 8½.. Three-vol. set.
Vol. I: 60088-2 Vol. II: 60089-0 Vol. III: 60090-4

GEOMETRY OF COMPLEX NUMBERS, Hans Schwerdtfeger. Illuminating, widely praised book on analytic geometry of circles, the Moebius transformation, and two-dimensional non-Euclidean geometries. 200pp. 5⅝ x 8¼. 63830-8

DIFFERENTIAL GEOMETRY, Heinrich W. Guggenheimer. Local differential geometry as an application of advanced calculus and linear algebra. Curvature, transformation groups, surfaces, more. Exercises. 62 figures. 378pp. 5⅜ x 8½. 63433-7

CURVATURE AND HOMOLOGY: Enlarged Edition, Samuel I. Goldberg. Revised edition examines topology of differentiable manifolds; curvature, homology of Riemannian manifolds; compact Lie groups; complex manifolds; curvature, homology of Kaehler manifolds. New Preface. Four new appendixes. 416pp. 5⅜ x 8½. 40207-X

TOPOLOGY, John G. Hocking and Gail S. Young. Superb one-year course in classical topology. Topological spaces and functions, point-set topology, much more. Examples and problems. Bibliography. Index. 384pp. 5⅝ x 8¼. 65676-4

Physics

OPTICAL RESONANCE AND TWO-LEVEL ATOMS, L. Allen and J. H. Eberly. Clear, comprehensive introduction to basic principles behind all quantum optical resonance phenomena. 53 illustrations. Preface. Index. 256pp. 5⅜ x 8½. 65533-4

ULTRASONIC ABSORPTION: An Introduction to the Theory of Sound Absorption and Dispersion in Gases, Liquids and Solids, A. B. Bhatia. Standard reference in the field provides a clear, systematically organized introductory review of fundamental concepts for advanced graduate students, research workers. Numerous diagrams. Bibliography. 440pp. 5⅜ x 8½. 64917-2

QUANTUM THEORY, David Bohm. This advanced undergraduate-level text presents the quantum theory in terms of qualitative and imaginative concepts, followed by specific applications worked out in mathematical detail. Preface. Index. 655pp. 5⅜ x 8½. 65969-0

ATOMIC PHYSICS (8th edition), Max Born. Nobel laureate's lucid treatment of kinetic theory of gases, elementary particles, nuclear atom, wave-corpuscles, atomic structure and spectral lines, much more. Over 40 appendices, bibliography. 495pp. 5⅜ x 8½. 65984-4

AN INTRODUCTION TO HAMILTONIAN OPTICS, H. A. Buchdahl. Detailed account of the Hamiltonian treatment of aberration theory in geometrical optics. Many classes of optical systems defined in terms of the symmetries they possess. Problems with detailed solutions. 1970 edition. xv + 360pp. 5⅜ x 8½. 67597-1

THIRTY YEARS THAT SHOOK PHYSICS: The Story of Quantum Theory, George Gamow. Lucid, accessible introduction to influential theory of energy and matter. Careful explanations of Dirac's anti-particles, Bohr's model of the atom, much more. 12 plates. Numerous drawings. 240pp. 5⅜ x 8½. 24895-X

ELECTRONIC STRUCTURE AND THE PROPERTIES OF SOLIDS: The Physics of the Chemical Bond, Walter A. Harrison. Innovative text offers basic understanding of the electronic structure of covalent and ionic solids, simple metals, transition metals and their compounds. Problems. 1980 edition. 582pp. 6⅛ x 9¼. 66021-4

HYDRODYNAMIC AND HYDROMAGNETIC STABILITY, S. Chandrasekhar. Lucid examination of the Rayleigh-Benard problem; clear coverage of the theory of instabilities causing convection. 704pp. 5⅜ x 8¼. 64071-X

INVESTIGATIONS ON THE THEORY OF THE BROWNIAN MOVEMENT, Albert Einstein. Five papers (1905–8) investigating dynamics of Brownian motion and evolving elementary theory. Notes by R. Fürth. 122pp. 5⅜ x 8½. 60304-0

THE PHYSICS OF WAVES, William C. Elmore and Mark A. Heald. Unique overview of classical wave theory. Acoustics, optics, electromagnetic radiation, more. Ideal as classroom text or for self-study. Problems. 477pp. 5⅜ x 8½. 64926-1

METHODS OF THERMODYNAMICS, Howard Reiss. Outstanding text focuses on physical technique of thermodynamics, typical problem areas of understanding, and significance and use of thermodynamic potential. 1965 edition. 238pp. 5⅜ x 8½. 69445-3

TENSOR ANALYSIS FOR PHYSICISTS, J. A. Schouten. Concise exposition of the mathematical basis of tensor analysis, integrated with well-chosen physical examples of the theory. Exercises. Index. Bibliography. 289pp. 5⅜ x 8½. 65582-2

RELATIVITY IN ILLUSTRATIONS, Jacob T. Schwartz. Clear nontechnical treatment makes relativity more accessible than ever before. Over 60 drawings illustrate concepts more clearly than text alone. Only high school geometry needed. Bibliography. 128pp. 6⅛ x 9¼. 25965-X

THE ELECTROMAGNETIC FIELD, Albert Shadowitz. Comprehensive undergraduate text covers basics of electric and magnetic fields, builds up to electromagnetic theory. Also related topics, including relativity. Over 900 problems. 768pp. 5⅝ x 8¼. 65660-8

GREAT EXPERIMENTS IN PHYSICS: Firsthand Accounts from Galileo to Einstein, edited by Morris H. Shamos. 25 crucial discoveries: Newton's laws of motion, Chadwick's study of the neutron, Hertz on electromagnetic waves, more. Original accounts clearly annotated. 370pp. 5⅜ x 8½. 25346-5

RELATIVITY, THERMODYNAMICS AND COSMOLOGY, Richard C. Tolman. Landmark study extends thermodynamics to special, general relativity; also applications of relativistic mechanics, thermodynamics to cosmological models. 501pp. 5⅜ x 8½. 65383-8

LIGHT SCATTERING BY SMALL PARTICLES, H. C. van de Hulst. Comprehensive treatment including full range of useful approximation methods for researchers in chemistry, meteorology and astronomy. 44 illustrations. 470pp. 5⅜ x 8½. 64228-3

STATISTICAL PHYSICS, Gregory H. Wannier. Classic text combines thermodynamics, statistical mechanics and kinetic theory in one unified presentation of thermal physics. Problems with solutions. Bibliography. 532pp. 5⅜ x 8½. 65401-X

Paperbound unless otherwise indicated. Available at your book dealer, online at **www.doverpublications.com**, or by writing to Dept. GI, Dover Publications, Inc., 31 East 2nd Street, Mineola, NY 11501. For current price information or for free catalogues (please indicate field of interest), write to Dover Publications or log on to **www.doverpublications.com** and see every Dover book in print. Dover publishes more than 500 books each year on science, elementary and advanced mathematics, biology, music, art, literary history, social sciences, and other areas.